# Agriculture's Ethical Horizon

# Agriculture's Ethical Horizon

*Second Edition*

***Robert L. Zimdahl***
*Professor Emeritus*
*Department of Bioagricultural Sciences and Pest Management*
*Colorado State University*
*Fort Collins, CO, USA*

ELSEVIER

AMSTERDAM • BOSTON • HEIDELBERG • LONDON • NEW YORK • OXFORD • PARIS
SAN DIEGO • SAN FRANCISCO • SINGAPORE • SYDNEY • TOKYO

Elsevier
32 Jamestown Road, London NW1 7BY
225 Wyman Street, Waltham, MA 02451, USA

First edition 2006
Second edition 2012



**British Library Cataloguing-in-Publication Data**
A catalogue record for this book is available from the British Library

**Library of Congress Cataloging-in-Publication Data**
A catalog record for this book is available from the Library of Congress

ISBN: 978-0-12-416043-9



This book has been manufactured using Print On Demand technology. Each copy is produced to order and is limited to black ink. The online version of this book will show color figures where appropriate.

# Contents

# Foreword—First Edition

In one sense, the current era of agricultural ethics began in the 1970s when Glenn L. Johnson, an agricultural economist known for his work on asset fixity, took a sabbatical at Oxford University to work with several philosophers there. The result was a series of papers calling for a new area of explicit and logically critical exposition of the values underlying applied and problem-solving research in the agricultural sciences (1976, 1982). One could also argue that there has been a continuous and unbroken string of ethical and philosophical reflections on agriculture that can be dated back at least to Xeonophon's *oeconomicus* in the fourth century BC. Here, one would trace a succession in the twentieth century that notes the writings of Liberty Hyde Bailey, Louis Bromfield, and Wendell Berry. Bailey (1858–1954), best known as a taxonomist and for his leadership as the Dean of Agriculture at Cornell University, was almost certainly the leading agricultural scientist of his generation. In addition to his many scientific publications, he contributed a number of reflective philosophical essays, including *The Holy Earth* and his work for the Theodore Roosevelt administration's Commission on Country Life.

Louis Bromfield (1896–1956) was a novelist and Hollywood screenwriter who turned his pen to farming after returning to his Ohio homeland in the 1940s. During the 1950s he became a potent spokesman for conservation and the values of rural life, but his writings are little appreciated today. Wendell Berry (born 1934) is the current generation's Bromfield. The Kentucky poet and novelist has perhaps become best known for his writings on farming and conservation. But Bailey was a leading agricultural scientist in his own right, and Bromfield worked closely with agricultural scientists such as Hugh Hammond Bennett (1881–1960), the father of soil conservation in the United States. Berry, in contrast, largely has been ostracized from the agricultural science establishment. His trenchant critique of land-grant university science and education in the 1977 book *The Unsettling of America* caused him to be perceived as an enemy by those agricultural scientists who were aware of him. For reasons that Robert Zimdahl makes clear in this volume, most faculty at agricultural universities in the 1970s and 1980s simply ignored Berry.

The transition from Bailey to Berry is thus significant, and the neglect that post-war agricultural science showed for ethics was a central point of analysis not only for Johnson's essays of the 1970s, but also for this extended and systematic study by Zimdahl. Johnson and Zimdahl both emphasize the rise of positivism as a philosophy of science within agricultural universities and research organizations. Positivism can be succinctly defined as the view that agricultural scientists should confine their activity to the collection of empirical data and to the analysis of quantifiable relationships among data. It is, on the one hand, clear that agricultural scientists have never so confined themselves. If Norman Borlaug is the prototypical agricultural scientist of the late twentieth century, one must note his tireless work to ensure that modern maize, rice, and wheat varieties responsive to fertilizer would be supported by government and accepted by farmers, not to mention his public advocacy on behalf of the green revolution (see Borlaug, 2000). On the other hand, it is undeniable that sustained, critical debate over the goals of agricultural science has been exceedingly rare during Borlaug's professional lifetime. Positivism is the philosophy that holds that such debate has no place in science. To the extent that scientists such as Borlaug campaign on behalf of their preferred vision of agriculture, it is considered to be an extra-scientific activity, a necessary evil, perhaps, but in no sense part and parcel of the scientific process itself.

As I have argued elsewhere (Thompson, 2004), this brand of positivism had its philosophical roots in a short-lived philosophical movement associated with the Vienna Circle philosophers Morritz Schlick (1882–1936), Rudolph Carnap (1891–1970), and Kurt Godel (1906–1978), among others. The Vienna Circle philosophers were active in the 1920s and early 1930s, but this philosophical movement may have had its greatest influence over post World War II through a single book, A.J. Ayer's *Language, Truth and Logic*, published in 1936. The key philosophical doctrine rests on a theory of meaning that had been promulgated by Gottlob Frege (1848–1925), according to which the "sense" of a word or sentence must be distinguished from the thing or state of affairs to which it refers. As Ayer (1910–1989) expressed it, statements are meaningful only if one of two conditions hold: they express purely conceptual relationships that arise in virtue of definitions (the mental "sense") given to terms, or they correspond to (that is, describe) possible states of the world. Ayer proposed a "verification principle" for determining whether sentences met the second, empirical criterion for meaning, to wit, that all empirically meaningful sentences, in principle, are capable of being determined true or false through the collection and analysis of data. One consequence of this view was that sentences expressing norms or values were deemed neither true nor false, but "meaningless." The positivists denigrated such talk, labeling it as "metaphysics," and implying that it was tantamount to the outdated superstition of a bygone era.

This form of positivism has had a profound impact on the history of science since World War II. It has vindicated countless decisions by journal editors, tenure, promotion, and review committees, not to mention individual scientists, who rejected and repressed themselves or their colleagues when they engaged in speculative, philosophical, and reflective exercises on the grounds that such activities are "not science." In fact, the Vienna Circle philosophers who survived the war and enjoyed distinguished careers in the United States had discovered a host of problems in the verification principle by 1950. Each had significantly modified their views, adopting a form of pragmatism that recognizes the value-laden character of knowledge, as did Ayer himself. Nevertheless, *Language, Truth and Logic* was assigned widely in classrooms well into the 1980s, and undoubtedly had a profound influence on the philosophical views of scientists who were educated in the 50-year period following its original publication in the 1930s.

The other book of philosophy that was especially influential was Karl Popper's *Logic of Scientific Discovery*, published in 1935 and translated into English in 1959. Here, Popper puts forth the view that science progresses not through verification, but through falsification, by eliminating hypotheses that are inconsistent with data collected through experiments. Popper's characterization of scientific logic also yielded the view that science properly is occupied with the formulation of hypotheses that predict specific outcomes. Although such hypotheses are not "proven" when the predicted outcomes materialize, the failure of a prediction falsifies the hypothesis in question. Crucially, hypotheses incapable of such falsifying tests cannot be characterized as properly scientific on Popper's view. Although the view that science properly is concerned with the collection of data and the analysis of relations among data rests jointly on the positivist rejection of metaphysics and on Popper's more sophisticated characterization of progress through falsification, Popper was one of the most severe critics of the verification principle from the very outset. In part, his criticism focused on his view that proving false was more important than proving true, but he also believed that one would never be able to actually conduct falsifying experiments without also engaging in philosophical arguments intended to frame and contextualize empirical research, including debates over *which* experiments to conduct and *how* to conduct them. There is a world of difference between Popper's belief that ethical norms cannot be subjected to logically decisive falsifying tests and the belief that they are wholly meaningless.

Popper was right. Science cannot be done without philosophy, and this philosophy includes ethics. In fact, the statement that scientists should confine themselves to the collection and analysis of data is an ethical norm, a norm for the conduct of inquiry. Such norms cannot become widely established in scientific practice without a significant amount of philosophical discussion and argument. Thus, in addition to the value-laden campaigning for which Norman Borlaug is so well known, there have been countless conversations and exchanges in which scientists have established the positivist tenet as an ethical principle for inquiry in the agricultural sciences. In fact, the journal editors and tenure committees who have imposed this norm of practice have not succeeded in eliminating metaphysics, ethics, and philosophy from scientific disciplines. They have succeeded only in expunging such philosophical reflection from the scientific record. The result is that a significant amount of the work that was necessary to make science possible in the last half of the twentieth century cannot be passed down to the present generation, nor can it be brought before a public anxious to believe that science is conducted according to a discipline of logic, honesty, and adherence to standards of rigor.

The actual philosophy of scientific practice for the period in which the agricultural disciplines took their present shape is thus as ephemeral as the casual remarks that the scientists who built these disciplines exchanged over coffee. Where are the books and articles in which the scientists of the 1950s and 1960s articulated the rationale for developing chemical pesticides, herbicides, and fertilizers? Where are the course syllabi in which instructors in the agricultural sciences discussed alternative approaches for understanding agriculture's impact on the broader environment? Where is the evidence that this generation of scientists debated and perhaps rejected the ideas of Albert Howard? The failure to record the considerations and deliberations that led scientists to undertake the studies that led to the rise of chemical and molecular technologies in the plant sciences, and to a mechanical revolution in animal husbandry, has left the current generation vulnerable to the charge that such developments were undertaken in secret by profit- and power-seeking individuals with little regard for farmers, farm animals, the environment, or the broader public.

Even under assault from authors such as Rachel Carson or Wendell Berry, the agricultural disciplines of the 1960s, 1970s, and 1980s displayed too little willingness to articulate the reasons for, values behind, and logic of their science. I have argued that an implicit and poorly articulated utilitarianism was integrated into the rationale for agricultural science. According to this view, the rising productivity of industrial agriculture leads to lower food costs for consumers (Thompson, 1995). The argument runs like this: Because food is essential, because everyone eats, and because expenditures for food are particularly critical for the poor, the net benefit from lower consumer prices for food offsets any cost experienced in the form of environmental impact, as well as farm bankruptcies and associated impacts on rural communities that may occur as farms become larger and fewer. Although a few agricultural economists, notably Luther Tweeten (1984), have accepted the importance of articulating the utilitarian rationale of this argument explicitly, this view, if indeed it is the view of mainstream agricultural scientists, remains wholly implicit within the biologically oriented agricultural sciences. Does the current generation of scientists see no reason to articulate the rationale for doing what they do, to engage in self-reflection, or for defending what they do against mounting criticism? Thankfully, the answer is, "Not entirely." The movement to embrace ethics has gone farthest and most quickly in the animal sciences, beginning not surprisingly with entomology. The controversy sparked by Carson's *Silent Spring* in 1962 did indeed result in substantive debate within this discipline. Robert Van Den Bosch offered a book-length ethical critique of his discipline in 1978, and entomologists were among the first faculty in agricultural science to publish in the journal *Agriculture and Human Values* shortly after it was launched in 1982. However, it is fair to note that this self-criticism within the discipline of entomology did not embrace the vocabulary and conceptual resources available within philosophical ethics and the philosophy of science. Livestock researchers, the last group of agricultural scientists I would have expected to break with positivism when I began my own work on agricultural ethics in 1980, were in fact the first to do so. In retrospect, it is not surprising that this group would move first because no other area of the agricultural sciences has been subjected to such sustained criticism from so many different directions. The lead issue, of course, has been the welfare of livestock in concentrated animal feeding operations, but the food animal industries have dealt with enormous issues with respect to environment, food safety, and changes within farm structure, as well (see Thompson, 2001). Here the charge has been led, perhaps, by my philosophical colleague, Bernard Rollin (1995), who had at least a 10-year head start on me in his work on ethical issues in agriculture. But many animal scientists have taken up the task of articulating, critiquing, and refining the key norms for their discipline. Here, one must note papers by David Fraser (1997, 1999) and Keith Schillo (1998, 1999), along with Ray Strickland's efforts to establish a standing bioethics section at the annual scientific meetings. Here also there are at least two book-length studies on ethical issues by animal scientists: Peter Cheeke's *Contemporary Issues in Animal Agriculture* (1998) and H.O. Kunkel's *Human Issues in Animal Agriculture* (2000).

Things are not as well developed on the plant side. There is, of course, Wes Jackson, but Wes long ago forsook the agricultural university/experiment station complex to become friends with Wendell Berry, and no one would take his ideas to represent reflective self-criticism on the part of mainstream crop production science. From the mainstream, the first hints of a call for ethics only now are being heard. Writing for the centennial issue of *The Journal of Agricultural Science,* L.T. Evans closes his reflections on the last 100 years by noting that "future agricultural scientists will be called on not merely to enhance agricultural production, but also to consider more explicitly the ethical as well as the environmental consequences of their research" (2005, p. 10). Maarten Chrispeels (2003) has coauthored an article entitled "Agricultural Ethics," and included a series on the topic as editor of *Plant Physiology.* But so far as I know, the volume that you hold in your hands is the first book-length study to incorporate sustained ethical and philosophical reflections by a mainstream agricultural scientist working with plants or crop production, at least since roughly the time that Liberty Hyde Bailey published *The Holy Earth,* in 1915. Zimdahl has also gone farther into the philosophical literature than any of the scientifically trained authors listed earlier. The book you are reading now contains sophisticated expositions of philosophical concepts such as utilitarianism and positivism, and develops a careful application of these concepts to the practice of agricultural science.

Some will read Zimdahl's analysis and react in anger. In itself, that is fine. There are certainly different philosophical positions that can be taken with respect to the philosophy of the agricultural sciences than the one Zimdahl develops in the following pages. What must happen next, however, is the translation of that reaction into words, then into arguments intended to show just where Zimdahl goes wrong in the reader's mind. These debates need to be aired at scientific meetings, and the reactions to this book need to find their way into print, if not in the major scientific journals then in outlets such as *Agriculture and Human values* or *The Journal of Agricultural and Environmental Ethics.* Finally, there must be significant portions of graduate education in the agricultural sciences given over to the philosophy of agricultural science. Only then and by such means will the public record of values and rationale for agricultural research be constructed and laid open to anyone who cares to look. What is more likely is just what Zimdahl anticipates. Those who disagree with his book will ignore it, just as they ignored Wendell Berry. Perhaps this will arise from the mistaken and ultimately self-serving belief that to cite and discuss literature that one disagrees with is to lend credibility to its conclusions; or perhaps it just is the continuing legacy of positivism that prevents scientists from engaging in the debate that so desperately needs to happen.

Even the brief and idiosyncratic list of works listed in this preface shows that agriculture was not wholly without any philosophical reflection during the last half of the twentieth century. The trouble is that this reflection was disjointed, each effort emerging *de novo* as if nothing had gone before. Here, too, Zimdahl's approach is an important departure from the norm. My challenge to the reader of Zimdahl's book, then, is to respond in kind. Even for those who are largely sympathetic with the main thrust of Zimdahl's argument, there is a responsibility to offer him the benefit of engaged criticism. It is only through the give and take that occurs when philosophical ideas are batted about that a literature in the philosophy of agricultural science can be built. It is only by authors being willing to articulate the reasons why they see things one way rather than another that we can have a public record of the values that underlie research and technology choices in agriculture. It is only through such a public record that we can have the ability to sharpen, refine, and reevaluate those choices from one generation to the next. And perhaps even more importantly, it is only through such a public record that we can have any confidence that something more than the most self-seeking and venal motives actually are guiding the key decisions that are made when articles are published or rejected, scientists are tenured or denied, grants are awarded, and technology is developed.

It is not in its emphasis on data, logic, rigor, or even quantification that positivist philosophy of science fails. All these values, perhaps better articulated by Popper than by Ayer, should be cherished. The error came in establishing a practice of silence among agricultural scientists when it comes to articulating, critiquing, and then defending various reasons and rationales for doing things one way rather than another. This practice is irresponsible because it fails future generations of scientists, who are deprived of the ability to survey those rationales, examining their strengths and possibly also finding places where they need to be adapted to changing circumstances. Silence is especially irresponsible within the land grant mission of public science, where the public has a reasonable expectation that research choices be consistent with a broad conception of the public interest. Because the public interest itself is an open-ended, evolving, and revisable ideal, it is doubly critical that science intended to serve the public interest be engaged in an ongoing and public process of evaluation and debate. Zimdahl has taken giant steps in the direction of restoring a practice nobly evident in the legacy of Liberty Hyde Bailey. Let us hope that his colleagues will not fail to honor both Zimdahl and Bailey with a considered response.

**Paul B. Thompson**
W.K. Kellogg Professor of Agriculture, Food, and Community Ethics
Michigan State University

## References

Ayer, A.J. (1936). *Language, Truth and Logic*. New York, Dover Publications (republished 1952).

Bailey, L.H. (1915). *The Holy Earth*. New York, Charles Scribner's and Sons.

Berry, W. (1977). *Thee Unsettling of America: Culture and Agriculture*. San Francisco, CA, Sierra Club Books.

Borlaug, N. (2000). *The Green Revolution Revisited and the Road Ahead*, http://nobelprize.virtuaJ.museum/peace/articles/borlaug!borlaug-lecture.pdf (accessed January 4, 2006).

Cheeke, P. (1998). *Contemporary Issues in Animal Agriculture*. Danville, IL, Interstate Publishers.

Chrispeels, M. and D.F. Mandoli. (2003). Agricultural ethics. *Plant Physiol.* 132:4–9.

Evans, L.T. (2005). The changing context for agricultural science. *Agric. Res.* 143:7–10.

Fraser, D., Weary, D.M., Pajor, E.A. and Milligan, B.N. (1997). A scientific conception of animal welfare that reflects ethical concerns. *Animal Welfare.* 6:187–205.

Fraser, D. (1999). Animal ethics and animal welfare science: Bridging the two cultures. *Appl. Animal Behav. Sci.* 64:171–189.

Johnson, G.L. (1976). Philosophic foundations: problems, knowledge and solutions. *Eur. Rev. Agric. Econ.* 3(2/3):207–234.

Johnson, G.L. (1982). Agro-ethics: Extension, research and teaching. *South. J. Agric. Econ.* July:1–10.

Kunkel, H.O. (2000). *Human Issues in Animal Agriculture*. College Station, TX, Texas A&M University Press.

Popper, K. (1959). *The Logic of Scientific Discovery*. New York, Basic Books.

Rollin, B. (1995). *Farm Animal Welfare*. Ames, IA, Iowa State University Press.

Schillo, K. (1998). Toward a pluralistic animal science: posdiberal feminist perspectives. *Animal Sci.* 76:2763–2770.

Schillo, K. (1999). An appropriate role for ethics in teaching contemporary issues. *Animal Sci.* 77(Suppl. 2):154–162.

Thompson, P.B. (1995). *The Spirit of the Soil: Agriculture and Environmental Ethics*. New York, Routledge Publishing Co.

Thompson, P.B. (2001). Animal welfare and livestock production in a postindustrial milieu. *J. Appl. Animal Welfare Sci.* 4(3):191–205.

Thompson, P.B. (2004). The legacy of positivism and the role of ethics in the agricultural sciences. Pp. 335–351 *in* C.G. Scanes, J.A. Miranowski (eds.). *Perspectives in World Food and Agriculture 2004*. Ames, IA, Iowa State University Press.

Tweeten, L. (1984). Food for people and profit: Ethics and capitalism. *The Farm and Food System in Transition: Emerging Policy Issues* No. 2. East Lansing, MI, Cooperative Extension Service, Michigan State University.

Van Den Bosch, R. (1978). *The Pesticide Conspiracy*. New York, Doubleday.

# Foreword—Second Edition

Writing a foreword for the revised version of a previously published book which is widely recognized as a well-documented and respected systemic study of agriculture and ethics, and one which already had a stellar and comprehensive foreword written by Paul Thompson for the first edition, is challenging to say the least. Since Paul already traced the history and philosophy of ethics and agriculture in such a comprehensive manner, there is certainly no point in addressing those issues further. So, I will concentrate on making a few brief comments about the revisions to this important work, and add a few observations of my own that may be useful to the reader.

The significant shift that everyone interested in agricultural sustainability, or for that matter a sustainable biosphere, must address, is the fact that the future we will all experience in the decades ahead will be significantly different from the world we lived in for the past century or more. Zimdahl quotes Jane Lubchenco (1998) in this regard. In her presidential address to the American Association for the Advancement of Science she asked whether the scientific community that had successfully met past challenges "is prepared for the equally crucial and daunting challenges that lie in our immediate future." The answer to her own question was, "No" since "the real challenges facing us have not been fully appreciated nor properly acknowledged by the community of scientists whose responsibility it is, and will be, to meet them." Lubchenco then went on to say that it was time for the "scientific community to take responsibility for the contributions required to address the environmental and social problems before us, problems that, with the best of intentions in the world, we have nonetheless helped to create."

These insights from Lubchenco perhaps best sum up the reason why Zimdahl's revised edition is so important. In this edition, he addresses many of the ethical and moral issues, which now impose themselves on the agricultural sciences as we attempt to address future challenges—issues related to biotechnology, organic agriculture, animal agriculture, etc.

Certainly our agricultural scientists are addressing future challenges facing agriculture, but they have rarely addressed Lubchenco's larger question. Most of us involved in sustainable agriculture, for example, still concentrate most of our efforts on "tweaking the edges" of existing systems, assuming, it seems, that the world itself will remain much the same. Rarely do we contemplate whether or not the production models that have been so successful in increasing the production of agricultural commodities in the past century can continue, or if radically new models need to be developed. As Zimdahl points out, this raises a host of ethical and moral questions. A standard phrase that characterizes much of today's industrial agriculture and which agricultural scientists often pose is, "How are we going to feed nine billion people?" Posing the question this way, of course, suggests a moral imperative and proposes a largely unquestioned strategy by which we must fulfill our responsibilities, as agriculturalists, to future generations. It assumes that feeding a growing human population is our only challenge and that we have a mandate to simply intensify the agriculture, which has been so successful over the past century —continue to develop new technologies, to increase the yields of a few commodities, and to continue "feeding the world."

On the other hand it may be that the real question we should be asking is "how are we going to restore and maintain the capacity for self-renewal of our entire biotic community (Leopold, 1949) so that we have healthy ecosystems to sustain productivity, and, what kind of food systems should we design to function effectively in such a resilient ecological environment?"

In this regard, anthropologist Schusky (1989) has raised some interesting questions that most agricultural scientists are failing to address today. As an anthropologist, Schusky asks the question of sustainability from the perspective of the long trajectory of human history. How have we fed ourselves successfully as a human species? At first, we, of course, fed ourselves as hunter gatherers, and he argues that from an energy perspective it was the most efficient food system we ever had since it required so little labor. Then about 10 thousand years ago we introduced agriculture. The first period of agriculture was the Neolithic Revolution and while it provided many benefits it required much more labor and probably provided a "less nutritious diet than food collecting and would lead to the first famines." It also enabled the beginning of "a population explosion." Then we introduced a second period of agriculture which Schusky calls the "neocaloric" period, which began in the mid-twentieth century. He calls it the neocaloric era because it is entirely based on old calories—fossil energy and all of the other old, stored, concentrated calories that made modern, industrial agriculture possible. Schusky goes on to argue that this Neocaloric era will of necessity be a very short period in the time-line of human history since we are rapidly using up all these old calories—fossil fuels, fossil water, rock phosphate, potassium, rich deposits of top soil, etc. And once these old calories are gone the neocaloric era must end. So, the question facing us now is, what kind of food system will we design for the next era of feeding ourselves as a human species?

Consequently, a key moral question facing us as a species is how are we going to maintain the capacity to feed the human population in the light of these future challenges? Without the "old calories" that served as our primary natural resource for the neocaloric era, what new designs can we imagine?

Fortunately we have an emerging group of scientists that are exploring some of the options. For example, Wes Jackson and the scientists at the Land Institute are demonstrating that we can replace annual crop plants with perennials that are much more energy efficient, require less irrigation and fertilizers, sequester more carbon, require less labor, and provide many other services that begin to address some of the challenges in our new future. Perfecto et al. (2010) imagine a radically different landscape for our agriculture of the future, one based more on agroecological models that use the science of adaptive cycles to design social, economic as well as ecological resilience into our agriculture future. Resilience thinkers, who are taking the lessons learned from the science of ecology, determine how landscapes and communities can be designed to absorb some of the disturbances we are likely to encounter as we move into the post-neocaloric era, and still maintain essential functions. They suggest how some of those insights may apply to agriculture (Walker and Salt, 2006).

Of course, once we begin to address these larger design issues we will have to ask far more questions than the simplistic "How are we going to feed nine billion people?" We will need to begin asking "what is the appropriate number of human species within the context of the larger biotic community to achieve sustainability"? Leopold, among others, has reminded us that the health of the entire biotic community is essential to our own survival, and suggested how we might maintain such a healthy "biotic pyramid" (1949). We will also have to ask how we can redesign a food system that wastes a lot less food. As many food system analysts have pointed out we are already producing enough food to feed 9 billion people, the problem is that we are wasting nearly half of it. And there is no compelling evidence to suggest that we can maintain the current system.

Consequently, as Zimdahl points out we have to begin addressing additional, far more complex questions than the ones our agricultural scientists have grown used to. We need to begin addressing all of the interrelated challenges that are on our doorstep, and begin devoting much more of our incredible brain power to research and education that begins to come up with much more complex and nuanced answers (answers that take deep ethical and moral problems into consideration) than the isolated "science-based" strategies that have occupied most of our scientists in recent decades. In this regard, the holistic manner in which Crib (2010), a science writer from Australia, addresses our future challenges, provides a nice complement to the challenging questions Zimdahl poses for us in his new edition of *Agriculture's Ethical Horizon*.

**Frederick Kirschenmann**
Distinguished Fellow
The Leopold Institute, Iowa State University
Ames, IA

## References

Cribb, J. (2010). *The Coming Famine: The Global Food Crisis and what we can do to avoid it*. Berkeley, CA, University of California Press.

Leopold, A. (1949). *A Sand County Almanac*. New York, Oxford University Press.

Lubchenco, J. (1998). Entering the century of the environment: A new contract for science. *Science* 279:491–497.

Perfecto, I., J. Vandermeer, and A. Wright. (2010). *Nature's Matrix: Linking Agriculture, Conservation and Food Sovereignty*. Washington, DC, Earthscan.

Schusky, E.L. (1989). *Culture and Agriculture: An Ecological Introduction to Traditional and Modern Farming Systems*. New York, Bergin and Garvey Publishers.

Walker, B. and D. Salt. (2006). *Resilience Thinking*. Washington, DC, Island Press.

# Preface

*My conduct this day, I expect, will give the finishing blow to my once great and now too diminished popularity . . . But thinking as I do on the subject of debate, silence would be guilt.*

***John Dickinson of Pennsylvania,***
***speaking against independence on July 1, 1776***
***during the Continental Congress in Philadelphia.***
***Cited in McCullough, D. (2001)***

This book is the result of the development of a career in agriculture. It is neither a Homeric story of great, heroic deeds nor a story that will change civilization or be told over and over again. It is a small story of the life of one mind that began in one agricultural subdiscipline, weed science, and ended elsewhere.

After completing a Master of Science degree at Cornell University in 1966 and a doctorate at Oregon State University in 1968, I arrived in Fort Collins, Colorado to begin a new life as an Assistant Professor of Botany and Plant Pathology at Colorado State University. The job required teaching a class, the Biology and Control of Weeds, and doing research on soil persistence of herbicides and weed control in agronomic crops. It was the long-desired opportunity and I was confident that I was ready to take full advantage of it.

In the beginning of my university career, life and work resembled a mobile my wife gave me some years ago.[1] It hangs in my home study and consists of a black paper circle and three porpoises made from red construction paper; each with a sharply contrasting black eye. Each porpoise hangs from a string at the end of a slim metal wire and they move alone or in unison, with frail elegance, grace, and beauty. I walked into my study one morning expecting to admire the porpoise's elegant, floating grace and they were gone. The supporting stick, fastened so carefully, had come loose and the mobile had fallen to the floor. The frail elegance was no more. As I reflect on my weed science career, its direction, and on what I thought and knew as fact when I began, I know my career has resembled my mobile.

In 1968, and for some years after, my life was fascinating and everything moved forward in order and harmony. I knew the Vietnam Tet offensive occurred on January 30, Martin Luther King was assassinated on April 4, Robert F. Kennedy was assassinated on June 5, and Neil Armstrong and Buzz Aldrin walked on the moon on July 20. The

[1] I am indebted to Dr. D. A. Crosby, Professor Emeritus of Philosophy, Colorado State University for this metaphor.

oldest building on the Colorado State University campus was destroyed by an arsonist on May 8, 1970. But these events, while very important, didn't affect me, my family, or my new career. Then the stories and facts about the use of the herbicide 2,4,5-T during the Vietnam war intervened. My career's supports began to loosen. I began to doubt if what I knew to be the foundational facts and the supporting myths of my science were adequate. It was, in a very real way, a crisis of faith; a crisis of faith in science.

In 1964, a study initiated by the National Cancer Institute suggested concern about the public safety of 2,4,5-T, an important herbicide for control of woody brush on rangeland and for weed control in forests. By 1950, 4.5 million kilograms (9.9 million pounds) of 2,4-D and 2,4,5-T were being applied annually in the United States (Wildavsky, 1995, p. 82). The National Cancer Institute study revealed the possibility that 2,4,5-T or one of its formulation's constituents might be a teratogen. Other allegations appeared over the next several years, many because an ester form of 2,4,5-T was half of Agent Orange, a defoliant used in Vietnam. By 1970, there was enough toxicological evidence to halt military use of 2,4,5-T and for the U.S. Environmental Protection Agency (EPA) to initiate administrative proceedings to suspend its registration. Throughout the 1970s, increasing attention was given to the dioxin contaminant in 2,4,5-T. Extensive studies confirmed that a dioxin[2] was the teratogen in 2,4,5-T. In 1979, following a still controversial study of human miscarriages after 2,4,5-T had been used in forests in the Alsea basin of Oregon, the EPA issued an emergency suspension of all uses of 2,4,5-T for forestry, rights-of-way, and pastures. Public sentiment against the herbicide grew in the 1970s. The manufacturers and EPA attempted to negotiate settlements to keep some uses, but discussions broke down in 1983 and the EPA canceled all U.S. uses in 1985.

In 1971, I presented a volunteer paper titled Human Experiments in Teratogenicity, in the ecology section of the Weed Science Society of America meeting in Dallas, Texas. The philosophical supports of my elegant, ordered, satisfying professional life began to crumble after that paper. The major objective of the paper was to question the role weed scientists played and ought to play in an increasingly polluted world. I was troubled and asked my colleagues to help me think about under what conditions it is possible to say that any pesticide is so necessary to our food production system that any risk of human harm is acceptable. The paper suggested pesticides were means to the desirable end of food production. I proposed that those who work with pesticides must ask and answer questions about whether means and ends are compatible. The paper argued that members of society must feel they are participants in determining the way things are ordered. They must think they have and actually have the power to choose. To make the sense of choosing and participation real, people must have the evidence required to judge possible alternatives. People must also have, beyond the evidence, a sense of general purpose that serves as a context into which particular judgments are fitted.

The room was partially full for my paper. The normally perfunctory applause was minimal. A group of senior colleagues spoke to me after the paper to tell me how

[2] There are several dioxins. The dominant teratogenic molecule in 2,4,5-T was 2,3,7,8-tetrachlorodibenzo-p-dioxin (TCDD).

wrong I was. The essence of the rather unpleasant encounter was that they wanted to know why I was so eager to bite the hand that fed me and much of the rest of the world. Their comments assured me that something was wrong, but it was something that was wrong with me and my thinking. In my colleagues' view, there was nothing wrong with agriculture, weed science, or with herbicides. They believed that weed scientists should continue the scientifically responsible quest for wise use of federally approved herbicides. I knew something was wrong but wasn't able to define it well, and I was beginning to doubt that the unquestioned development of herbicides for agriculture was *a priori* good.

A 1972 paper (Zimdahl, 1972) elaborated the oral presentation and continued the quest to decide what I thought and see if anyone cared. The issues didn't go away. I continued to read and think and tried to learn more about the issues when I wasn't doing the teaching and research my job required. A second paper (Zimdahl, 1978) was published later in the same journal. It included two fundamental propositions.

1. Some species are pests and it is necessary to control their populations to produce food.
2. Pesticides are the primary means to control pests but there may be an unnecessary dependence on them.

The paper argued that special knowledge and the highly trained mind produce their own limitations. They tend to breed an inability to accept views from outside the discipline, usually owing to a deep preoccupation with the discipline's conclusions.

After doing weed science research and teaching for 20 years and making another attempt to clarify my thoughts (Zimdahl, 1991), it was time to reflect on what had been learned and plan my future. This led to study and thought about the values and ethics of agriculture and particularly of weed science. It required learning how to do things I didn't know how to do; an old dog that wanted to learn new things. Exploring the ethical foundation of a science that had been my professional life was what I wanted to do.

Such decisions and changes, especially radical changes, don't come without costs. The costs have been personal and financial. It had taken a long time to realize that solving weed problems was a very important task, but not the one I wanted to spend the rest of my career working on. The personal cost included loss of colleagues and friends who didn't understand my quest and assumed the worst. In the minds of many, I *was* biting the hand that fed me. The costs also included the financial and intellectual difficulty of venturing into philosophy—a new, unknown area. The financial cost was just that; ethical reflection does not provide opportunities for research grants offered by the study of the kinetics of herbicide degradation in soil. Learning how to reflect on the history and values of agriculture and weed science has been a difficult challenge.

Freeman Dyson (1988) reflected on physics, his discipline, and said it was "passing through a phase of exuberant freedom, a phase of passionate prodigality." Weed scientists have always been exuberant and prodigious in their scope of work and ambition for the future. During weed science meetings, others, especially the young, revel in their exuberant freedom. They walk confidently along trails that, to me, are almost invisible, discussing ideas I don't even have the vocabulary to discuss. They are not wandering aimlessly. They are explorers, mapping the territory, finding the way to a new weed science with new scientific terminology and, perhaps, a new territory. However, this book suggests that these capable explorers lack an understanding of the ethical foundation of agriculture and of weed science. Indeed, in my opinion, they often lack a concern for discovering the ethical foundation of their work.

I often feel like a foreigner at weed science meetings even though I share a language with my colleagues. As I overhear conversations or listen to papers, I often feel as I have felt when lost in a foreign city. I don't know the language and can't find the way home. When wandering the streets in a foreign city, I often don't know the best route to my destination. I frequently can't understand the signs or the spoken words. I don't know the language. There are people all around, speaking, laughing, eating, trading ideas—the stuff of life. There is an intricate profusion of activity, and, as I walk, the surrounding activity seems to grow in complexity and abundance and my confusion, my sense of direction worsens, at the same, or at a faster, pace. Those who are native to the place are, of course, at home. They understand and only I am confused.

I am puzzled by the new directions of agricultural science and seek direction and confidence as I explore the ways of ethics, a route I have learned, but one my colleagues have not trod. Predictions about the future of agricultural science by scientists say that it is good, essential, and going to get better. When I was a student, I don't recall hearing the word sustainable, and the environment was acknowledged but not endangered. Weeds invaded crops but they were not invasive species, a new and growing area of study. Genetic engineering of species was unknown. All of these are now powerful ideas with powerful constituencies and they are changing agriculture's direction and its foundational ideas. Hopes will be dashed, new elites and new ideas will consolidate power and privilege and frustrate the dreams of others. Dreams of more equitable and just societies may be at risk. It is wise to remember, as we change, that agriculture and the technology of its sub-disciplines can affect and be affected by the development, direction, and future of the greater society.

The central norm, the primary moral stance of agronomy, plant pathology, weed science, entomology, indeed of agricultural science is that the scientific research that is done should benefit humanity by aiding the production of food and fiber. Agriculture and its technological disciplines are primary moving forces behind many social changes. Agriculture is one of the few production activities that takes pride in and seeks public adulation for reducing its labor force and weed science has been a major contributor. What becomes of the people displaced is someone else's problem.

My quest, what I have learned in a new language, albeit slowly, is to think about and then try to engage students and my colleagues in thought about what constitutes an appropriate ethical basis for making judgments about value differences in agriculture. I have learned that among my agricultural colleagues, "concern about moral questions is often relegated to the realm of private anxiety, as if it would be awkward or embarrassing to make it public" (Bellah et al., 1985, p. vi). I have also learned that we must take the risk of appearing awkward and being embarrassed as we discuss:

- What are the goals of agricultural science?
- What should the goals of agricultural science be?

- What should the goals of weed science be?
- How do and how should the practitioners of agriculture address complex ethical questions?

The aim of my ethical quest and the goal of the learning process is not what many have assumed. Many think what I want is to tell them that they are ethically wrong because they have no ethical foundation for their work. They are wrong. It is not a matter of sorting things out to a final, definitive truth that I understand and others do not. The aim is to construct a more harmonious and mutually acceptable view from which to address existing and future value conflicts. The aim is not unanimity, but a functional, mutually respectful plurality; not a solo, a practicing chorus. And, as is true for all good choruses, the practice must continue. Discussion of foundational values, of why we practice agriculture as we do should become a central rather than peripheral or absent part of agricultural practice and education.

One of the important things I have learned is that the persistence of moral conflict, of value questions, is an inevitable and important part of the human condition. Engaging in the ensuing debate stimulates the full development of the intellect and of humanity. Such discussions normally occur in a time of political and cultural imponderables. Calm discussion and rational thought may be impeded by irrational anger. Dyson (1988, p. 6) said that calm discussion is akin to holding a small candle in a hurricane to see if there are any paths ahead where people who share goals can walk together, while thinking about and planning their future. A fear, perhaps a fact, is that if agricultural scientists do not venture forth to understand and shape the ethical base of the future, it will just happen or be imposed by others.

The *primary* purpose of this book is to continue the discussion of agricultural ethics begun by others (Blatz, 1991; Lehman, 1995; Mepham, 1998; Thompson, 1995, 1998; Thompson et al., 1994) each of whom will be cited as we proceed. The 2004 compilation of manuscripts produced from a series of articles that appeared in *Plant Physiology* from 2001 to 2003 is important (Anonymous, 2004). The goal of this book is to explore agriculture's ethical horizon; the boundary line that separates and delineates one's outlook and knowledge. There will be special focus on weed science. The perpetuation and improvement of agriculture and weed science is my goal:

> *Never will it teach us all we need to know. Never will it provide us with final answers, and since none exist, then science's weakness becomes science's strength. Never will it cease its controversies, and that too is just as well if truth, like infinity, is to be eternally sought, though never captured. So it is that I must prefer the informed to the convinced, the demonstrated to the revealed, the observed to the imagined the probable to the impossible, the unalterable fact to the evanescent wish, the reasoned conclusion—however offensive—to the unquestioned assumption—however pleasing.*
>
> *Ardrey, R. (1976)*
> *The Hunting Hypothesis, p. 71.*

Ardrey's is a plea for reasoned conclusions. It is reason—the ability to think, form judgments, draw conclusions coherently and logically—that guides one in the ethical realm. It is reason, which is not the moral equivalent of scientific facts or evanescent wishes about the way things ought to be that will be the most reliable guide to the future.

In this day when my university emphasizes strongly that the ability to attract external support will be a major factor in hiring decisions for new faculty, the role of reason, central to ethics, seems to be losing its primacy to market interests. The role of higher education institutions now seems to be shaped almost exclusively by the wants of students seeking educational credentials, and businesses and government agencies seeking research outcomes (Zemsky, 2003). "When market interests totally dominate colleges and universities, their role as public agencies significantly diminishes—as does their capacity to provide venues for the testing of new ideas and agendas for public action" (Zemsky, 2003). We may be losing the understanding that as good and as powerful as our science is, knowledge has more than just instrumental value (Zemsky, 2003). What is lost is the ability to recognize that "what we should know, pretend that we know, and wish that we knew, we don't. Worse still, we do not know, without risk of embarrassment, how to ask about what we need to know" (Gomes, 1996, p. 4). Ideas and values are important whether or not they have marketability or confer personal advantage. Universities are the traditional and best places to generate ideas and discuss values. They are the places to begin to ask about what we need to know. One of the things those engaged in agriculture continually need to explore is how to ask and respond to the many ethical challenges agriculture faces.

This book is written by one whose career has been in agriculture. It is not a philosophical text, although it uses philosophical language and concepts and focuses on ethics. The book does not attempt to deal with all of agriculture's scientific and ethical challenges. Several important topics have not been discussed: agriculture and world population growth, urbanization and loss of agricultural land, and the implications of limited water supply for future food production. Foreign food aid is an important ethical question that affects agriculture, but because of its complex policy and political dimensions, it has not been included (see Aiken and LaFollette, 1996; Lappé, et al., 1981). The ethical questions raised by the kinds of food we eat and those who provide it are well covered elsewhere (e.g., Cook, 2004; Pollan, 2006, 2008; Roberts, 2008; Schlosser, 2002) and have been omitted.

The first chapter explores the differences between scientific and experiential truth and discusses why that difference is important to thought about the ethics of agriculture. *Chapter 2* asks what kind of agricultural research ought to be done and questions the past orientation of agricultural science, which has tended to keep the domain of scientific inquiry dissociated from the rest of the world and from human experience. This has resulted in affirmation of the incorrect thought that science is value free. Because of the frequent negative and well-known effects of agricultural technology and the fact that agriculture is the single largest and most ubiquitous human interaction with the environment, its science is feared by many, and disparaged by some. Therefore, the chapter concludes with advocacy of several appropriate changes in agriculture. *Chapter 3* explores the problems that have developed due to agricultural technology and examines some of the responses when things go wrong and bad results occur. The discussion questions the dominant view that whatever problems technology creates will be solved by new technology. The importance of recognizing the limits of scientific knowledge and the need to adopt a broader integrative approach to managing complexity, one that allows for surprises and values ethical considerations is emphasized. This is especially important in societal debates about whether or not to interpret evidence in a precautionary framework that seeks to minimize false judgments that no hazard exists when in fact one does. *Chapter 4* is an introduction to ethics intended for nonphilosophers. It is not intended to be an in-depth exploration. The chapter claims that ethics is regarded by agricultural scientists as something that is peripheral to the conduct of their science, rather than central. It is regarded as something that is purely academic. It is not. References are provided for those who want to explore ethics in greater depth. *Chapter 5* suggests that those engaged in agriculture possess a definite but unexamined moral confidence or certainty about the correctness of what they do. The chapter examines the origins of that confidence and questions its continued validity. The chapter argues that the basis of the moral confidence is not obvious to those who possess it or to the public. In fact, the moral confidence that pervades agricultural practice is potentially harmful because it is unexamined. Suggestions for re-moralizing agriculture are made to approach the questions of where moral values originate and what are or ought to be the moral standards for agriculture in our post-industrial, information age society. The chapter advocates analysis of what it is about our agriculture and our society that thwarts or limits our aspirations for agriculture and needs modification. *Chapter 6* attempts to establish the relevance of ethics to modern agriculture. It is directed at agricultural people not professional philosophers. The agricultural relevance of five moral theories is discussed and utilitarianism is proposed as the theory that dominates agricultural thought, although it is largely unexamined within agriculture. *Chapter 7* discusses sustainability, including its definition, the moral case for achieving sustainability, and why it must be achieved. *Chapter 8* tackles the still evolving, controversial topic of agricultural biotechnology. The chapter outlines the debate and describes the current regulatory situation here and elsewhere. Arguments in favor of and opposed to biotechnology are presented followed by some moral arguments. The potential effects on family farms, academic–industry relationships, and transgenic pharming are discussed. Animal ethics (*Chapter 9*) was not covered in the first edition. It is one of the most carefully explored areas of agricultural ethics. Significant changes have been made in the raising, housing, and slaughter of animals raised for human food. The issues have been discussed well by others (e.g., Cavalieri, 2001; Rollin, 1995; Singer, 1975; Thu and Durrenberger, 1998; Varner, 1998). Their work will be summarized in this edition. *Chapter 10* will present the ethical issues unique to alternative/organic agriculture and how these may affect its evolution. The *final chapter (11)* asks how should we proceed? It begins with a discussion of seven agricultural problems and how agricultural/scientific mythology interacts with them. The chapter concludes with presentation of the imperative of responsibility and the task of finding partners as we proceed into the future.

A *secondary* but important purpose of the book is to seek the counsel of agricultural colleagues who care to offer comments and guidance. I ask, *Quid vobis videtur?*—How does it seem to you? It was in 1971 when I began to question if the foundational facts and supporting myths of my science were adequate. The challenge that confronted me then and with which I still struggle, as I write this edition, is the ethical paradox described by Steinbeck (2001). He said:

> *There is a strange duality in the human which makes for an ethical paradox. We have definition of good qualities and of bad; not changing things, but generally considered good and bad throughout the ages and throughout the species. Of the good, we think always of wisdom, tolerance, kindliness, generosity, humility; and the qualities of cruelty, greed, self-interest, graspingness, and rapacity are universally considered undesirable. And yet in our structure of society, the so-called and considered good qualities are invariable concomitants of failure, while the bad ones are the cornerstones of success. A man—a viewing-point man—while he will love the abstract good qualities and detest the abstract bad, will nevertheless envy and admire the person who through possessing the bad qualities has succeeded economically and socially, and will hold in contempt that person whose good qualities have caused failure. When such a viewing-point man thinks of Jesus or St. Augustine or Socrates he regards them with love because they are symbols of the good he admires, and he hates the symbols of the bad. But actually he would rather be successful than good.*

A reasonable conclusion is that we can be accused of being moral hypocrites. In agriculture, we are confident that we are acting virtuously even when what we do is something we would condemn in others. Therefore, as we explore agriculture's ethical horizon, it is my hope that it will facilitate navigation through complex issues, including thought about moral hypocrisy, and that the exploration will serve as a guide to ways to construct common ground for resolution of agriculture's many ethical issues.

## References


Aiken, W. and H. LaFollette. (1996). *World Hunger and Morality*, 2nd Ed. Upper Saddle River, NJ, Prentice Hall.

Anonymous, (2004). *Agricultural Ethics in a Changing World.* Rockville, MD, American Society of Plant Biologists.

Ardrey, R. (1976). *The Hunting Hypothesis: A Personal Conclusion Concerning the Evolutionary Nature of Man.* New York, Bantam Books.

Bellah, R.N., R. Madsen, W.M. Sullivan, A. Swidler, and S.M. Tipton. (1985). *Habits of the Heart: Individualism and Commitment in American Life.* New York, Harper & Row.

Blatz C.V. (ed.). *Ethics and Agriculture: An Anthology of Current Issues in World Context.* Moscow, ID, University of Idaho Press.

Cavalieri, P. (2001). *The Animal Question—Why Nonhuman Animals Deserve Human Rights.* New York, Oxford University Press.

Cook, C.D. (2004). *Diet For a Dead Planet: How the food industry is killing us.* New York, The New Press.

Dyson, F. (1988). *Infinite in All Directions.* New York, Harper and Row, Publishers.

Gomes, P.J. (1996). *The Good Book: Reading the Bible with mind and heart.* New York, W. Morrow and Co.

Lappé, F.M., J. Collins, and D. Kinley. (1981). *Aid as obstacle—Twenty Questions about our Foreign Aid and the Hungry.* San Francisco, CA, Institute for Food and Development Policy.

Lehman, H. (1995). *Rationality and Ethics in Agriculture.* Moscow, ID, University of Idaho Press.

McCullough, D. (2001). *John Adams.* New York, Simon & Schuster.

Mepham, B. (1998). Agricultural ethics. Encyclopedia of Applied Ethics. San Diego, CA, Academic Press, Pp. 95–110.

Pollan, M. (2008). *In Defense of Food: An Eaters Manifesto.* New York, The Penguin Press.

Pollan, M. (2006). *The Omnivore's Dilemma: A Natural History of Four Meals.* New York, The Penguin Press.

Roberts, P. (2008). *The End of Food.* New York, Houghton Mifflin Co.

Rollin, B.E. (1995). *Farm Animal Welfare—Social, Bioethical, and Research Issues.* Ames, IA, Iowa State University Press.

Schlosser, E. (2002). *Fast Food Nation—The Dark Side of the All-American Meal.* New York, Houghton Mifflin Co.

Singer, P. (1975). *Animal Liberation.* New York, Random House.

Steinbeck, J. (2001). *The Log from the Sea of Cortez.* London, Penguin Books.

Thompson, P.B. (1998). *Agricultural Ethics: Research, Teaching, and Public Policy.* Ames, IA, Iowa State University Press.

Thompson, P.B. (1995). *The Spirit of the Soil: Agriculture and Environmental Ethics.* New York, Routledge.

Thompson, P.B., R.J. Matthews, and E.O. van Ravenswaay. (1994). *Ethics, Public Policy, and Agriculture.* New York, Macmillan Publishing Co.

Thu K.M., E.P. Durrenberger (eds.). *Pigs, Profits, and Rural Communities.* Albany, NY, State University of New York Press.

Varner, G.E. (1998). *In Nature's Interests?—Interests, Animal Rights, and Environmental Ethics.* New York, Oxford University Press.

Wildavsky, A. (1995). *But is it True? A Citizens Guide to Environmental Health and Safety Issues.* Cambridge, MA, Harvard University Press.

Zemsky, R. (2003). Have we lost the 'public' in higher education? *The Chronicle of Higher Education* May 30, Pp. B7–B9.

Zimdahl, R.L. (1972). Pesticides—a value question. *Bull. Ent. Soc. Am.* June:109–110.

Zimdahl, R.L. (1978). The pesticide paradigm. *Bull. Ent. Soc. Am.* 24:357–360.

Zimdahl, R.L. (1991). *Weed Science—A Plea for Thought.* Washington, DC, USDA/CSRS.

# Acknowledgments—First Edition

Dr. James W. Boyd, Professor of Philosophy, Colorado State University, has been a friend and mentor for more than 30 years. He was the first to acquaint me with the intricacies of philosophy and the importance of continuing to ask difficult questions. I am indebted to him for showing me how to explore the ethics of weed science and agriculture and for leading me to new approaches to what could be done to solve problems of weed management in agriculture.

Mr. R. Lee Speer, Senior Lecturer, Department of Philosophy, Colorado State University, has been a friend and teaching companion. He taught me as we tried to teach students how to think about agriculture's ethical dilemmas. This book could not have been created without Lee's guidance and patience. He has been my co-teacher and I his student.

Dr. Thomas O. Holtzer, Professor of Entomology and Department Head, Department of Bioagricultural Sciences and Pest Management, Colorado State University, has made this work possible. He permitted and supported my change in scholarly emphasis from weed science to philosophy and agricultural ethics. His administrative support allowed me to learn a new field while significantly diminishing my routine research efforts in weed science.

Acknowledgments are, by custom, brief. To fully acknowledge the contributions of the three colleagues above, I would have to make them co-authors.

I thank those who reviewed all or parts of the manuscript and whose comments improved it. They include Cynthia S. Brown, Department of Bioagricultural Sciences and Pest Management, Colorado State University; Henry A. Cross, Professor Emeritus, Department of Psychology, Colorado State University; Ruth A. Hufbauer, Department of Bioagricultural Sciences and Pest Management, Colorado State University; Jack Paxton, Retired, Department of Crop and Soil Sciences, University of Illinois; Edward E. Schweizer, Retired, USDA/ARS, Fort Collins, CO. The insightful comments of five anonymous reviewers improved the manuscript significantly.

Special thanks are owed to David Belles, who provided regular, prompt, and accurate research assistance.

The continued support and love of my wife, Pam, have been essential to this effort.

# Acknowledgments—First Edition

[illegible]

[illegible]

[illegible]

[illegible]

[illegible]

[illegible]

# Acknowledgments—Second Edition

Many of the thoughts and arguments in both editions of this book were created and nurtured over many years in classes I have taught and others I have attended, during seminars, professional meetings, and in numerous conversations with colleagues. All of these took place over several years and with far too many people to list them all. I am indebted to all who have given me the privilege of sharing their thoughts, even when we knew we did not agree. I must offer my special appreciation to Dr. James W. Boyd, Professor Emeritus of Philosophy, Colorado State University, who remains a valued friend and philosophical mentor. Dr. Thomas O. Holtzer, Professor of Entomology and Head, Department of Bioagricultural Sciences and Pest Management, Colorado State University, has supported and advocated my work and provided office space and departmental administrative support. His careful reviews of portions of the manuscript have been especially helpful. Dr. Bernard E. Rollin's and Dr. Terry Engle's comments on Chapter 10 were especially helpful.

I express my gratitude to Ms. Maggie Hirko and Ms. Janet Dill, who have made my task more pleasant and easier by their courtesy, regular assistance, and tolerance of my (perhaps too frequent) requests for assistance.

My wife, Pamela J. Zimdahl, encouraged my writing and offered comments and criticism when she thought it was appropriate (it usually was).

# 1 The Horizon of Agricultural Ethics

*We should be on our guard not to overestimate science and scientific methods when it is a question of human problems; and we should not assume that experts are the only ones who have a right to express themselves on questions affecting the organization of society.*

**Albert Einstein**

There are many differences in the words used and in the understanding of their meanings as one moves from the scientific to the experiential realm; from the laboratory where life's processes are studied to the world where life is experienced. The words of scientific language are necessarily precise and understandable to other scientists, whereas the words of experiential language rarely have the same meaning to all.[1]

For example, a scientific description of the common synthetic organic herbicide (2,4-dichlorophenoxy)acetic acid (2,4-D) might use these words: 2,4-D is a herbicide composed of a benzene ring with chlorine atoms in the ortho (2) and para (4) positions. An acetic acid moiety is in the 1 position via a phenoxy (oxygen) link. The herbicide's mode of action is that of a persistent auxin whose concentration cannot be controlled by susceptible plants. Most broadleaf species (dicotyledons) are susceptible (i.e., their growth is severely reduced or they may die) and most grasses (monocotyledons) are not. Several formulations of 2,4-D are available and they can be used for selective weed control in wheat, barley, oats, rye, sorghum, and field and sweet corn. The molecular formula is $C_8H_6Cl_2O_3$ and the molecular weight of 2,4-D acid is 221.04. Epinastic symptoms in susceptible plants occur within a few days after application and absorption occurs through roots or shoots. Susceptible plants die within 3–5 weeks. It is translocated in the symplast and metabolism occurs slowly. 2,4-D has a field half-life ($t_{1/2}$) in soil of 10–12 days. It leaches in soil but rapid microbial degradation in soil and plant uptake prevents leaching below 6 in. in most soils. Volatility occurs for some ester formulations, but is typically negligible for acid, salt, and low volatility ester formulations.

Another example of the precision of scientific language is a description of the common simple perennial weed, dandelion. I suspect that all plant scientists and most homeowners know the common dandelion. The plant scientist (taxonomists in particular) properly calls it *Taraxacum* (the genus) *officinale* (the species) Weber in Wiggers (the authority). The authority is the name or designation of the person or persons given credit for unequivocally identifying and naming the species.

[1] I am indebted to my colleague Dr. J.W. Boyd, Professor Emeritus of Philosophy, Colorado State University, who guided me toward an understanding of the importance of language and models of truth.

**Agriculture's Ethical Horizon. DOI: 10.1016/B978-0-12-416043-9.00001-5**

Dandelion is a member of the lettuce tribe of the sunflower or *Asteraceae* family. It was introduced to the United States from Europe. It is a deep-rooted simple perennial that reproduces by seed and, if cut, asexually reproduces from its tap root. The plant has a bitter, milky latex in all parts. Leaves are all basal, 2–12 in. long, and are lightly pubescent especially beneath and on the midvein. They sometimes form a flattened rosette, and other times are more or less erect. They are oblong to spatulate and deeply and irregularly cut. Leaves are coarsely pinnatified, sinuate-dentate, and rarely subentire. The paired lobes or divisions are somewhat acute. The inflorescence is bright golden yellow to orange, 1–2 in. across, containing 150–200 ray florets. Involucral bracts are not glaucous but the outer ones are elongated and conspicuously reflexed. Each composite flower is borne on a hollow stalk, 2–18 in. tall. At maturity they form white, fluffy, seed-bearing blowballs, about 1½ in. in diameter. Achenes are gray to olive-brown, 1/8 in. long, ridged, oblong, bluntly muricate, and bear a silky white pappus. Dandelion is distributed throughout the world's temperate and tropical zones.

A child and most adults who read such descriptions of 2,4-D or dandelions would probably regard them as nearly incomprehensible. Both descriptions are correct statements of scientific truth. That is to say, both are rational, publicly verifiable, falsifiable, literally true, definitive, and specific. These characteristics describe the language of science. Rationality, based on or derived from experiment and observation, is a cornerstone of scientific language. Often the language is mathematically based and precise. The language and the truth it represents are publicly verifiable, a hallmark of good science. Scientific findings, the result of research, are published in open, accessible journals and can be verified or denied by others. The meaning of the words in research reports is precise, when one understands them, and that understanding is available to anyone with a glossary of terms, a dictionary, or the right textbook. The language is definitive in that the words define 2,4-D, but not any of several other herbicides, and dandelions, but not other common dicotyledonous plants.

The language of rationality is the ideal model of objective scientific truth. But what of the child who picks the pretty, yellow dandelion flower or blows the pieces (the pappus) from the gray-white puffball of the mature flower and watches them float away and settle on the ground? What of the adult who has heard of 2,4-D and may even have used it to kill dandelions in the lawn, but is concerned about it and all pesticides and their possible effects on human and environmental health? Scientifically rational language may speak to them, but usually does not address what they see and feel. The objective of science is to understand phenomena, not judge them (Pinker, 2008). The language of rationality, the model of scientific truth, is not adequate to describe the child's experience or the adult's attitude. The flower's attraction, its beauty, the fun one can have with it, and one's concern, perhaps fear, about a herbicide and its possible side effects, require a different language—the language of experiential truth.

The language of experiential truth is personal and subjective. It is purposefully vague, and because it is so personal, it is not subject to public verification. My granddaughter told me as I was using 2,4-D to kill dandelions in my lawn, that she thought dandelion flowers were really pretty. Words such as pretty, playful, concern, and possible effects are imprecise. The language of experiential truth is rich in meanings because it is nonliteral, symbolic, and dependent on the personal subjectivity of the speaker, which scientific truth wants to diminish, if not eliminate. Subjective, personal opinions are least worthy of consideration in a model of scientific truth, but have the highest importance in a model of experiential truth.

When my granddaughter picked and showed me "the pretty dandelion flower," I realized quickly that my rapport with her in the midst of the flowers (or were they just weeds?) would have been damaged by the scientific response—"Well, actually what you think is a flower is not a flower at all. It is a complex inflorescence composed of several ray florets, etc." My rational, precise, literal, publicly verifiable words would have fallen on deaf ears or on no ears at all as she wandered off to pick more pretty flowers. My relation to her is durable, but my relation with her at that moment would not have been improved. My focus on correct, scientific, exterior data would have clashed with her focus on her interior consciousness about dandelion flowers.

Among the models of how truth can be perceived, the scientific model is valued by the scientific community, all of whom also know experiential truth, but many of whom have not considered the differences, place, and value of each model of truth. The order of value in the scientific model is:

1. *Rational truth*: Can be defined mathematically, is publicly verifiable, literal, definitive, falsifiable, and precise.
2. *Relational truth*: Exterior data take precedence over one's interior consciousness of the relationship of one observation to another.
3. *Personal truth*: The realm of subjectivity is least worthy of being called scientific truth.

A model of experiential truth reverses the order and importance of the two models of truth.

1. *Personal truth*: The language is often vague, imprecise, nonliteral, symbolic, descriptive, and highly subjective. It speaks of what is most important; what has the highest value.
2. *Relational truth*: Interior consciousness determines what one sees and how it is described and valued. Exterior data concerning the relation of one observation to another are interpreted subjectively.
3. *Rational truth*: Is present, but has the lowest value as a determinant of what is true.

## Scientific Truth and Myth

Many citizens of the world's developed countries are very well connected to their work. One sees examples everywhere: Blackberrys, I-pads, Kindles, cell phones with Internet access and more "apps" than many can use or know how to use, watches that tell time and connect to e-mail, etc. Cell phones may indicate status, are fashion statements and cameras as well as links to the daily grind (Coleman, 2000). Those who possess these marvelous technological achievements assume they lead to greater efficiency, productivity, and perhaps even more importantly, greater happiness. Another view says that we are so connected that we never can be disconnected. Proximity and constant connection reduce the time available to disconnect. Such time is required to think and reflect and to see where we have been so we can determine where we ought to go (Coleman, 2000). Most agricultural scientists are well-connected models of efficiency and productivity. However, they are often so busy being productive that direction becomes secondary or lost. Gallopín et al. (2001) suggest that there is a growing feeling (not a scientific certainty) that in spite of the marvels of communication and the appearance of efficiency and productivity, agricultural science is not responding adequately to the challenges of our time. Many of those engaged in agriculture are aware of the critique. However, they operate within the usually unexamined (frequently because it is unknown), guiding myth, the paradigm, that increasing production and profit is the proper (perhaps the only) goal for agriculture. They adhere to the paradigm while the real world reveals new realities (Kirschenmann, 2010). People frequently are so committed to the old ideas that they learned represented the world as it is that they resist change, even though they know the world is changing. They illustrate the fallacy of misplaced concreteness. In Whitehead's (1997) terms, the fallacy occurs when an abstract belief, opinion, or concept (the old paradigm) about the way things are, is reified. That is, one treats something which is not a real thing but an idea as if it were a physical or concrete reality. As Kirschenmann points out, the trick is determining which idea(s) really reflect what is happening in the real world and distinguishing them from unrealistic ideas.

It takes effort for any group to become aware of its guiding myths and then to gain sufficient intellectual distance from the myths so they can be examined dispassionately. The difficulty is compounded by the fact that groups believe strongly in the value of the governing myth, even though it is generally unexamined and the fact that in science, admission of the existence of a guiding myth is so foreign that scientists dismiss discussion of such things because they are not scientific and inherently subjective. This view helps explain why agriculture's practitioners dismiss those who criticize agriculture and agricultural science because they do not understand their importance; the essentiality of agriculture's mission. Critics question the paradigm, the operative myth, which is to say that such discussions lie in the arena of personal as opposed to rational truth. Asking agricultural scientists to describe the myths that guide their science is like asking a fish to describe water. The myths (the guiding paradigm) for the scientists, like the water for the fish, are just there. It is the nature of a myth that those who hold it do not believe it to be a myth (Bronowski, 1977, p. 21). Myth and science are like first cousins who strongly resemble each other and passionately hate the resemblance (Alexie, 2003).

Agriculture's practitioners seem to be so preoccupied with the vision of the necessity, indeed the responsibility, of continuing to increase production so the world's people will be fed that they do not pause to reflect on means (Midgley, 2002, p. 36). To properly criticize alternative visions of agriculture's present and its future:

> *we need to compare those visions, to articulate them more clearly, to be aware of changes in them, to think them through so as to see what they commit us to. This is not itself scientific business, though of course scientists need to engage in it. It is necessarily philosophic business (whoever does it) because it involves analysing concepts and attending to the wider structures in which those concepts get their meaning*
>
> *(Midgley, 2002, p. 36)*

The philosophic process of analyzing concepts will lead toward a just and realistic balance among competing visions of agriculture's future. The process will include consideration and analysis of scientific and experiential truths. The scientific view will, of course, not be hostile to science, but the point of view that includes experiential truth should also not be regarded as hostile. It is potentially a wider point of view from which science and our scientific myths arise and that provides support for them. The purpose is to strive for rational analysis to achieve what Midgley (p. 37) calls "a just, a realistic balance among our various assumptions and ideals."

> *The scientific point of view is itself an abstraction from it. The scientific angle is the one from which we attend only to certain carefully selected abstractions which are meant to be the same for all observers. When we move away from that specialized angle to the wider, everyday point of view we are not 'being subjective' in the sense of being partial. Instead we are being objective—ie. realistic—about subjectivity, about the fact that we are sentient beings, for whom sentience is a central factor in the world and sets most of the problems that we have to deal with.*
>
> *(Midgley, 2002, p. 101)*

Agriculture and all its subdisciplines (e.g., soils, animals, breeding of plants and animals, economics, entomology, plant pathology, weed science, etc.) are guided by a core mythology—an arena of experiential truth, which I claim is usually unknown and unexamined. Such mythologies are not myths in the sense of lies or in the colloquial sense of a false tale, but imaginative visions or pictures that express a belief and appeal to the deepest needs of our nature (p. 200)—our need for myth (May, 1991). They are essential. In agriculture and in life, we cannot live without myths. A lack of myths would break our required links to the past; we would become uprooted from the past and from our own society. It is our myths that may or may not be founded on fact, that capture human imagination so powerfully. They are one way we order our and other's experiences. It is an essential way we use to order our world that is not exhaustive (Midgley, 2002, p. 101). It is best when considered with other views, other ways of ordering and interpreting the world.

Scientific truth, spoken in empirical language, refers to objective facts, whereas myth refers to experiential things, the quintessence of the human experience that gives meaning and significance to life (May, 1991, p. 26). When we examine our myths, we automatically move away from the realm of scientific truth, but that does not mean one dismisses scientific truth. The examination of guiding myths often compels questions that cannot be answered easily and may not be answerable. It is asking that is cathartic (p. 284).

Part of our knowledge about scientific agriculture includes some level of certainty about the ability of technology to continue to solve problems as it has in the past. Technology, the knack of so arranging the world that we do not experience it (p. 57), can tell us what it is possible to do and perhaps how to do something but not why. Technology deals with the "what" of human existence rather than the "why" and it is the latter for which we are famished (p. 57). There is no question that scientific agriculture has solved many production problems. Part of the prevailing mythology of agricultural science is that the problems that some identify as being caused by the science lie in the way the science and its associated technology are used and misused (what to do), but not with the scientific approach to problem definition or problem solving (why to do something) (Gallopín et al., 2001). No thoughtful agricultural scientist denies that soil erosion, soil salinization, pesticide resistance, pesticide presence in groundwater, and a host of other problems are real problems caused or exacerbated by agricultural technology. Few go on to the possibly cathartic question about "the existing rules of enquiry, and to what extent (and in which situations) the scientific rules themselves have to be modified, or even replaced" (Gallopín et al., 2001). That is, few go on to question the myth of the objectivity of the scientific method and how science is done. Science is criticized because of its use and misuse, but the model of scientific enquiry is not usually questioned. Gallopín et al. suggest it is necessary to consider modifying or replacing the fundamental rules of scientific inquiry in some situations, especially when it comes to study of agricultural sustainability, which requires integrating economic, social, cultural, political, and ecological factors. Sustainability (see Chapter 7) is not simply a scientific question.

Agricultural science has defined its domain as solving agricultural production problems. It is what scientists and technologists do. The world is a vast array of problems, many known and many unknown. The job of the scientist is to work on and solve the problems the world presents (Gallopín et al., 2001). In close association has been what Gallopín et al. call a strong "privileging of the intended purpose" of the scientific enterprise. That is the intended outcome, the desired solution is consistently seen as good and likely, while the unintended side effects are ignored or dismissed as externalities (Gallopín et al., 2001).[2] There may be inconvenient or undesirable effects but they are relegated to another domain and are not the responsibility of the scientists who developed the technology or those who apply it. For example, herbicides were not designed or intended to leach to groundwater and their presence there is unfortunate. But removing them or paying the costs created by their presence is not regarded as the responsibility of those who develop, study, apply, or benefit from the herbicides used to control weeds in crops. The problems are external to agricultural science, which strives to eliminate future problems but does not emphasize solving or apologizing for the problems created. The accepted view within weed science has been that the benefits of weed control and herbicides exceed the negative costs, including the externalities. The view is reinforced by economic analysis. Possible negative effects (soil, water, and air pollution; resistance; loss of biodiversity; poisoning or physical or mental impairment of humans and other species; etc.) are difficult, if not impossible, to evaluate by standard cost/benefit analysis because determining a monetary value is unavailable and some are priceless (see Ackerman and Heinzerling, 2004). That is, cost/benefit analysis does not clarify, it

[2] An externality is a cost that is not reflected in price, or more technically, a cost or benefit for which no market mechanism exists. In the accounting sense, it is a cost that a decision maker does not have to bear, or a benefit that cannot be captured. From a self-interested view, an externality is a secondary cost or benefit that does not affect the decision maker. It can also be viewed as a good or service whose price does not reflect the true social cost of its consumption.

can confuse. Simple cost/benefit analysis cannot and it is unacceptable to determine the value of a child's life, a fragile forest, or the view into the Grand Canyon. It can be especially inappropriate in developing countries (Atreya et al., 2011).

Tegtmeier and Duffy (2004) suggest, with adequate supporting data, the negative external cost effects of crop and livestock agriculture in the United States are between \$5.7 and \$16.9 billion each year. Crop production had negative effects between \$4,969 and \$16,561 million annually, while livestock's negative externalities were \$714 to \$739 million per year. Their work was based on 417 million US acres cropped in 2000.

Is a system that yields very high external costs one that should remain unexamined for its defects or means of change? Is the method of scientific inquiry that contributed to the production of these external costs above question? The obvious answer is no. The complexity of the problems faced by agriculture and agricultural science is clear to all involved in agriculture. It is not a simple enterprise. The approach and the answer to many of the questions agriculture faces require value judgments. Determining whether something is good or bad, right or wrong, decent or indecent is frequently complex and requires more than scientific truth. Such judgments are subjective and experiential and although they may be supported by reason, they are not totally dependent on scientific evidence. Scientific reasons alone are a poor guide to matters of value and judgment (Ehrenfeld, 1978, p. 223). Consensus about goodness may be reached, but it is not subject to proof or verification by science.

The problems of agriculture seem to multiply faster than the solutions. Gallopín et al. (2001) offer three reasons why things have become more complex. The *first* reason is ontological or human-induced changes in the nature of the real world. This is not just a twentieth-century concern (see Marsh, 1864; Turner et al., 1990). Humans are a new force of nature (Lubchenko, 1998) that modifies "physical, chemical, and biological systems in new ways, at faster rates, and over larger spatial scales than ever recorded on Earth." Humans stand in sharp contrast to all other species that must adapt to the environment. Man and nature have become separate. Man is master now, and it was meant to be so. Man's power yields dominion and the ability to subdue nature even though we strive to obtain goals that are ecologically unsound and unsustainable. Those engaged in agriculture can be justly accused of being moral hypocrites. Their actions yield results they would condemn in others (e.g., air pollution from cars, water and soil contamination from industrial sites, cruelty to animals, etc.). Nature and natural things are judged by what they can do for man, not by any value judgment about intrinsic natural patterns that control us and are affected by our actions. For example, carbon dioxide ($CO_2$) emitted from the fossil fuels burned (mostly in the north) combined with carbon dioxide produced by deforestation (mostly in the south) increased atmospheric $CO_2$ levels by about 20% over the pre-industrial background (Turner et al., 1990, p. 6). $CO_2$ and methane, whose atmospheric concentration has doubled since the mid-eighteenth century (Turner et al., 1990), have become primary drivers of global climate change. Data from Mauna Loa show that atmospheric, $CO_2$ concentration increased an average of 1.8 parts per million (ppm) per year from 1995 to 2009. Since 2001 the average annual increase has been 2 ppm. World $CO_2$ levels are the highest they have been in 650,000 years. In 2010 the atmospheric $CO_2$ level at the Mauna Loa Observatory was 387 ppm up almost 40% since the industrial revolution (Adam, 2008).

Soil erosion caused by human and natural activity continues. "The overwhelming impression is that transfer of materials is changing the face of the earth at a faster rate than that at which the world's population is growing" (Douglas, 1990). More atmospheric nitrogen is fixed by humans than by all natural terrestrial sources combined (Vitousek et al., 1997). The high productivity of modern agriculture is dependent on modifications of the Haber-Bosch process for synthesis of nitrogen fertilizer. More than one-half of all the nitrogen fertilizer used in all of human history has been used since 1990 (Clayton, 2004) and as much as half of that ends up in the atmosphere or local waterways releasing 2.1 billion tons of carbon dioxide equivalent as nitrous oxide (World Watch, 2008), a potent greenhouse gas. As much as 1.5 billion pounds of nitrogen fertilizer is applied, primarily to US corn fields each year. About 50–60% is used by the corn and the rest is free in the environment. This massive use has contributed to the growing hypoxic ($O_2$ concentration <2 mg/L), dead zone in bottom water on the Louisiana–Texas coast. The sediment load in the Missouri/Mississippi river basin is about 616 million tons (2,000 lb/t) (or approx. 550 million metric tons; 2,200 lb/t) annually. Much of the nitrogen that does not fertilize corn and other crops reaches the Gulf of Mexico in sediment and creates the growing hypoxic zone. The ultimate cause of hypoxia is excess nutrient loading from human activities which causes algal blooms. The algae sink to the bottom and use oxygen to decompose at a rate faster than it can be added back into the system by physical mixing. The lack of oxygen (anoxia) kills bottom-living organisms and creates dead zones.

The US area begins where the Mississippi River enters the Gulf of Mexico. In 2010 it extended east to Alabama and west to Galveston, Texas. The area in mid-2010, 7,722 $mi^2$, was 10% less than the area of Massachusetts (8,721 $mi^2$). Sewage effluent contributes but the primary cause is application (or over application) of nitrogen fertilizer in the Missouri/Mississippi river drainage basin. There is a growing consensus that corn grown for ethanol production in the United States exacerbates the problem due to high nitrogen fertilizer use and the substitution of corn for soybeans, which do not require nitrogen fertilizer. The combination of increasing corn acreage, nitrogen fertilizer use, the quest for ever-higher production, and government subsidies for ethanol production creates human-induced change (Goolsby et al., 2001; Rabalais et al., 2002). Mean annual nitrate N concentrations at St. Francisville, LA from 1980 to 1996 were more than double the average concentration from 1955 to 1970 (Goolsby et al., 2001). Hypoxia is not limited to the United States. It has spread rapidly in recent decades. There are at least 146 areas in the world (Postel, 2005, p. 23). The largest hypoxic dead zone is in the Baltic Sea (northern Europe). The Gulf of Mexico is 7.5 times larger than the Baltic but the hypoxic area in the Baltic is 14% larger. Hypoxia is "the most widespread anthropogenic induced deleterious effect in estuarine and marine environments" (Diaz, 2001). More than half of all accessible freshwater is used by humans (Postel et al., 1996); most to irrigate crops and hypoxia is a common outcome.

However, one must acknowledge that doubling food production by 2050 will require increasing nitrogen application 2–2.5 times, which will exacerbate its well-documented negative effects (Myers, 2009, p. 24).

The *second* reason agriculture problems multiply more rapidly than solutions (Gallopín et al., 2001) is epistemological change. Epistemology is the branch of philosophy concerned with the nature of knowledge. It is the study of the origin, nature, and limits of knowledge. Essentially it is study of the foundation of knowing. Gallopín et al. assert that our understanding of the world has changed because modern science has made us aware of the behavior of complex systems, especially of their unpredictability. Surprise is part of the world's reality at the microscopic and macroscopic level. Scientists are coming to understand that the mysteries of ecology, in all its grand complexity, are more important (albeit more difficult) science than economics (Midgley, 2002, p. 188). Agricultural economics has a role to play in measuring agriculture's future, but limiting definition of that role to economic analysis based on efficient use of resources is too limited because it ignores the human dimension of agriculture (Dundon, 2003). The focus of economic analysis is on developing a better society, but economics often limits the purview of better to price and profit. Madden (1991) suggests that focus must be expanded to "ethics and values far beyond those embodied in current market prices." This, of course, makes things more complex, less scientifically precise, and increases the significance of personal truth.

Much of what we need to know about agriculture is related to the behavior of complex ecological systems, about which we know little. Ecosystem services operating on generally unappreciated and unknown large and small scales are impeded by human activities and cannot be replaced by technological advances in agriculture as they have been in the past (Daily et al., 1997). The weed scientist who asks what herbicide will control weed X in crop Y is asking a good but incomplete question. It is a technical question that leads to ignoring or assuming that it is someone else's responsibility to ask questions such as:

- What happens to the herbicide after it is applied?
- What are the effects of attempts to remove the weed on the system?
- Are weeds an inevitable concomitant of agriculture or is the weed there because of the way agriculture is practiced?

All involved in agriculture are aware of the *third* reason for added complexity offered by Gallopín et al. (2001): changes in the nature of decision making. A more "participatory style of decision making" is gaining and "technocratic and authoritarian" decision making is less in favor. The ecocentric as opposed to technocentric view often prevails. Other decision criteria (gender, human rights, the environment) are gaining credibility as nongovernmental organizations (NGOs) and multinational corporations expand the dimensions that define issues and solutions. In general, while changes in the nature of decision making are known and often lamented in agriculture, that knowledge has not led to changes in agricultural practice. Change in the ways agriculture is practiced have been imposed from outside. It is reasonable to posit that changes resulting from environmental concern, gender issues, human rights, and animal rights have initially been resisted within agriculture.

Everyone is for agricultural sustainability (see Chapter 7). It has achieved the universally good status of God and motherhood. Even though all do not agree on what it is, there seems to be agreement that a sustainable agriculture must be economically successful. It also has to be ecologically, socially, culturally, and politically acceptable. Lubchenko (1998) said that the goal of obtaining a more sustainable biosphere means obtaining that which is ecologically sound, economically feasible, and socially just. She, as President of the American Association for the Advancement of Science, asked if the scientific enterprise that "had met these past challenges is prepared for the equally crucial and daunting challenges that lie in our immediate future." Her answer was, No, science is not prepared to meet future demands because "the real challenges facing us have not been fully appreciated nor properly acknowledged by the community of scientists whose responsibility it is, and will be, to meet them." Lubchenko firmly says that it is time for the "scientific community to take responsibility for the contributions required to address the environmental and social problems before us, problems that, with the best intentions in the world, we have nonetheless helped to create."

The agricultural community knows that our modern agricultural system is very productive but not always profitable for those who produce. It has been quite profitable for corporations that create and sell agricultural technology and for many large farms. More than 40 years ago, Berry (1970, p. 78) noted the condition of the American farmer in an era of unparalleled affluence and leisure. His view is still valid:

> *... the American farmer is harder pressed and harder worked than ever before; his margin of profit is small, his hours are long; his outlays for land and equipment and the expenses of maintenance and operation are growing rapidly greater; he cannot compete with industry for labor; he is being forced more and more to depend on the use of destructive chemicals and on the wasteful methods of haste and anxiety. As a class, farmers are one of the despised minorities. So far as I can see, farming is considered as marginal or incidental to the economy of the country, and farmers, when they are thought of at all, are thought of as hicks and yokels whose lives do not fit into the modern scene.*

The modern agricultural system created by the cooperative research of colleges of agriculture in the nation's land grant universities and by agribusiness companies has done at least seven things worthy of note. They are:

- Food and fiber production have increased,
- The long-term health of soil and groundwater has declined,
- Plant and animal genetic diversity have been reduced,
- The political and economic climate have reduced crop and livestock choice,
- The US diet favors animal over plant products,
- The creation of a capital, energy, and chemically intensive production system that to survive requires high production volume at low cost, and
- The system has driven small- and medium-sized farms out of business.

Many college of agriculture faculty members will claim that their work was not intended to create this kind of system and in fact did not create it. This may be true and, if it is, one must ask, what these faculty members were doing? Perhaps their work was irrelevant to the creation of the modern agricultural system the above characteristics describe. One cannot be sure. Therefore, we must ask, as Lubchenko (1998) did, if the challenges "facing us have not been fully appreciated nor properly acknowledged by the community of scientists whose responsibility it is, and will be, to meet them."

We must continually ask the cathartic questions. What should we do? What is the agricultural research task? What are the questions we ought to be asking? Maintenance of production and, presumably, profit have been the premier, perhaps the only, goal of agricultural research and of colleges of agriculture. Production has been maintained and even increased for most crops, grower profit has not, except for some large farms. We ought to explore whether this has been a proper and sufficient goal, and if it is the proper goal for the future.

Those engaged in agriculture must begin to examine and expand agriculture's ethical horizon. Most people think of a horizon as the apparent line where the sky meets the earth. A horizon can also be regarded as a limit or the extent of one's outlook, experience, interest, knowledge, etc. In the same sense as the earth–sky horizon, our intellectual horizon is what separates, divides, binds, and defines us. An intellectual horizon is the full range or widest limit of our perception, interest, appreciation, knowledge, and experience. It is the intellectual horizon that those engaged in agriculture must examine and it is a major purpose of this book to explore agriculture's intellectual horizon, particularly as our collective, yet unexamined, ethical position, may limit what agriculture's ethical horizon defines.

Lubchenko (1998) concludes with a Calvin and Hobbes cartoon (Watterson, 1992). Watterson, through Calvin and Hobbes, has been a perceptive commentator on our society. His observations apply to our agricultural and general scientific dilemma.

Calvin and Hobbes are careening through the woods in their red wagon.

*Calvin:* "It's true, Hobbes, ignorance is bliss!
Once you know things, you start seeing problems everywhere...
... and once you see problems, you feel like you ought to try to fix them...
... and fixing problems always seems to require personal change...
... and change means doing things that aren't fun!
I say phooey to that!"

Moving downhill, they begin to go faster.

*Calvin:* (looking back at Hobbes): "But if you're willfully stupid, you don't know any better, so you can keep doing whatever you like!
The secret to happiness is short-term, stupid self-interest!"

*Hobbes:* (looks concerned): "We're heading for that cliff!"

*Calvin:* (hands over his eyes): "I don't want to know about it."They fly off the cliff: "Waaaugghhh!"

After crash landing,

*Hobbes:* "I'm not sure I can stand so much bliss."

*Calvin:* "Careful! We don't want to learn anything from this."

Another comic strip we all know well is Peanuts by Charles Schulz, now presented as Classic Peanuts®.[3] The comics often comment succinctly and incisively on

[3] Peanuts, a creation of C.M. Schulz, is published by United Features.

a fundamental truth. It is wise counsel as we proceed to discuss agriculture's ethical horizon.

*Charlie Brown asks Lucy—"What are you looking for?"*

*Lucy:* "A tennis ball."
*Charlie:* "How did it get way out here?"
*Lucy:* "I threw it at Linus. He ducked and it flew into these weeds."
*Charlie:* "You know what?"
*Lucy:* "What?"
*Charlie:* "Perhaps this is the punishment you get for losing your temper."

Lucy then slugs Charlie and knocks him over.

*Charlie:* "I always moralize at the wrong time!"

In contrast to Calvin and Hobbes, we bear a responsibility to ask: What do we know and what must we learn from the agricultural experience and the limits of agriculture's ethical horizon? What are we responsible for that we can be proud of and what are we responsible for that we regret? We must learn how to ask as Eliot did:

*Where is the wisdom we have lost in knowledge?*
*Where is the knowledge we have lost in information?*
*Eliot, 1934 Choruses from "the Rock"*

Finally, Charlie Brown wisely tells us that there is a time to moralize and a time to be quiet.

## References


Ackerman, F. and L. Heinzerling (2004). *Priceless: On knowing the Price of Everything and the Value of Nothing*. New York, The New Press.

Adam, D. (2008). World carbon dioxide levels highest for 650,000 years. *The Guardian*, May 13. http:www.Guardian.co.uk/environment/2008 (accessed December 23, 2010).

Alexie, S. (2003). Washington: Coyote's unauthorized guide to Washington State. Pp. 454–463 *in* John Leonard (ed.). *These United States: Original Essays by Leading American Writers on Their State within the Union*. New York, Thunder's Mouth Press.

Atreya, K., B.K. Sitaula, F.H. Johnsen, and R.M. Bajracharya. (2011). Continuing Issues in the Limitations of Pesticide Use in Developing Countries. *J. Agric. Environ. Ethics* 24:49–62.

Berry, W. (1970). Think little. A Continuous Harmony. Harcourt Brace, New York & Co. Pp. 71–85.

Bronowski, J. (1977). *A Sense of the Future*. Cambridge, MA, MIT Press.

Clayton, M. (2004). 'Dead zones' threaten fisheries. *Christ. Sci. Monitor* May:13, 16.

Coleman, J. (2000). Is technology making us intimate strangers? *Newsweek* March:12.

Daily, G.C., S. Alexander, P.R. Ehrlich, L. Goulder, J. Lubchenko, P.A. Matson, H.A. Mooney, S. Postel, S.H. Schneider, D. Tilman, and G.M. Woodwell. (1997). Ecosystem services: benefits supplied to human societies by natural ecosystems. *Issues Ecol.* 2:2–15.

Diaz, R.J. (2001). Overview of hypoxia around the world. *J. Env. Qual.* 30:275–281.

Douglas, I. (1990). Sediment transfer and siltation. Pp. 215–234 *in* B.L. Turner, W.C. Clark, R.W. Kates, J.F. Richards, J.T. Mathews, W.B. Meyer (eds.). *The Earth as Transformed by Human Action: Global and Regional Changes in the Biosphere Over the Past 300 Years*. Cambridge, Cambridge University Press.

Dundon, S. (2003). Agricultural ethics and multifunctionality are unavoidable. *Plant Physiol.* 133:427–437 and Pp. 7–17 in M. J. Chrispeels, (ed.). *Agricultural Ethics in a Changing World*. Plant Physiology Society of America.

Ehrenfeld, D. (1978). *The Arrogance of Humanism*. New York, Oxford University Press.

Eliot, T.S. (1991). Collected poems 1909–1962. Centenary Edition. Harcourt Brace Jovanovich. New York. P. 221.

Gallopín, G.C., S. Funtowicz, M.O. Connor, and J. Ravetz. (2001). Science for the twenty first century: from social contract to the scientific core. *Int. J. Social Sci.* 168:219–229.

Goolsby, D.A., W.A. Battaglin, B.T. Aulenbach, and R.P. Hooper. (2001). Nitrogen input to the Gulf of Mexico. *J. Environ. Qual.* 30:329–336.

Kirschenmann, F. (2010). Anticipating changes. *Leopold Lett.* 22(2):5.

Lubchenco, J. (1998). Entering the century of the environment: a new social contract for science. *Science* 279:491–497.

Madden, P. (1991). Values, economics and agricultural research. Pp. 285–298 *in* C. Blatz (ed.). *Ethics and Agriculture*. Moscow, ID, University of Idaho Press.

Marsh, G.P. (1965 [1864]). *Man and Nature; Or, The Earth as Modified by Human Action*. Cambridge, MA, Belknap Press of Harvard University Press.

May, R. (1991). *The Cry for Myth*. New York, A Delta book, Dell Publishing.

Midgley, M. (2002). *Science and Poetry*. New York, Routledge.

Myers, S.S. (2009). Global environmental change: the threat to human health. Worldwatch report no. 181. Worldwatch Institute, Washington, DC.

Pinker, S. (2008). The moral instinct. *The New York Times*, 13 pp. (http://www.nytimes.com/2008/01/13/magazine/13Psychology-t.html?scp=1&sq=%22the+moral+instinct.) (accessed September 2009).

Postel, S. (2005). Liquid assets: The critical need to safeguard freshwater ecosystems. Worldwatch Paper 170. Worldwatch Institute. Washington, DC.

Postel, S.L., G.C. Daily, and P.R. Ehrlich. (1996). Human appropriation of renewable fresh water. *Science* 271:785–788.

Rabalais, N.N., R.E. Turner, and D. Scavia. (2002). Beyond science into policy: Gulf of Mexico hypoxia and the Mississippi river. *BioScience* 52(2):129–142.

Tegtmeier, E.M. and M.D. Duffy. (2004). External costs of agricultural production in the United States. *Int. J. Agric. Sustainability* 2(1):1–20.

Turner B.L., W.C. Clark, R.W. Kates, J.F. Richards, J.T. Mathews, W.B. Meyer (eds.). *The Earth as Transformed by Human Action: Global and Regional Changes in the Biosphere Over the Past 300 Years*. Cambridge, Cambridge University Press.

Vitousek, P.M., J.D. Aber, R.W. Howarth, G.E. Likens, P.A. Matson, D.W. Schindler, W.H. Schlesinger, and D.G. Tilman. (1997). Human alteration of the global nitrogen cycle: sources and consequences. *Ecol. Appl.* 7:337–750.

Watterson, B. (1992). Calvin and Hobbes. May 17. Distributed by Universal Press Syndicate.

Whitehead, A.N. (1997 [1925]). Science and the Modern World. New York, Free Press, Simon & Schuster.

World Watch (2008). Adjustments in agriculture may help mitigate global warming. May/June. P. 4.

# 2 The Conduct of Agricultural Science

> *The mistrust of science—the fear that people with an impenetrable language of their own are tinkering with things that are better left alone—has always run deep. In the best of cases scientists have responded to this fear head on, encouraging discourse, publicly exploring the limits and the unknowns.*
>
> ***Cohen, 2002***

All educated people know something about science, about what it does and what it is. Exactly what science is and what it does is not elusive, but it is not easy to find a single definition that satisfies all. Below are a few of the many definitions available.

> *Science among us is an invented cultural institution, an institution not present in all societies, and not one that may be counted upon to arise from human instinct. Science exists only within a tradition of constant experimental investigation of the natural world.*
>
> *Eiseley, 1973, pp. 18 & 19*

> *Science is the organization of knowledge in such a way that it commands more of the hidden potential of nature.*
>
> *Bronowski, 1956, p. 7*

> *Science is a method rather than a body of knowledge.*
>
> *Sagan, 1974, p. 82*

> *Science is that branch of pure learning which is concerned with the properties of the external world of nature.*
>
> *Campbell, 1952*

> *Science does not and cannot pretend to be true in any absolute sense. It does not and cannot pretend to be final. It is a tentative organization of mere working hypotheses.*
>
> *Campbell, 1972, p. 15*

> *Science is the attempt to make the chaotic diversity of our sense experience correspond to a logically uniform system of thought.*
>
> *Einstein, 1940*

> *Science is the systematic observation of natural events and conditions in order to discover facts about them and to formulate laws and principles based on these facts. It is the organized body of knowledge that is derived from such observations and that can be verified or tested by further investigation.*
>
> *Academic Press Dictionary of Science & Technology*

**Agriculture's Ethical Horizon. DOI: 10.1016/B978-0-12-416043-9.00002-7**

*Science involves more than gaining knowledge. It is a systematic, organized inquiry into the natural world and its phenomena. Science is about gaining a deeper and often useful understanding of the world.*

*The Multicultural History of Science, Vanderbilt University*

*Science alone of all the subjects contains within itself the lesson of the danger of belief in the infallibility of the greatest teachers in the preceding generation. As a matter of fact, I can also define science another way: Science is the belief in the ignorance of experts.*

*Feynman, 1999*

*Science: systematized knowledge derived from observation, study and experimentation carried on in order to determine the nature and principles of what is being studied. A branch of knowledge or study, especially one concerned with establishing and systematizing facts, principles, and methods, as by experiments and hypotheses.*

*Webster's New World Dictionary of the American Language. 2nd College Ed. 1970.*

For Lewis Thomas (1973), one of our age's most careful observers of science, especially medical science, the process was far more interesting than the definition:

*Scientists at work have the look of creatures following genetic instructions; they seem to be under the influence of a deeply placed human instinct. They are, despite their efforts at dignity, rather like young animals engaged in savage play. When they are near to an answer their hair stands on end, they sweat, they are awash in their own adrenalin. To grab the answer, and grab it first, is for them a more powerful drive than feeding or breeding or protecting themselves against the elements. It sometime looks like a solitary activity, but it is as much the opposite of solitary as human behavior can be. There is nothing so social, so communal, so interdependent. An active field of science is like an immense intellectual anthill; the individual almost vanishes into the mass of minds tumbling over each other, carrying information from place to place, passing it around at the speed of light. There is nothing to touch the spectacle. In the midst of what seems a collective derangement of minds in total disorder, with bits of information being scattered about, torn to shreds, disintegrated, reconstituted, engulfed, in a kind of activity that seems as random and agitated as that of bees in a disturbed part of the hive, there suddenly emerges, with the purity of a slow phrase of music, a single new piece of truth about nature. In short, it works. It is the most powerful and productive of the things human beings have learned to do together in many centuries, more effective than farming, or hunting and fishing, or building cathedrals, or making money.*

Thomas (1974) describes the planning and conduct of medical science and his description is, in many ways, precisely analogous to the agricultural experience. Agricultural scientists, farmers, and ranchers generally think of agriculture as an essential, immensely productive, successful enterprise. But one can argue that it is not successful, because while productivity has been maintained, profit has declined, and many small-scale producers have suffered and disappeared. Busch and Lacy (1986, p. 77) point out that "the core agricultural science community remains surprisingly homogenous, despite new entrants from divergent backgrounds." The core community represents "an impressive array of traditional American values: respect for independence, initiative, cooperation, hard work, usefulness, pragmatism, education, science, nature, age, experience, and tradition" (Busch and Lacy, 1986, p. 77). The agricultural community is a community of shared values that are rarely questioned from within. These values have enabled the community to withstand, perhaps even ignore, external questions about agricultural technology that has improved lives and worked in the service of all humanity by feeding people. This is true even though much of the technology has generated private profit and undesirable external effects, the costs of which are not borne by the developers or adopters of the technology (Busch and Lacy, 1986, p. 271; Tegtmeier and Duffy, 2004).

Within agriculture, the green revolution of the 1960s is generally regarded as a great success. New high-yielding varieties of wheat and rice increased yields around the world and more people were fed; massive starvation was avoided. A doubly green revolution has been proposed (Conway, 1997) to address the world's continuing problem of feeding a growing number of people. All have a demand for food and some are not fed neither because they do not have a need for food nor because too little food is produced. People are not fed because they have ineffective demand. They are poor, without money to buy food or land to produce it. Conway (1997, p. 42) suggests reversing the goals of the first green revolution. The doubly green revolution should begin with the socioeconomic demands of poor households and then seek to identify appropriate research priorities. The goal would not be to produce more food, although that is not a trivial goal, but to create food security and sustainable livelihoods for the poor. It is too soon to know if Conway's shift in emphasis has been accepted or if it will be effective. In developed countries, the public view, if agriculture is thought of at all, is that agriculture must be successful, because the grocery store is always full of food. That is not the view in the world's developing nations.

A more nuanced view expresses uncertainty about agriculture's past and future technology (e.g., pesticides, genetic modification, animal confinement), its results (loss of small- and medium-sized farms and rural communities, soil degradation and erosion, diminishing natural resources, groundwater mining, pesticide contamination of food and water, and continued hunger in the United States and the world), its opaqueness, and its complexity. An example of its complexity is the lack of a clear connection between the technological achievements of agricultural research and the producer's income. The price received by the farmer for producing basic food commodities (e.g., wheat, corn, soybeans) is a small and diminishing part of the cost of these commodities in the supermarket. Productivity increases have not increased producer's income. The farmer's income from the wheat in a box of Wheaties™ is minuscule compared to the income from the wheat received by General Mills. It is reasonable to claim that there is only limited evidence that agricultural research has contributed to solving human and environmental problems. There is no question that production increased. It is true that farmers feed most, albeit not all, of the world's 6.9 billion people (in mid-2010). In 1950 when the world had 2.5 billion people, most people, not just pessimists, found it difficult to conceive of the possibility of feeding 7 billion people. Now we face the fact, barring nuclear war or pandemic disease, that 7.9 billion must be fed in 2030 and 8.9 in 2050.[1] The grocery stores may be full but people are still hungry here and elsewhere. Developed-country agriculture is enormously productive but its techniques and their results compel questions about its goodness. The US Congress and the European Union approve large agricultural subsidies but "the rural home and rural life"[2] suffer. The productive, environmental, and social challenges to agriculture cannot be answered by science alone, but they should not be ignored by agricultural scientists.

Critiques of the perceived focus of agricultural science and modern agricultural practice are often superficial, one sided, and lacking in any obvious attempt to understand its complexity. However, it is that very complexity that fosters the critique. Some think that modern agriculture has become so complex that it tends to establish its own conditions, create its own environment and draw us, unknowingly, into it. Agriculture performs an essential function. It produces food and we cannot survive without our daily bread. But we do not live by bread alone. We all need a reason or several reasons for living. Humans want more than just biological existence once our needs for food, clothing, and shelter are met. We want a good life that enables us to realize our potential for human development. We want purpose, hope, and a way to make sense of the meaning of things. The fundamental evolutionary process is survival of the fittest, but far more basic is survival. We have an urge for survival (Cobb, 1993, p. 223). We want to live. Modern agriculture provides abundant food but works against these other things because man is not included in its design. Agriculture has moved from a position of terrifying ignorance and dependence to a place of knowledge and power (Morison, 1966). We are the master manipulators of the environment. We live where we want and eat what we want when we want it. But the modern, very successful agricultural system is beyond the comprehension of most citizens. They feel removed from the source of their food, which has become just another commodity. Do farmers grow food or just a commodity to sell? Many agriculturalists will say it is the latter and, if that is so, is it wrong to ask if people eat food or just another commodity they buy? Our technological triumphs in agriculture have produced an environment and a food supply we no longer trust.

Similar to Thomas' (1974) claim for medical research, in agriculture it is understood that specifically targeted research goals are required. The problem or problems to be solved must be identified clearly and it will be best if the problem can be addressed and solved in a 2- or 3-year funding period. It is good to have clear targets—well-defined problems that can be solved in a short time—so one can move on to other clear, well-defined problems. A critical aspect of this approach is that it assumes that basic biological and agricultural knowledge has "a much greater store of usable information, with coherence and connectedness, than actually exists" (Thomas, 1974, p. 116). The tendency in agricultural research has been to proceed without the basic knowledge and the approach has been quite successful.

[1] World population never doubled for anyone that died by 1930. Anyone born after 2050 may not live to experience the result of another doubling (Jackson, 2007).

[2] This is the language used in the Hatch Act, United States Code, Section 361b.

For example, in weed science most weeds (not all) can be controlled in most crops. Complete life histories of weeds are rare (they are generally just annuals, biennials, or perennials). Why a weed grows where it does is less important than how to control it. The fact that the empty ecological niche created by control will be filled by another weed, which keeps the cycle of control going, is not regarded as even a minor problem. Weed scientists have demonstrated over and over again that controlling one weed well, means that another will begin to dominate and it must be controlled, but they persist in applying the same technological control methods without ever really acknowledging the problems the approach creates. Technological application rules weed science, which continues to advocate applying herbicides to solve weed problems while expecting better results and fewer problems from each new herbicide. Thomas (1974) noted that early medical science was hoaxed by bleeding, cupping, and purging and, more recently, by overuse of antibiotics. All were used with the best of good intentions and all failed to some degree because of lack of understanding of the etiology of disease.

The great achievement of weed science and other pest control disciplines has been the development of diverse technologies to control weed/pest infestations in most crops. That ability in weed science began with the development of selective organic herbicides subsequent to World War II. It was not preceded by extensive, painstaking, laborious, demanding research on weed biology and ecology. This history stands in sharp contrast to medical research where years of work on disease etiology demonstrated [in Thomas' words (1974)] "bits of information being scattered about, torn to shreds, disintegrated, reconstituted, engulfed, in a kind of activity that seems random and agitated." All of this necessarily preceded the advent of antibiotics.

Weed science research proceeded with a limited basic biological and ecological foundation. The primary approach to solving today's important, unsolved weed problems (e.g., parasitic weeds, perennial weeds) remains application of the chemical technology that has been used successfully to solve annual weed problems. The environmental, human health, and nontarget species effects remain largely unaddressed, ignored, or externalized, and the crucial biological and ecological information to address these things is unavailable or appears only slowly. Medical science has demonstrated repeatedly that the direct, frontal approach to disease does not work.[3] Freedman (2010) claims that "much of what medical researchers conclude ... Is misleading, exaggerated, or flat-out wrong." He recognizes that "the scientific enterprise is probably the most fantastic achievement in human history." But he cautions that one should not assume a right to overstate accomplishments. Weed science illustrates the agricultural penchant for the direct frontal approach and continues to rely on it while extolling its accomplishments. Medical science (Freedman, 2010) and weed science are equally subject to "an intellectual conflict of interest that pressures researchers to find whatever it is that is most like to get them funded." Thus, we return to the reasonable conclusion mentioned in the preface. Those in agriculture can be accused

[3] An example is the tendency to favor pharmaceutical remedies over physical or psychological therapies that may be better solutions to chronic pain (See Wallis, 2005. The right and wrong way to treat pain. TIME Feb. 28, pp. 46–57).

of being moral hypocrites. They are confident they are acting virtuously and morally correctly even when what they do is something they would condemn in others.

Thomas (1974, p. 118) notes that the element of surprise is what marks the difference between applied and basic science. Agriculture, especially in the pest control disciplines, has been well organized to establish targets, apply knowledge, and produce a usable product. This procedure requires a high degree of certainty. The facts on which experimental protocols are based must be accurate with unambiguous meaning. Pesticide development is a good example. The work is planned and organized so the result (a product that works) will be obtained. There is a central authority, elaborate time schedules, and a reward system based on speed and perfection. These are development projects. They do not test and do experiments to explore a scientific hypothesis. Much pest control research has been of this kind.

Basic research is the opposite. It begins with a high degree of uncertainty, otherwise Thomas (1974) asserts, "it isn't likely to be an important problem." The initial facts are incomplete and ambiguous. The challenge may be to find the threads of connection between unrelated bits of information. Experiments are most often planned on probability rather than certainty and results are uncertain, surprise is common. If one did a survey of agricultural research over the last several decades, the data would show a strong emphasis on applied studies. This is an observation of fact, not a criticism. The results of the volume of applied work have been impressive. For example, as mentioned above, the world now feeds more people a better diet than ever before. In the mid-1960s about 1 billion people (50% of the population) in developing countries did not get enough to eat. Today less than 20% of the people in developing countries do not get enough to eat, but that is 1.4 billion people.

I agree with Thomas' (1974, p. 119) assessment of medical research and suggest his thoughts are directly applicable to agricultural research—"the majority of important research to be done is in the class of basic science. The mass of knowledge is formless, incomplete, lacking the essential threads of connection, displaying misleading signals at every turn, riddled with blind alleys" (Thomas, 1974, p. 189). If the assessment is correct for agricultural research, and if the orientation changes, the future will be difficult and exciting.

It is correct to posit that the primary emphasis of agricultural research for many decades has been to increase grower's productivity and profit. It is also reasonable to assume that agricultural scientists and farmers have other goals. Applied research has had a higher priority, a higher value, than basic research (Busch and Lacy, 1986, p. 15). The value is increasing production. They and most people want to create a good world,[4] a world that is just, peaceful, generally prosperous, democratic, free of prejudice of all kinds, and humane. It may also be other things. To create a good world, agriculture has specific responsibilities. These include:

- Achieving sustainable production practices.
- Decreasing pollution.

[4]The following definition of good is not intended to be precise or all inclusive. Good is an experiential term not a precise, scientifically definable one.

- Eliminating soil erosion.
- Eliminating harm to other plant and animal species.
- Ending habitat destruction (see Green et al., 2004).
- Ending the detrimental effect of agriculture on other species.
- Ending water pollution and mining water for irrigation.

There are other values one could include to create a good world (e.g., ending war, empowering women and minorities, creating more equitable social arrangements, etc.), but these are not things that fall directly within agriculture's purview. To begin to work on the problems that agriculture can work on, we must learn how. Maxwell (1992) argues that the academic research of which agriculture is a part, is "devoted by and large to improving knowledge and technological know-how"—which from "the standpoint of helping us create a good world, is grossly and damagingly irrational." What is produced merely increases our power to act without affecting our power "to act humanely, cooperatively, and rationally" (Maxwell, 1992). Modern technology that permits one person to do the work of a thousand also permits that person to wreak the environmental havoc of a thousand (Peterson, 1978).

New goals for agriculture require a new kind of inquiry, which pursues scientific truth but simultaneously has humane goals. It is an inquiry that is specifically value-laden. Its basic aim is to improve knowledge and "personal and global wisdom" (Maxwell, 1992). Wisdom in this context is "the capacity to realize what is of value in life, for oneself and others." Wisdom includes knowing how to do X, knowing why it is good to do X, and why it is good to do X rather than Y or Z.

The philosophy of knowledge as it has been inherited from Sir Francis Bacon, the scientific revolution of the sixteenth and seventeenth centuries, and the Enlightenment of the eighteenth century represents the unspoken scientific, intellectual creed adopted whole by agricultural scientists. It is a widely held ideal of science as it is practiced in the academic and industrial worlds (Maxwell, 1992). In its simplest form, one defines a problem, does the appropriate research, and arrives at a management solution that can be applied within a policy framework (Walton et al., 2002). The essential problem with this philosophy of knowledge is that the fundamental methodological prescription is one that keeps the domain of scientific inquiry dissociated from the rest of the world and from human experience (Maxwell, 1992). It results in affirmation of the incorrect thought that science is value free. Only scientific truth can enter the domain of inquiry. Politics, religion, values, emotions, and desires (the realm of experiential or personal truth; see Chapter 1) are excluded because they must be. Economic considerations may be part of the management decision because the scientific solution proposed must, after all, be profitable or it won't be employed.

Maxwell (1992) proposes that the reason this is the dominant view is that "the intellectual aim of inquiry is to improve knowledge" of scientific truth. This can only be achieved when "we allow only those factors relevant to the assessment" of scientific "fact and truth to influence ... choice of results and theories." When the scientist allows an argument with a spouse, the misbehavior of children, strong feelings, the desire for promotion, political objectives, the availability of funding, etc., to influence the choice of experiments and theories, scientific truth will be corrupted.

Therefore, in this view, the scientist *qua* scientist must in the scientific intellectual domain absolutely exclude all human value considerations. "Empirical success or failure alone must decide the fate of scientific theories" (Maxwell, 1992) and of each experiment. What a scientist chooses to study can be influenced by personal or societal factors, but the evaluation of the data, the justification of the results, must never be so influenced.

Rudner (1953) claims the contention that scientists make value judgments is supported by the argument that doing science at all involves a value judgment. Selection among alternative scientific approaches to a problem involves a value judgment. Scientists cannot escape their humanity, which influences all activities, including scientific ones. The criticism of these arguments is that while they all may be true, they also are extra-problematic (Rudner, 1953) in that they "form no part of the procedures involved in the scientific study of a particular problem." Such procedures, the essence of science, have not been shown to include any value judgments. *Ergo*, the scientific method is (must be) free of value judgments (a value claim). Scientists, when engaged in the scientific enterprise, strive diligently to omit personal idiosyncrasies and value judgments from the conduct of science (Rudner, 1953). However, Rudner argues persuasively that scientists *qua* scientists make value judgments, although they may not be recognized as such. His first claim is that scientists accept or reject hypotheses and that this includes a value judgment. No scientific hypothesis is ever completely verified. Scientists use statistical inference to accept or reject hypotheses and the statistical evidence enables judgment about how strong the evidence is. How strong the evidence must be is often determined by the importance (the value) of the question being studied. If a scientist asks if a pesticide will harm human health, the evidence must be stronger than if one asks if a new crop variety has a greater yield. In agriculture (as in most of the scientific world), two levels of confidence (1% or 5%) have become accepted by frequent use. These values represent the risk one is willing to assume and that is clearly a value judgment. Rudner (1953) notes that some object to this line of reasoning because it is "the function of the scientist *qua* member of society to decide whether a degree of probability associated with the hypothesis is high enough to warrant its acceptance." However, the task of the scientist *qua* scientist is only the determination of the degree of probability or the strength of the evidence for a hypothesis and not the acceptance or rejection of that hypothesis. Rudner argues that the scientific method requires making value judgments and for the scientist to omit this fact does not in any way contribute to scientific objectivity. To ignore the fact that value decisions are an inevitable part of science and to make them intuitively, haphazardly, or unconsciously is "to leave an essential aspect of the scientific method scientifically out of control" (Rudner, 1953).

Omission of value considerations and maintenance of the myth that science is value free has been a problem in agricultural science. I agree with Rudner, who argues that scientific objectivity should include thought about what value judgments are made or might be made and even about those that ought to be made. Full awareness of ethical matters is an essential part of scientific progress toward objectivity (Rudner, 1953). An operative philosophy, which excludes all but empirical success, has enabled agricultural scientists to produce scientific solutions rapidly to many production problems while simultaneously ignoring the problems the solutions create. The global problems mentioned in the introduction and the continuing agricultural problems mentioned above have not been solved and have become worse. It is undeniably true that agricultural production feeds more people than ever before, but too many people are still hungry in a world with serious environmental problems, many exacerbated by the practice of agriculture (Green et al., 2004). Many people would not be fed, but many environmental problems would not exist were it not for modern agricultural practice that enables feeding people. Solutions to the problems involve science, but the continued dissociation of science from human problems and from questions of value will assure that the "priorities of scientific research will come to reflect, not the interests of those whose needs are greatest, the world's poor, but the interests of the powerful and wealthy—First World rather than Third World interests" (Maxwell, 1992). In agriculture, this is exactly what one finds—First World interests dominate.

## What Research Ought to Be Done?

The challenge of research is to find that "single new piece of truth about nature" (Thomas, 1973). The challenge of agricultural research includes discovering truth about nature and discovery, development and employment of new practices, techniques, and machines to improve agricultural productivity and profit. Both approaches to science are worthy of praise and support.

As noted in Chapter 1, a major purpose of this book is to explore agriculture's ethical horizon, particularly as our collective, yet unexamined, ethical position may limit what that horizon defines. Given this goal, it is clear that scientific objectivity in discussions of what agriculture's ethical horizon is or ought to be, should include thoughts about what value judgments are made or might be made and about those that ought to be made. Part of the scientific enterprise is critical inquiry. That means doing experiments well with proper procedures and appropriate analysis of results. It also means examining what we think is known, to learn about what is not known, what is accepted as true, and what is opinion. "It is far easier to label than to understand, and intellectual laziness undermines our studies with deadly inversion of the scientific method: 'I'll believe it when I see it!' becomes 'I'll see it when I believe it'" (Homerin, 2003).

It is a certainty that the ideas of science are remaking the world (Bronowski, 1977, p. 3). But there is nothing absolute about the ideas or concepts of science. They form a flexible framework that is always being built and rebuilt. The only thing the framework must fit, or adapt to, is the facts—the scientific facts. Bronowski (p. 211) identifies a "tyranny of facts that distresses even intelligent people, who fear that the spread of science is robbing them of some freedom of judgment."

Because of its technology and the fact that agriculture is the largest, most ubiquitous human interaction with the environment, its science is applauded by those directly involved, feared by others, and disparaged by some. Therefore, Maxwell (1992) advocated a move from a philosophy of knowledge to a philosophy of wisdom. In his view, a move toward a philosophy of wisdom demands several things that seem germane to appropriate changes in agriculture. Of the 15 changes he recommended seven seem particularly applicable to agriculture. They are:

1. Academic problems must include the problems of living as intellectually more fundamental than the problems of knowledge.
2. Proposals for academic ideas need to change so they become proposals for action as well as claims to knowledge.
3. The definition of intellectual progress should be expanded from progress in knowledge to include progress in ideas relevant to achieving a sustainable, wise world.
4. The intellectual domain of science which has consisted of evidence and theory should be expanded to include research aims. Scientific discussion should be expanded to include discussion of the effects of scientific achievements on life; all life, not just human life, for example, the advent of pesticides in agriculture and their relationship to the state of the environment and the health of other creatures.
5. Social inquiry and natural (agricultural) science must be more integrated. The social effects of agricultural technology that appear after adoption should be considered before adoption.
6. The academic enterprise should not be intellectually dissociated from the world but "constantly learning from, speaking to, and criticizing society" (Maxwell, 1992) as they move together toward cooperative rationality and social wisdom.
7. Philosophy must cease to be a specialized discipline and become an integral part of all inquiry that is concerned with the most fundamental problems. Many fundamental problems are essentially agricultural. How can all people be fed? How should we practice agriculture? Is there a human right to food?

Philosophers will help people in agriculture seek answers to the "inordinate number of paradoxes, puzzles, and ironies" encountered in explorations of the agricultural enterprise (Perkins, 1997, pp. 3–4). These are not simple puzzles with clear, single solutions. They are complex issues that require the best science and the best philosophy.

Perkins identified several questions raised by the practice of agriculture. The questions are clearly not scientific or technical and the answers won't be found in any set of scientific data.

- Agriculture feeds the world, but few people understand it or care much about how it works.
- People think of agriculture as an activity that creates a landscape that is "alive, verdant, lush, and redolent of wholesome naturalness," but agricultural practice often destroys natural ecosystems and wildlife habitat (Green et al., 2004).
- Most Americans think, as Thomas Jefferson did, that the practice of agriculture produces honorable farmers who are the backbone of the nation and a primary source of its cultural richness. In spite of this political mythology, farmers and ranchers are as noted in Chapter 1 (Berry, 1970, p. 78)—a despised minority and, if they are thought of at all, are thought of as hicks and yokels who probably couldn't make it in the modern world of American business.
- Agriculture is a business but it is also a human-created ecosystem generating a food-web of which we are an integral part and without which most of us could not survive.
- Agriculture is not usually regarded as something that is essential to the security of any nation. It is, however, as important as any activity in guaranteeing national independence because it is the activity on which all others depend.
- Agriculture is usually perceived as a "romantic, tranquil refuge from the relentless blight of industrial civilization." It is far from tranquil as it is continually changed by the relentless advance of technology. Agriculture is "the foundation upon which the machinery of urban industry was built and is maintained."

It is one among many definitions, but technology can be defined as a set of predetermined operations that yield predictable results. For example, it is true that when one uses the right pesticide at the right time at the proper dose, susceptible pests die. When the right amount of the right fertilizer is applied correctly, crop yield will increase. But such simple relationships overlook the fact that using any agricultural technology has social, economic, and political ramifications that should be identified and discussed. Agricultural science has focused on results that increase productivity, the end of research. Ends may be analyzed to determine if they are good. But philosophical analysis obligates comparing ends to means to reveal their compatibility or lack thereof. The means contain the ends. Emerson (1937) said:

> *cause and effect, means and ends, seed and fruit, cannot be severed; for the effect already blooms in the cause, the end pre-exists in the means, the fruit in the seed.*

The means result in the natural end, not necessarily the end predicted or advocated by strong advocates.

Agricultural scientists have assumed that as long their research and the resultant technology increased food production and availability, they and the end users were somehow exempt from negotiating and re-negotiating the moral bargain that is the foundation of the modern democratic state (Thompson, 1989). It is a moral good to feed people and agriculture does that. Therefore, its practitioners assume that anyone who questions the morality of the ends or its technology simply doesn't understand the importance of what is done. It is assumed that researchers are technically capable and that the good results of technology make them morally correct. Berry and Wirzba (2002, p. 20) questions that assumption and reminds us of the obligation it entails.

> *We have lived by the assumption that what was good for us would be good for the world ... We have been wrong ... For I do not doubt that it is only on the condition of humility and reverence before the world that our species will be able to remain in it.*

I conclude that while agricultural scientists are ethical in the conduct of their science (they don't cheat, don't fake the data, give proper credit, etc.) and in their personal lives (they earn their wages, take care of family, respect others, are responsible for their actions, etc.), they do not extend ethics into their work. Agricultural scientists are not only reluctant revolutionaries that Ruttan (1997) identified, but also realists. Realists run agricultural research and the world; idealists do not. Idealists attend academic conferences and write thoughtful articles (Kaplan, 1999). The action is elsewhere. The reality may be publish or perish in academia, but it is produced profitably or perish in the real agricultural world. Realism rules, and neither philosophical nor ethical correctness, are necessary for useful scientific work (Rorty, 1999).

I find that true, but I want more. I want all to accept the difficult task of analyzing scientific results. We must strive for an analysis of what it is about weed science, agriculture, and our society that limits our aspirations and needs modification. We must strive to strengthen features that are beneficial and change those that are not. We must be sufficiently confident to study ourselves and our institutions and, when necessary, dedicated to the task of modifying both. It will not be easy because people don't want assumptions about their science, its results, or their lives challenged. People believe their assumptions are correct and they want to use them, not question them, especially when the questions come from others.

Those engaged in agriculture need to examine the ethical foundation and its associated values. In short, agriculture needs a new ethic that does not ignore the importance of increasing food production and availability, but does not end there. It will involve seeking answers to what Perkins (1997) called the "inordinate number of paradoxes, puzzles, and ironies" of agriculture. Full analysis demands exploration of the philosophical dimensions of agricultural science and technology. That is, in the classical sense, we must begin to analyze the principles underlying conduct, thought, and knowledge of the agricultural enterprise rather than just its productive results.

A comment by the Russian author Leo Tolstoy[5] about art is relevant. Tolstoy urged questioning and debate about the correctness of scientific and moral assumptions:

> *I know that the majority of men who not only are considered to be clever, but who really are so, who are capable of comprehending the most difficult scientific, mathematical, philosophical discussions, are very rarely able to understand the simplest and most obvious truth, if it is such that in consequence of it they will have to admit that the opinion which they have formed of a subject, at times with great effort,—an opinion of which they are proud, which they have taught others, on the basis of which they have arranged their whole life,—that this opinion may be false.*

To preserve what is best about modern agriculture and weed science, to identify the abuses modern technology has wrought on our land, our people and other creatures, and begin to correct them will require many lifetimes of work (Berry, 1999). Agriculture should be seen in its many forms—productive, scientific, environmental, economic, social, political, and moral. A new ethic will acknowledge the value of increased production. Other criteria, many with a clear moral foundation, must be

**Highlight 2.1**

"We need a corpus of people who consider that it is important to take a serious and professional crude look at the whole system (Friedman (1999) quoting M. Gell Mann). It has to be a crude look, because you will never master every part of every interconnection ... Unfortunately, in a great many places in our society, including academia and most bureaucracies, prestige accrues principally to

[5] Tolstoy, L. (1904). What is Art? The Christian Teaching. Page 274 *in* L. Wiener (ed.) Resurrection Vol. II. Boston, Dana Estes & Co. Pub.

those who study carefully some (narrow) aspect of a problem, a trade, a technology, or a culture, while discussion of the big picture is relegated to cocktail party conversation. That is crazy. We have to learn not only to have specialists but also people whose specialty is to spot the strong interactions and the entanglements of the different dimensions, and then take a crude look at the whole. What we once considered the cocktail party stuff—that's the crucial part of the real story."

included. We live in a postindustrial, information age society, but we do not and no one ever will live in a post-agricultural society. All nations have an agricultural foundation within their borders or elsewhere. Those in agriculture must strive to ensure all that the foundation is secure.

## References


Berry, W., Wirzba, N. (2002). The art of the common-place: the agrarian essays of Wendell Berry. *Washington*, DC.: Counterpoint. P. 330.

Berry, W. (1999). Distrust of movements. The land report 65 (Fall), 3–7. The Land Institute, Salina, KS.

Berry, W. (1970). Think little. In: A Continuous Harmony. Harcourt Brace & Co., New York, Pp. 71–85.

Bronowski, J. (1977). *A Sense of the Future: Essays in Natural Philosophy*. Cambridge, MA, MIT Press.

Bronowski, J. (1956). *Science and Human Values*. New York, Harper and Row.

Busch, L. and W.B. Lacy. (1986). The Agricultural Scientific Enterprise—a System in Transition(1986). *Boulder, CO*. Westview Press.

Campbell, J. (1972). *Myths to Live By*. New York, The Viking Press, Inc. Pp. 287.

Campbell, N. (1952). *What is Science?* New York, Dover Publications. (first pub. 1921). Pp. 186.

Cobb, J. (1993). Ecology, ethics, and theology. Pp. 211–227 *in* H.E. Daly, K.N. Townsend (eds.). *Valuing the Earth: Economics, Ecology, Ethics*. Cambridge, MA, MIT Press.

Cohen, J. (2002). Designer bugs. *Atlantic Mon.* July/August:113–124.

Conway, G. (1997). *The Doubly Green Revolution: Food for All in the 21st Century*. Ithaca, NY, Comstock Publication Associates division of Cornell University Press.

Einstein, A. (1940). Conversations concerning the fundamentals of theoretical physics. *Science* 9:487.

Eiseley, L. (1973). *The Man Who Saw Through Time—Francis Bacon and the Modern Dilemma*. New York, C. Scribner's Sons.

Emerson, R.W. (1937). Compensation. Essays and English traits 90 *in* C.W. Eliot (ed.). *Vol. 5 of the Harvard Classics (51 vols.)*. New York, P.F. Collier & Son Corp.

Feynman, R. (1999). The pleasure of finding things out. *Am. Sci.* 87:462.

Freedman, D.H. (2010). Lies, dammed lies, and medical science. The Atlantic, November. Pp. 76–78, 80–82, 84–86.

Friedman, T.L. (1999). *The Lexus and the Olive Tree*. New York, Farrar Straus Giroux. Pp. 394.

Green, R.E., S.J. Cornell, J.P.W. Scharlemann, and A. Balmford. (2004). Farming and the fate of wild nature. *Sci. Expr.* December 23, 2004. 6 Pages. (accessed January 12, 2004).

Homerin, T.E. (2003). Homerin offers wisdom to new ΦBK members. *The Key Reporter* 68(3):6–7.

Jackson, W. (2007). The next 49 years. The land report Summer (18), 19. The Land Institute, Salina, KS.

Kaplan, R.D. (1999). Kissinger, Metternich, and realism. *Atlantic Mon.* June: 73, 74, 76–78, 80–82.

Maxwell, N. (1992). What kind of inquiry can best help us create a good world? *Sci. Technol. Human Values* 17:205–227.

Morison, E.E. (1966). Men, Machines, and Modern Times. Cambridge, MA, MIT Press.

Perkins, J.H. (1997). *Geopolitics and the Green Revolution: Wheat, Summer 3Genes, and the Cold War.* New York, Oxford University Press.

Peterson, R. (1978). *Technology: Its Promise and its Problems.* Boulder, CO, Colorado Association of University Press.

Rorty, R. (1999). Phony science wars: a review of Hacking, I. 1999. The social construction of what? *Atlantic Mon.* November:120–122.

Rudner, R. (1953). The scientist *qua* scientist makes value judgments. *Philos.Sci.* 20:1–6.

Ruttan, V. (1991). Moral responsibility in agricultural research. Pp. 107–123 *in* P.B. Thompson, B.A. Stout (eds.). *Beyond the Large Farm: Ethics and Research Goals for Agriculture.* Boulder, CO, Westview Press.

Sagan, C. (1974). *Broca's Brain—Reflections on the Romance of Science.* New York, Random House.

Tegtmeier, E.M. and M.D. Duffy. (2004). External costs of agricultural production in the United States. *Int. J. Agric. Sustainability* 2(1):1–20.

Thomas, L. (1973). Notes of a biology watcher. *New Engl. J. Med.* 288:307–308. Also in— Thomas, L. (1974). Natural science *in The Lives of a Cell: Notes of a Biology Watcher.* New York, Viking Press. Pp. 100–103.

Thomas, L. (1974). The planning of science(1974). *The Lives of a Cell: Notes of a Biology Watcher.* New York, Viking Press. Pp. 115–120.

Thompson, P.B. (1989). Values and food production. *J. Agric. Ethics* 2:209–223.

Wallis, C. (2005). The right and wrong way to treat pain. *TIME,* Feb. 28, pp. 46–57.

Walton, C., C. McGaw, and P. Mackey. (2002). Problem + Research ≠ Management Solution. In: 13th Australian Weeds Conference, Perth, Western Australia. Abstract, p. 528.

# 3 When Things Go Wrong—Balancing Technology's Safety and Risk

*Feeding antibiotics to livestock helps the animals grow faster (for some unknown reason) and provides a preventive measure of protection against some health problems. On the other hand, this widespread use of antibiotics leads to "super bugs" through genetic selection, that can infect both livestock and humans. When a person contracts a drug resistant strain of a bug, standard methods of treatment are usually ineffective. The consequences are a more severe illness and possibly death. Thus, the continued feeding of antibiotics to livestock is not good for society. Any amount of benefit-risk assessment can be done to quantify the number of people who might die as a result of this agricultural practice. But, the scientific approach fails to address whether any number of deaths is acceptable and thus right. Philosophy doesn't dispute the power of antibiotics. Philosophy addresses the concept of acceptable risk using reason about the morality of causing death in an otherwise healthy person because the treatment has been pre-empted for scientific, economic, and material gains of a few.*

***See: http://www.missoulian.com/articles/2003/11/09/opinion/opinion1.txt***

My first experience with herbicides for weed control was in Columbia County, in the Hudson Valley of New York State. Dr. S.N. Fertig, Professor of Weed Science in Cornell University's Department of Agronomy had given me, the assistant county agricultural agent, a small Hudson sprayer and several small, brown paper bags with differing amounts of white to gray, odd smelling, powdery material in them. My instructions were to find a corn field, mark off four adjacent corn rows each 20 ft long, with four similar areas (four replications) for each bag of material. Then I was to put the material in each bag in the Hudson sprayer, mix it thoroughly with water, pump up the sprayer, and spray the mixture on each of the four small plots before the corn emerged. Each numbered plot, marked with a white, wooden stake was in a corn field on the farm where I lived and I was able to observe them frequently. I do not recall how many different materials were included or the specific rates used for each material. Given the application method, precise rates were probably not achieved. I do recall that some of the results of this simple experiment were absolutely astonishing. As the corn emerged, two sets of four plots were completely weed-free. Only corn grew and it grew well with no obvious detrimental effects from the herbicide. The year was 1958, the herbicide applied at two rates, separately to each set of four plots was among the first of the triazines—simazine™. The results, in my view, were magic. I had done field research, for the first time that achieved marvelous results that I thought could eliminate the need to weed; a task no one liked. I was convinced

**Agriculture's Ethical Horizon. DOI: 10.1016/B978-0-12-416043-9.00003-9**

of the potential of chemical weed control and similar to most others, in that time, I did not consider possible disadvantages. In view of the magic of total weed control observed, I was not inclined even to try to think of any disadvantages. My memory is that no one I knew asked about possible disadvantages. The agricultural magic I had facilitated, would inexorably lead to my future, although I did not understand how that simple experiment would affect my future.

## The Development of Herbicides

The chemical era of agriculture really began after 1945, when new fertilizers and pesticides were developed rapidly and became widely available. Increases in crop production and labor productivity were caused by mechanization, the use of agricultural chemicals, increased education of farmers, improved crop varieties, and improved farming practices. My simazine experiment was not at the beginning of the rapid chemicalization of agriculture, but it was close to the beginning.

The chemical era developed rapidly after World War II, but it began much earlier (see Committee on the Future Role of Pesticides in US Agriculture (2000) and for a more complete history, Zimdahl, 2010). In 1000 B.C. the Greek poet Homer wrote of pest-averting sulfur. Theophrastus, regarded as the father of modern botany (372–287 B.C.), reported that trees, especially young trees, could be killed by pouring oil, presumably olive oil, over their roots. The Greek philosopher Democritus (460–370 B.C.) suggested that forests could be cleared by sprinkling tree roots with the juice of hemlock in which lupine flowers had been soaked. In the first century B.C., the Roman philosopher, Cato, advocated the use of amurca, the watery residue left after the oil is drained from crushed olives, for weed control (Smith and Secoy, 1975).

History tells us of the sack of Carthage in 146 B.C. by the Romans, who spread salt on the fields to prevent crop growth. Later, salt was used as a herbicide in England. Chemicals had been used as herbicides in agriculture for a long time, but their use was sporadic, frequently ineffective, and lacked any scientific basis (Smith and Secoy, 1975, 1976).

Bordeaux mixture, a combination of copper sulfate, lime, and water was applied to grapevines to control downy mildew. Someone (perhaps several people) in Europe (no one knows who it was) in the late nineteenth century noted that when the mixture was used to control downy mildew, yellow charlock = wild mustard [*Brassica kaber* (DC.) L.C. Wheeler = *Sinapis arvensis* L.] leaves turned black. That led Bonnet, in France in 1896, to show that a solution of copper sulfate would selectively kill yellow charlock growing with cereals. In 1911, Rabaté demonstrated that dilute sulfuric acid could be used for the same purpose. The discovery that salts of heavy metals might be used for selective weed control led, in the early part of the twentieth century, to research on the use of heavy metal salts for weed control by the Frenchmen Bonnett, Martin, and Duclos, and the German, Schultz (cited in Crafts and Robbins, 1962). Nearly concurrently, in the United States, Bolley (1908) studied iron sulfate, copper sulfate, copper nitrate, and sodium arsenite for selective control of broadleaved weeds in cereal grains.

Bolley, a plant pathologist who worked in North Dakota, is widely acknowledged as the first in the United States to report on selective use of salts of heavy metals as herbicides in cereals. Succeeding work in Europe observed the selective herbicidal effects of metallic salt solutions and acids in cereal crops (Zimdahl, 1995, 2010).

Use of the inorganic herbicides developed rapidly in Europe and England, but not in the United States. Weed control in cereal grains is still more widespread in Europe and England than in the United States. The primary reasons for slow development in the United States were lack of adequate equipment and frequent failure because the heavy metal salts were dependent on foliar uptake that did not occur readily in the low humidity of the primary grain-growing areas of the United States. Other agronomic practices such as increased use of fertilizer, improved tillage, and new varieties increased crop yield in the United States, without weed control. US farmers could choose to move on to what many thought of as the endless frontier of new land and farmers therefore were not as interested, as they would be later, in untried, yield-enhancing technology.

Petroleum oils were introduced for weed control along irrigation ditches and in carrots in the early twentieth century. Field bindweed (*Convolvulus arvensis* L.) was controlled successfully in France in 1923 with sodium chlorate, which is now used mainly as a soil sterilant in combination with organic herbicides. Arsenic trichloride was introduced as a product called KMG (kill morningglory) in the 1920s. Sulfuric acid was used for weed control in Britain in the 1930s. The first synthetic organic chemical for selective weed control in cereals was 2-methyl-4,6-dinitrophenol (DN or Dinitro), introduced in France in 1932 (King, 1966, p. 285). It was used for many years for selective control of some broadleaved weeds and grasses in large-seeded crops such as beans.

Pokorny (1941) first synthesized 2,4 dichlorophenoxy acetic acid (2,4-D). Accounts vary about when the first work on growth-regulator herbicides was done (Akamine, 1948). Zimmerman and Hitchcock (1942) of the Boyce-Thompson Institute (now part of Cornell University) first described the substituted phenoxy acids (2,4-D is one) as growth regulators, but did not report herbicidal activity. They were the first to demonstrate that these chemicals had physiological activity in cell elongation, morphogenesis, root development, and parthenocarpy (King, 1966). A Chicago carnation grower's question about the effect of illuminating gas (acetylene) on carnations led to the eventual discovery of several plant growth-regulating chemicals by Boyce-Thompson scientists (King, 1966).

E.J. Kraus was Head of the University of Chicago Botany Department and had studied plant growth regulation for several years. He supervised the doctoral programs of J.W. Mitchell and C.L. Hamner, who in the early 1940s were plant physiologists with the US Department of Agriculture Plant Industry Station at Beltsville, Maryland. Kraus thought these new, potential plant growth regulators that often distorted plant growth when used at higher than growth-regulating doses, and even killed plants, might be used beneficially to kill plants selectively. He was the first to advocate purposeful application in toxic doses for weed control. Because of World War II, much of this research was done under contract from the US Army and was directed toward discovery of potential uses for biological warfare against an enemy's crops. Similar work for similar reasons was done in Great Britain (Kirby, 1980; Peterson, 1967; Troyer, 2001).

Hamner and Tukey (1944a, 1944b) reported the first field trials with 2,4-D for successful selective control of broadleaved weeds. They also worked with 2,4,5-T as a brush killer. At nearly the same time, Slade et al. (1945), working in England, discovered that naphthaleneacetic acid at 25 lbs/acre would selectively remove yellow charlock from oats with little injury to oats. They also discovered the broadleaved herbicidal properties of the sodium salt of MCPA, (4-chloro-2-methylphenoxyacetic acid) (King, 1966), a compound closely related to 2,4-D. Slade et al. (1945) confirmed the selective activity of 2,4-D. Marth and Mitchell (1944), also former students of Kraus, first reported the use of 2,4-D for killing dandelions and other broadleaved weeds selectively in Kentucky bluegrass turf. These discoveries were the beginning of chemical weed control. All previous agricultural use of chemicals was just a prologue to the rapid development that occurred following discovery of the selective activity of the phenoxyacetic acid herbicides. The first US patent (No. 2,390,941) for 2,4-D as a herbicide was obtained by F.D. Jones of the American Chemical Paint Co. in 1945 (King, 1966). In 1943, 2,4-D was patented (No. 2,322,761) as a growth-regulating substance (King, 1966). Jones patented only its herbicidal activity and made no claim about selective action (King, 1966).

The great era of herbicide development came at a time when world agriculture was involved in a revolution of labor reduction, increased mechanization, and new methods to improve crop quality and produce higher yields at reduced cost. Herbicide development built on and contributed to agriculture's change. Crop growers were ready for and welcomed improved methods of selective weed control. As weed control technology developed, the importance of methods changed, but no methods have been abandoned. The need for cultivation, hoeing, etc. has not disappeared and they are still used in small-scale agriculture. They are less important in developed world agriculture because of the rising costs, reduced availability of labor, and narrow profit margins.

Rapid development of pesticides occurred after World War II. In 1994, there were over 180 different selective herbicides in use in the world and several experimental herbicides in some stage of progress toward marketability (Hopkins, 1994). If proprietary labels are considered, there may be over 1,000 chemical and biological compounds used for pest control in the world (Hopkins). The EPA reports 16,263 pesticides registered under Section 3 of the Federal Insecticide Fungicide and Rodenticide Act (FIFRA) and 2,394 under section 24c. This includes 1,252 active ingredients and 2,505 inert ingredients.[1] The EPA data do not show the number of herbicides or other pesticides. It is accepted that herbicides for weed control are an important and growing part of world agriculture as well as a major part of the agrochemical industry. Their development as primary tools for weed control in the United States is shown by the fact that the number of herbicides included in the Herbicide Handbook of the Weed Science of America has grown from 97 representing 27 mode-of-action groups in 1967 (1st ed.) to 222 from 19 groups in 2007 (9th ed.). The total pounds of herbicides used decreased in recent years probably due to widespread planting of herbicide resistant crops, although no careful study

[1] Personal communication—Claire Gesalman, EPA, July 11, 2011.

**Table 3.1** World and US Herbicide Data for 1990, 1995, 2000, and 2007[a]

| Year | *World* Herbicides % Total Pesticide Market | *World* Herbicide Sales Billion $ | *US* Herbicides % Total US Pesticide Market | *US* Herbicide Sales Billion $ (pounds used) |
|---|---|---|---|---|
| 1990 | 44 | 12.6 | 45 | 4.5 |
| 1995 | 43 | 16.2 | 55 | 6.4 (5.6 mill) |
| 2000 | 44 | 14.1 | 57 | 6.3 |
| 2007 | 39 | 15.5 | 47 | 5.9 (5.3 mill) |

[a]Data from Grube, A.D. Donaldson, T. Kiely, and L. Wu. (2011). Pesticide industry sales and usage report—2006 and 2007 market estimates. http://www.epa.gov/oppfead1/Publications/catalog/subpage1.htm (accessed March 8, 2011).

of the reasons for decreased use has been done.[2] The global pesticide market was 5.7 million pounds (43% herbicides) in 1999 with a value of $33.6 billion (Donaldson et al., 2002). In 2007, 5.2 million pounds of pesticides (47% herbicides) were used in the world with a value of $39.4 billion. About one-third of the world market for pesticides is in the United States. Japan is the next largest national consumer. When the European market is considered as a whole, it is second largest, with France, the largest single pesticide user in Europe (Hopkins, 1994).

Worldwide herbicides (Table 3.1) have been 40% or more of total pesticide sales since 1990 (Donaldson et al., 2002; Hopkins, 1994; NASS, 2010).[3] More than 70% is used in agriculture. The United States exports about 2 billion pounds of pesticides each year, including legal export of about 25 million pounds that are banned for use in the United States.[4] The number of companies engaged in synthesis, screening, and developing herbicides was 46 in 1970. Now 10 companies (3 European, 3 American, 1 Israeli, 1 Australian, and 2 Japanese) control 80% of the world pesticide market of $39.4 billion, which is projected to grow by 2.9% annually through 2014.[5] The top six companies accounted for $28.8 billion, or 75% of the total market in 2007.[6] Monsanto, DowAgro Sciences, and DuPont are the American companies. With the advent of genetic engineering, several of these companies have also become major seed companies:

- Bayer (Germany), the world's biggest agrochemical company, is the seventh biggest seed company.
- Syngenta (Swiss), the world's second largest agrochemical company, is the third largest seed company.
- Monsanto (US), the world's biggest seed company, is the fifth largest agrochemical company.
- DuPont (US), the world's second biggest seed company, is the sixth largest agrochemical company.[7]

[2] L. Gianessi, Crop Life Foundation, Washington, DC, personal communication, March 23, 2011.
[3] USDA/National Agricultural Statistics Service (NASS).
[4] Data from Grube, A.D. Donaldson, T. Kiely, and L. Wu (2011). Pesticide industry sales and usage report—2006 and 2007 market estimates. http://www.epa.gov/oppfead1/Publications/catalog/subpage1.htm (accessed March 8, 2011).
[5] www.fredoniagroup.com/WorldPesticides.html (accessed December 2010).
[6] www.gmwatch.org (accessed December 2010).
[7] www.gmwatch.org (accessed December 2010).

The cause is not clear but seed prices have been rising rapidly as farmers make greater use of genetically engineered (GE/GMO) seeds. Since 1999, prices paid for seeds have risen 146%, with 64% of that rise occurring during 2007–2009. Between 1999 and 2009, seed expenses increased $7.9 billion (110%). During the last 3 years, seed expenses rose $4.5 billion (41%). The price increases were forecast to end, at least temporarily, in 2010 (see ers.usda.gov).

Managers of large companies in any industry are charged with maintaining and improving the company's market position. It is reasonable to suggest that development of new products, increasing market share, and improving profitability are important goals. Advertising, in its many forms, is a primary means of improving sales and profit. An interesting analysis of the role of advertising that is consonant with this book's exploration of agriculture's ethical horizon is Kroma and Flora's (2003) historical analysis of the social construction of pesticide advertising. This advertising, found in a wide variety of agricultural media (e.g., magazines, radio, company brochures), creates and maintains the message that pesticides and other agricultural products are an essential part of modern agriculture. The ads stress the scientific basis of productivity enhancing products while ignoring environmental and social risks of their use. Kroma and Flora show how pesticide advertising co-opts cultural values through use of names that "reflect strategic repositioning of brands through their changing names." The dominant theme of pesticide advertising is portrayal of nature as something to be conquered and adapted to human needs, rather than "an asset to be utilized." To make their point about co-optation of cultural values, Kroma and Flora illustrate changes in the dominant theme of pesticide advertising over time (Table 3.2). They identify a consistent pattern of strategic repositioning of products through their changing names and the societal values they express. None of this is evil or *prima facie* morally suspect. It is good business practice oriented toward achieving the desirable goals of improving sales and profit. Kroma and Flora argue that the "social, cultural, and technical messages symbolically embodied" in advertising of agricultural products have made it more difficult to distinguish between what agriculture and agricultural industrialization have done *for* society and what they have done *to* society.

**Table 3.2** Themes in Pesticide Advertising over Time (Kroma and Flora, 2003)

| Time Period | Advertising Theme |
|---|---|
| 1940s | Orientation toward military needs, progress through overcoming nature. |
| 1950s | Science will solve inadvertent negative side effects of use. |
| 1960s | Pesticides are scientifically sound, productivity enhancing products. |
| 1970s | More powerful technology, control, aggression, competition, dominance of nature. Nature is an adversary. |
| 1980s | Control of natural processes. Ability to achieve desirable environmental interactions and be sustainable, products that support the natural world. |
| 1990s | In harmony with nature, but dominant. Pesticides are part of the natural order. They are in affinity with nature. Emphasis on profitability for user. Beginning of the greening era of the pesticide industry. |
| 2000s | Genetic engineering is consistent with achieving a sustainable agriculture. Ads begin to shape the sustainable agriculture debate. |

## Progress of Weed Science

In 1958 when I saw the magic of simazine for weed control in corn, I was aware of a few people who questioned the advantages of chemical weed control, but those I knew were not among them. I thought I was joining what I saw as the march of progress toward reduced labor for weed control and increased agricultural productivity. Among those I knew and respected, there was an obvious quest for the development and use of herbicides in agriculture. There was no compelling reason even to consider the possible disadvantages of herbicides or other pesticides and those who mentioned such thoughts were dismissed as Luddites or environmentalists (an epithet) who did not understand agriculture and its needs.

Research on pesticide use in agriculture, that is applied research, went forward for the usual reasons noted by Reiss (2001). Scientists then as they do now regarded themselves as autonomous with respect to their scientific work. That is not to say that scientists were or are absolutely free to do whatever they want to do. The pursuit of science is constrained by required foundational knowledge (e.g., no work on genetically modified organisms was done in the 1950s because the foundational knowledge was lacking) and by the availability of funds, equipment, and help. However, when research begins, most scientists feel quite free of constraining value or moral questions in the conduct of their work. They are doing what they want to do in a way they decide is best. Scientists also believe that what they are doing is interesting (at least to them) and perhaps innovative because it has not been done before, or has not been done in quite the same way. In the 1950s, herbicides were new and development of new chemical families and modes of action was rapid. Pesticide research was one of the many bandwagons on the highway to increased yields and profit for agriculture. Agricultural researchers saw pesticide research as interesting and the right thing to do. Ethical questions that asked why it was right were not asked within the agricultural community. It was an easy way to obtain funding and gain publishable results in the right journals. Publication and funding were and remain important drivers for academic scientists. Pesticide research also had the glamour of being new. The results were often nearly magical and those who developed the technology to achieve improved, safe, inexpensive pest control first, received the adulation of peers and farmers.

Another important reason many agricultural scientists studied the practical application of pesticides is that the knowledge gained was rapidly adopted by farmers. The clear utilitarian argument (albeit unstated and unexamined) is that such research ought to be done because it is highly likely to increase the total amount of human happiness and reduce suffering. If weeding could be done with a simple application of what was thought to be a safe, highly efficacious, selective chemical, then agricultural scientists had a clear moral obligation to pursue research that would reduce or eliminate the arduous, undesirable task. In short, agricultural scientists ought to do whatever increases total human happiness (Reiss, 2001). Reducing the burden of weeding or management of other pests with pesticides was and remains, in the minds of agricultural scientists, an unquestionably good thing; the right thing to do.

Although there were a few voices that questioned the wisdom of widespread pesticide use in agriculture, the agricultural community stoutly rejected arguments against pesticide development. As Reiss points out, the reason was "the standard liberal one that one needs strong arguments before banning things." Banning or restricting the development or deployment of new technology should only be done in the face of compelling scientific evidence of harm. The most compelling, persuasive evidence would be harm to people. Pesticides were a technology that promised to reduce labor requirements and increase crop yield. Herbicides offered the promise of reduced weed competition, no harm to people, elimination of the need for people to weed, and increased profit to farmers. The case for going forward was compelling.

The case against pesticide development was largely absent in the 1950s and for several years after, as the chemicalization of agriculture proceeded. Reiss offered four reasons commonly used to argue that a particular research project should not proceed. The *first*, that it would be wrong to even want to do the research, was not regarded as relevant to agricultural research. A counter example Reiss uses is research designed to develop ways of more effectively torturing people. That would be morally reprehensible and nearly all members of civilized societies would consider it wrong even though reasonable arguments can be constructed to support such research. For example, if a person knows the location and time of detonation of a hidden bomb that will kill thousands of people, one could argue that it is permissible to develop and use any means, including torture, to prevent the death of thousands. Harm to one pales in comparison to possible harm to many. It is an argument that has been used. Torture is accepted in some quarters, but largely rejected, as justification to counter the current terrorist threat.

There is universal condemnation of the World War II Nazi regime's research to develop effective ways of killing people with deadly gas. One must assume there would be universal condemnation of research by weed scientists to develop ways to destroy another nation's crops through application of herbicides. Unfortunately, this has been done (Agent blue = cacodylic acid was used for defoliation of jungle vegetation and to kill rice in Vietnam), but the knowledge that made it possible was already available from what was known about the herbicide's activity. There was no research project to develop herbicides for the purpose of killing rice. However, the moral justification for using herbicides to destroy crops in wartime (winning the war is a high value) is weak, at best, and perhaps absent. It is always morally wrong to harm noncombatants, although it is common.

The *second* reason Reiss (2001) offers is that the process of the research would have undesirable consequences. Those doing the research could be harmed physically by inherent toxicity or harmed psychologically. Direct physical harm is possible in medicine, microbiology, chemistry, etc. It also is avoidable by proper isolation and protective clothing. Psychological harm may be more subtle and less easily observed but no less real. One can imagine a case where someone engaged in the development of herbicides for biological warfare (destroying crops) could be deeply disturbed by the moral implications of the work. Fortunately, as far as I know, this kind of work is not being conducted. It seems reasonable to assume that agricultural research may not lead to similar moral dilemmas.

If the net consequences of research lead to harm, one may question the justification for the research (Reiss). Thus, the *third* argument is common among those opposed to pursuit of genetic modification in agriculture. The potential for ecological and human harm is assumed to be real, therefore the research should not proceed until these potentials have been explored and eliminated. It is hard to understand how such things can be explored if the research on genetic modification is stopped or constrained, but the argument is common. Presumably researchers will recognize undesirable consequences before they occur. That is one of the purposes of research, albeit one that may be overlooked if the primary purpose is to create and use the new technology rather than to examine its consequences dispassionately.

Reiss also suggests that in a time when research funds are limited, one can claim that some research (usually someone else's) should not be done. The research may have merit, but with limited funds its merit is less than other work that must be done. This is not immediately a moral argument. It is, or at least it can begin as, a debate about the need for research project A as opposed to project B. Need can be justified economically, politically, or socially. However, in support of each of these arguments there is always a moral foundation. Agricultural researchers commonly justify their work with a moral argument that goes like this:

- *Premise*: Food is necessary for human survival.
- *Premise*: Agricultural research leads to greater food availability.
- *Conclusion*: Because agricultural research ensures human survival, it ought to continue.

The validity of this argument is unexamined within the agricultural scientific community. Other research communities make similar claims and decisions on research funding that undoubtedly reflect the values (the moral judgments) of those who make them.

Reiss' *final* argument concerns possible undesirable consequences of research. It is compelling in the case of herbicides. The cited advantages of herbicides include low cost, safety, high efficacy, labor saving, selectivity (weeds killed and crops not affected), persistence (weed control over time), energy efficiency, profit to growers, and increased crop yield. Each of these has been and is regarded as a good thing. It was only with the passage of time that questions were raised about each of these presumed advantages. None is a clear advantage because most can also be framed as a disadvantage. For example, herbicides clearly can be toxic to humans and nontarget species. Labor is saved, but that is not always good when labor is abundant and a job, even an undesirable, hard job is better than no job, in a developing country with endemic poverty. Persistence is good for weed control over time but may lead to undesirable soil residues, injury to a succeeding crop, harm to nontarget species, and movement from the site of application. Weed scientists know that valid counter arguments can be made for every claimed advantage. Undesirable consequences were not only possible, they were common. There is harm to nontarget species, residues are present in soil and water, the cost of weed control to users is often decreased, but environmental costs are commonly externalized. Advantages, while real, were nearly always short-term for the benefit of this year's crop and this year's profit. Disadvantages nearly always reflect a long-term view. They tend to shift focus from short-term self-interest to long-term human and environmental welfare. A long-term view demands a planning horizon an order of magnitude beyond the next election or 2–3-year funding for a research project. It also demands questioning the dominant view that whatever problems technology creates will be solved by new technology (Peterson, 1978).

## Challenges

Challenges to the presumed advantages of agricultural technology and its uses, and the quest for a long-term view began with Rachel Carson's *Silent Spring* (1962) and have continued. Agricultural scientists, especially those in pest management, were in van den Bosch's view (1989, p. 21):

> *Sucked in to the vortex, and for a couple of decades became so engrossed in developing, producing, and assessing the new pesticides that they forgot that pest control is essentially an ecological matter. Thus, virtually an entire generation of researchers and teachers came to equate pest management with chemical control.*

The ecological nature of pest control was affirmed by Julian Huxley in his preface to *Silent Spring* and is now a consistent, but not yet dominant, theme of weed science research.

For decades, agricultural scientists have been and largely remain technological optimists: new technology will solve the problems old technology created. Weed science began with the development of herbicides, in an era of forward-looking scientific optimism. The journal *Weed Science* first appeared in 1951 and the Weed Science Society of America was founded in December 1954 in Fargo, North Dakota. The society's first meeting was in New York City in January 1956. The magic of herbicides appeared to be able to solve one of humankind's oldest problems—how to reduce or eliminate weed competition in crops and reduce the need for human labor to weed crops. Early weed scientists felt that they were on the cusp of a marvelous new agricultural revolution and the burden of weeding, the arduous labor of the hoe, might disappear from earth. There was no pause to consider lessons from the history of science that reveal how early claims of spectacular advances in human welfare and benefits have, with time, been shown to demonstrate unexpected highly negative consequences (New Zealand's Parliamentary Commission, 2001). Agricultural education emphasized chemistry and physiology while ignoring questions raised by the history of science and philosophy and those raised by the public. A few of the glaring past errors of the agricultural community are illustrated in comments made by James Davidson (Emeritus Vice President for Agriculture and Natural Resources, University of Florida)[8]:

> *With the publication of Rachel Carson's book entitled Silent Spring, we, in the agricultural community, loudly and in unison stated that pesticides did not contaminate*

[8] Davidson's comments were made several years ago and were cited by Kirschenmann (2010).

*the environment—we now admit that they do. When confronted with the presence of nitrates in groundwater, we responded that it was not possible for nitrates from commercial fertilizer to reach groundwater in excess of 10 parts per million under normal productive agricultural systems—we now admit they do. When questioned about the presence of pesticides in food and food quality, we reassured the public that if the pesticide was applied in compliance with the label, agricultural products would be free of pesticides — we now admit they're not.*

Weed scientists confronted with the teratogenic effects of the dioxin contaminant of 2,4,5-T after its use in Vietnam challenged the findings—they were wrong. When some scientists found that very small concentrations of atrazine acted as an endocrine inhibitor in the early phases of the development of frogs,[9] weed scientists questioned the validity of the research—we now know that some herbicides do act as endocrine inhibitors. The research was correct, although the agricultural and weed science communities questioned its real world significance. Similar work with the same result has been done with glyphosate (Roundup™).[10] The most common claim is it is the dose that makes the difference. That argument was brought into question by Colburn et al. (1996) who formulated the hypothesis that many environmental pollutants, pesticides prominent among them, inhibited endocrine function and were particularly devastating if present in very low doses during fetal development. Within the agricultural community the challenge was quickly dismissed and has disappeared since it was presented in the early 1980s. It has not been dismissed by all (see Shulevitz, 2011).

The report of the New Zealand Parliamentary Commission examines a wide range of scientific and technological achievements. It discusses examples from the past century: transportation, chemicals and materials, energy systems, military initiatives, medicine, and ecology. Agricultural examples include antibiotic use, pesticides, prions, and the social costs of the green revolution. The report shows that "early claims of benefits are dangerously devoid of an adequate understanding of the underlying mechanisms." Agricultural scientists acted and continue to act like all proponents of new technologies; they are slow to respond to criticism and quick to raise defensive arguments when negative surprises emerge or are suggested. Science was used to reinforce defense of agricultural research. Science was used to counter objections to its results rather than to explore or enlighten the debate (Kirschenmann, 2005). This inevitably weakens public trust in science and scientists. The New Zealand report advocates "the need to recognize the limits to scientific knowledge and the need to adopt a broader integrative approach to managing complexity, one that allows for surprises and values ethical considerations."

Tenner (1996) raises similar issues for those who see only benefit in scientific progress. As reported by Dibbell (1996) Tenner posits that "every technological

[9] Weed killer deforms frogs in sex organs, study finds. NY TIMES, April 17, 2002, p. A-19. The principal researcher, T.B. Hayes (U.Cal-Berkeley), said "I'm not saying it's safe for humans. I'm not saying it's unsafe for humans. All I'm saying is that it makes hermaphrodites of frogs."

[10] Hoffman, K. (2005). Roundup™ highly lethal to amphibians, finds University of Pittsburgh researcher. http://www.eurekalert.org.pub_releases/2005-04/uopm-rh040105.php (accessed May 19, 2009).

endeavor is riddled with 'solutions' that backfire." Several negative consequences of agricultural endeavors are pointed out. The dust bowls of the "dirty thirties" and the "filthy fifties" are clear agricultural failures that could have been prevented (see Montgomery, 2007; Egan, 2006). Tenner mentions the promotion of pests, which was documented by van den Bosch (1989), who reported development of pest resistance from repeated use of the same pesticide and the creation of new pests after repeated pesticide use. When pesticide resistance first appeared among insects, weed scientists argued that it was a phenomenon that would be restricted to insects. The basis of the claim was that insects have short life cycles and weeds are annuals or perennials and therefore, if resistance appeared, it would take a long time to do so. The reasoning was wrong; 292 weed biotypes representing 174 species (104 dicots and 70 monocots) were resistant to herbicides from 18 mode-of-action groups, in more than 270,000 fields in 2005 (Heap, 2005). In 2010, 348 weed biotypes, 194 species (114 dicots and 80 monocots) in 400,000 fields were resistant to herbicides from 20 mode-of-action groups (http://www.weedscience.org/In.asp, accessed December 2010). One year later, the problem had grown to 357 biotypes, 197 species (115 dicots and 82 monocots) in 430,000 fields resistant to herbicides from 20 mode-of-action groups. Herbicide resistance, first discovered in 1968 and reported in 1970 (Ryan), is now known and accepted by weed scientists for nearly all classes of herbicides. Research proceeds to maintain use while minimizing the inevitable expression of resistance.

The invasion of animal and plant pest species into new areas is incompletely understood. Scientists don't understand why some species succeed and others fail. Among invasive plants, those that succeed are prolific seed producers or asexual propagators, that have escaped natural enemies, disperse easily and widely, and are good competitors, especially in habitats disturbed by humans. The aggressive annual weed, kudzu [*Pueraria lobata* (Willd.) Ohwi], once promoted by the US Department of Agriculture for soil erosion control, is an excellent agricultural example of a surprise, an unintended consequence.

In contrast to much of the environmental defense literature that objects to many technological innovations, Wildavsky (1979) suggests that being too cautious about technological developments may paralyze scientific endeavor and make society less safe. He suggests that "Chicken Little is alive and well in the richest, longest lived, best-protected, most resourceful civilization, with the highest degree of insight into its own technology" that has ever existed. In the past, society relied on expert opinion to determine what was right and what ought to be done. Now experts disagree about what the data mean and on what data count, and the public is left to decide without really understanding the issues. Wildavsky covers a wide range of topics (acid rain, ozone depletion, global warming, 2,4,5-T [Agent Orange and dioxin], asbestos, saccharin, and others). He uses these to illustrate that uses and abuses of science are widespread and anti-science bias is pervasive in the public mind. Many of the same issues were mentioned in the report of the New Zealand Parliamentary Commission (2001) with the same beginning assumption—surprise is inevitable in scientific endeavors and the opposite conclusion—caution is always appropriate.

Wildavsky asks if such claims about science and technology are true and if more caution is required or if the demand for caution impedes scientific progress? His answer in all the cases he examined, other than ozone depletion from use of chlorofluorocarbons (CFCs) is, No, more caution is not required! Nearly all claims about harm are not true; the technology is not harmful. The CFC example is worthy of special notice because the scientific community was sure that these simple, inert chemical compounds could not cause any environmental harm. Many of the examples Wildavsky examined are included in the New Zealand report and it is worthy of note that the two reports reach opposite conclusions.

Wildavsky's Chapter 3, "Dioxin, Agent Orange, and Times Beach," describes the dioxin, 2,4,5-T story in detail. He concludes that 2,3,7,8-tetrachlorodibenzo-p-dioxin (TCDD) is the one dioxin, among many, that has serious human health effects. It was found as a contaminant in trichlorophenoxy acetic acid products such as 2,4,5-T, but was never found, for clear chemical reasons, in dichlorophenoxy acetic acid products, such as 2,4-D. These are chemical facts. Wildavsky reports that such facts paled in comparison to the political and nonscientific debates that swirled around the dioxin issue, a debate that continues (Isaacson, 2007). Millions of dollars were spent to discover that dioxin has serious human health effects, but only at extremely high doses, and these occur only with unprotected, prolonged occupational exposure. There is a threshold: "below a certain level, little or no harm would occur; thus some body level might be harmless." The US government and the chemical industry paid millions of dollars to people who were not injured and equal amounts to regulate inconsequential exposures. Wildavsky concludes the discussion, as he concluded so many others in the book, with the question "Why expend so many resources in the name of public health with so little to show for it?"

The dioxin story is used by Wildavsky as one among many to illustrate that our society has become one that assumes most chemical substances and new technologies are dangerous and only a few can even be used safely. He proposes that it is culturally significant that we fear what is not harmful and therefore, by refusing risk we encourage danger. When everything chemical is regarded as potentially or actually dangerous, a society risks losing the advantages that new technologies may offer. A society may value safety so much that risk is minimized (Wildavsky, 1979).

Wildavsky concludes by objecting to the dominance of the precautionary principle in technological decisions (see Chapter 8). He suggests the precautionary principle should be rejected because "there are no health benefits from regulation of small, intermittent exposures to chemicals." The concluding section of the book provides an outline for what he calls the "case for modern technology endangering humans—the environmentalist paradigm." The paradigm is based on four guiding assumptions that are, or seem to be, the basis of evaluative processes. The assumptions illustrate, in his view, the paradigm's dominance which has resulted in excessive caution about technology and growing public fear of technology. The assumptions are:

- Possibility should replace probability as a criterion for regulation to protect human health and safety. The mode of risk assessment that increases prediction of harm by the largest amount should be adopted in the absence of proof to the contrary.

- No cause, however weak, is incapable of producing substantial harmful effects. That is, a chemical that causes effects at high doses can be assumed to be harmful at exceedingly low doses. In short, dose does not make the poison, exposure does. *Note*: This is opposite to the commonly accepted toxicological view attributed to Paracelsus, that dose alone makes the poison.[11]
- The purpose of risk regulation is to prevent health detriments, not to secure health benefits.
- What is not explicitly permitted is forbidden; substances or processes must be demonstrated to be benign before they can be used. The burden of proof rests on a conclusive demonstration that a substance or a process does not cause cancer or other, commonly unspecified, harm.

Wildavsky's (1997) optimistic view of technology is based on extensive research to ask if claims against technology are true, and he concludes that most claims have not been true. His work stands in sharp contrast to the majority of environmental literature and to the report of the New Zealand Parliamentary Commission (2001) that strongly recommends that society should "interpret evidence in a precautionary framework that seeks to minimize false judgments that no hazard exists when in fact one does." This recommendation values the precautionary principle and does not want to abandon or modify it as Wildavsky does. Scientific testing "can only deal with the scientifically tractable problem of known uncertainties." When there is scientific ignorance or when science asks the wrong questions or studies the wrong problem, the importance of the unknowns will remain. The dilemma within science is stated well by Holling (1998) who contrasts what he calls analytical science or the science of parts with integrative science or the science of integration of parts. The first is characterized by molecular biology and genetic engineering, which promise to give us health and economic benefits but whose progress is plagued by changing social values. Analytical science is "essentially experimental, reductionist, and disciplinary in character" (Holling). Integrative science, represented by evolutionary biology and ecology, deals with resource and environmental management where uncertainty and surprises are expected. In Holling's view, the latter is interdisciplinary and combines historical, comparative, and experimental approaches at scales appropriate to the issues to be addressed. Integrative science relates well to social sciences and is the bridge between analytical science and public policy. Analytical science is reductionistic and certain, while the other is integrative and uncertain. Holling claims that analytical science can be trapped by providing precise answers to the wrong question while integrative science runs the risk of providing useless answers to the right questions.

Wildavsky (1979, 1997) proposes that properly conducted, analytical science can and will answer the important human health questions about any technology. Scientific rigor and the demands of intellectual honesty are presumed to be adequate to answer all important questions. In contrast, the New Zealand report (2001) notes

[11] Paracelsus (Phillippus Aureolus Theophrastus Bombastus von Hohenheim, 1493–1541) often called the father of toxicology, wrote (in German) "All things are poison and nothing is without poison, only the dose permits something not to be poisonous." Or, more commonly "The dose makes the poison." That is to say, substances considered toxic are harmless in small doses, and conversely an ordinarily harmless substance can be deadly if over consumed.

that scientific inquiry too often omits integrative and ethical questions (see Chapter 1, experiential truth) precisely because the latter are not scientific and are often regarded as emotional and value laden, which, of course, they may be. This accusation does not mean that such questions are not important or lack substantive content. The ethical realm asks, by definition, different but no less important questions, which analytical science, by definition, is not prepared or designed to answer and which integrative science approaches but may not answer completely.

Disaster is not the inevitable consequence of good analytical or integrative science and the resulting technological innovations. If one views science and technology over the last hundred years, it is clear the optimists have had it right (Tenner, 1996). Nevertheless, public fear is real and not irrational. People are regularly notified of fallible technology and agriculture has not escaped public scrutiny. To wit:

Agricultural examples:

- The mid-1960s controversy over the real and suspected hazards of 2,4,5-T, a component of Agent Orange used in Operation Ranch Hand, a vegetation control program during the Vietnam war. It was the first major public debate that challenged the intellectual foundation of weed science and its dependence on herbicides (see Chapter 2).
- On December 3, 1984 a poisonous cloud of methyl isocyanate, used in the manufacture of pesticides, escaped from Union Carbide's plant in Bhopal, India killing 14,000 and permanently injuring 30,000 people.
- There were few immediate, reasonable solutions when weed resistance developed in glyphosate-resistant cotton and other crops used by many US farmers.
- Pesticide poisoning: no one knows for sure, but it is estimated that between one and five million cases of pesticide poisoning occur every year in the world, resulting in 20,000 deaths. Developing countries use 25% of pesticides, but experience 99% of the deaths (Goldman, 2004).
- The debate within and outside the agricultural community over the risks and ultimate beneficiaries of genetic modification of crops has raised legitimate economic, social, and biological concerns. The concern for weed science is the widespread adoption of herbicide resistant technology (see Chapter 8).
- Air and water pollution and animal suffering from Confined Animal Feeding Operations (CAFOs).
- Mad cow disease, swine flu, bird flu, meat recalls, and antibiotic resistance are all of concern or have been of major societal concern in the past.
- The ecological dead zone extending in the Gulf of Mexico from the Mississippi terminus.

Other examples:

- The 1979 US nuclear plant problem at Three Mile Island.
- On April 26, 1986 unit four at the nuclear power plant in Chernobyl, Ukraine exploded, causing approximately 6,000 deaths and injuring more than 30,000 people.
- January 28, 1986: 73 seconds into its flight, the space ship Challenger exploded, killing seven astronauts.
- The 1989 spill of 11 million gallons of oil from the Exxon Valdez.
- In 2005 Dell recalled 4.1 million notebook computers when it was discovered that their batteries were a fire risk.

- The failure of over 50 levees in August 2005 in and around New Orleans, Louisiana during Hurricane Katrina.
- The collapse of the eight lane, steel-truss bridge on Interstate 35 over the Mississippi River on the evening of August 1, 2007.
- On August 10, 2010 the Deepwater Horizon drilling platform caught fire in the Gulf of Mexico. There have been several other oil spills.
- The huge car recall by Toyota in 2010.
- The fear and uncertainty of the nuclear crisis at the Fukishima Dai-ichi nuclear plant in Japan that began after the earthquake and tsunami in March 2011.

The agricultural view has been that the eight examples are true and what was reported, happened. But the view always includes the question of whether or not the particular agricultural technology or practice is really dangerous. Of course, the answer depends on how "dangerous" is defined. As mentioned above, reports of the hazards of pesticides are frequently countered with the assertion that dose alone makes the poison (see footnote 12). Nearly everything can be toxic if enough is consumed or exposure is very high. Carbon monoxide is present on city streets and while automobile exhaust is unpleasant, exposure while stuck in traffic is not harmful. Its toxicity is easily expressed if one sits in a running car in a closed garage. The dose makes the difference. An ancillary aspect of public concern is that independent of the event there is rarely an ameliorative technology on a par with what has failed. The counsel of the English writer and dramatist Douglas Adams is worthy of consideration: "the major difference between a thing that might go wrong and a thing that cannot possibly go wrong is that when a thing that cannot possibly go wrong goes wrong it usually turns out to be impossible to get at or repair."[12]

On the other hand, there is a plethora of examples that affirm that agricultural science and technology have improved food production and human health and made life better. Disasters have occurred but they "mobilize the kind of human ingenuity that technological optimists believe exists in unlimited supply" (Tenner, 1996). Technological disasters will continue to occur and agriculture will cause some, but, on balance, because of analytical science, life will continue to improve for most people. The balance between protecting public safety and the risk of new technologies often takes subtle form. Pesticide residues in or on fruits and vegetables may, it is argued, be harmful to the health of those who consume them, especially children. If a new system or a new technology was developed that reduces or eliminates such residues it will be good, except that the poor may consume less fruit and vegetables because the requirements of a system to reduce pesticide residues on fruits, nuts, and vegetables might make them too expensive.

Ordinary citizens do not know what to do or who to believe and tend to ignore both the cranks who speak of gloom and doom and think everything leads to disaster of some kind and the optimists who speak of boom and zoom to ever-higher levels of technological achievement and human happiness, while acknowledging the possibility of greater disasters in the future. As Thompson (2010, p. 62) points out, ordinary citizens are gradually becoming more aware of and concerned about "the

[12]Adams, D.N. (2002). *The ultimate hitch hikers guide to the galaxy, book 5—Mostly Harmless.*

contemporary reality of American agriculture." They are plagued by recalls the problems mentioned above (Thompson). Those engaged in agriculture must acknowledge that their ability to envision the good or the bad of technology, prior to introduction, is surprisingly limited (Tenner, 1996) and science, unhelpfully, gives us only ranges of probability, not certainty. Because science cannot prove that things (life) will get worse or better, ordinary folks tend to adopt the Schlitz™,[13] "the beer that made Milwaukee famous" philosophy: "You only go around once in life, so you've got to grab for all the gusto you can."

## The Continuing Debate

### Highlight 3.1

"When R. Given Harper set out to understand why North America's migratory birds were declining, he set a unique course. While other researchers zeroed in on habitat loss as the key problem, he decided to look at an old culprit—the insecticide DDT—and its specific effects on songbirds."

"The effects were intriguing. Traces of DDT and other related chemicals were showing up in the birds." But the real shock came when Harper compared his results with DDT levels in nonmigrating songbirds. These year-round residents of North America—including the northern cardinal, black-capped chickadee, and dark-eyed junco—had more kinds of chemicals and dramatically higher levels of them than the migrating species (Clayton, 2005).

Agricultural people know that DDT has not been used in the United States for several decades. Residues are present but declining. The abandonment of chlorinated hydrocarbons was fought by most agricultural people but is now regarded as a difficult but correct decision. The problems associated with their use (long environmental persistence, lipid solubility, harm to nontarget species, bioconcentration) have been recognized. The more important lesson is that we must always be careful with what is put in to the environment because many things are not easily, and may be impossible, to retrieve. All effects are not known (cannot be known) at first use.

The possibility of disruption of endocrine function (Colburn et al., 1996) was not even considered when the chlorinated hydrocarbons were used. Now the phenomenon is known, but many regard it as a remote possibility because, in their view, dose alone determines the poison and environmental doses are very low. But that is the point. Low doses can act as endocrine disruptors when they are present at certain stages of embryo development (Colburn et al., 1996).

[13] Schlitz brewing was founded in 1849, purchased by Stroh in 1982, and the combined company was purchased by Pabst in 1999. At one time, Schlitz was the largest brewery in the world.

The arguments between the technological optimists (those who some call Pollyanas or Cornucopians) and the pessimists (the Cassandras or Jeremiads[14]) will continue. Final resolution is not highly likely because technology advances, the issues for debate and argument change, and, therefore the nature of the argument changes. The advent of agricultural biotechnology has altered the debate. Questions about agricultural technology used to focus on three questions:

1. Did the new technology work better than the old one? Was it more efficient, and/or more profitable for the user?
2. Was the new technology safer for users?
3. Was the new technology better than its predecessor? That is, was it more efficacious or cheaper?

Biotechnology has introduced a new question—Do we need it? The question might also be framed as, Ought we to do it? Thus, it becomes a moral rather than just a scientific or economic question. The results of analysis of the production and social effects of agricultural technology are not all negative (Wildavsky, 1997); much of it has been needed. The world now feeds more people an adequate diet than ever before. There are still too many hungry people but if we relied on past production methods we could not feed all that are now fed. It is correct to affirm that agricultural technology has been of great benefit. It is also correct to recognize that the technology has caused great problems.

A 1994 CAST report (Waggoner, 1994) claimed, with supporting analysis, that by harvesting more per unit area through the use of modern and improving agricultural technology, farmers could feed a future human population of 10 billion *and* spare land for nature. If farmers did not or could not take advantage of improving technology more land would be required to produce food and fiber. Farmers and the agricultural technology they use to produce food and fiber could be (should be) "at the hub of sparing land for nature." Waggoner's report is an optimistic appraisal of the value of agricultural technology.

Avery (1994) has shown, with careful analysis, that modern agricultural technology has saved thousands of acres that would have been required to produce today's yields with 1950s technology. He claims (1997) that modern agricultural technology, especially crop protection technology is "one of mankind's greatest environmental achievements." The land not required for production can be used for wildlife, urban development, or left alone because it is not needed to produce food and fiber. Because agriculture is the most widespread and most significant human interaction with the environment, only by reducing agriculture's need for land, can land be saved for nature. The biggest danger to wildlife and by implication to the environment is potential conversion to agriculture of land that is now in tropical rain forests so the land can be used to produce low yields of crops and livestock. In Avery's view, to eliminate world hunger quickly, society will have to do two things with far

[14] I set watchmen over you, saying, Hearken to the sound of the trumpet. But they said, We will not hearken. Jeremiah 6:17. Cassandra was a daughter of Priam, King of Troy. Apollo granted her the gift of prophecy but added a provision—however much she knew, no one would believe her. Pollyana was the eternally optimistic heroine of a 1913 novel by Eleanor H. Porter.

more effort than has been devoted to them to date. First, more investment will have to be made in high-yield farming technology for the third world. It is Avery's view that the necessary investments have not been made because what he calls eco-activists have successfully crusaded against "high-yield seeds, fertilizers, and pesticides that will need to be part of the high-yield packages." Avery (1994, 1997), Waggoner (1994), and Wildavsky (1997) agree on this point. Second, there must be a simultaneous upgrading of "the skills of third-world workers to instill the necessary concern for honesty and human rights in the governments of third-world countries." Avery claims that world hunger can be eliminated by those things that increase production—technology and human skill.

Sen's (1981) incisive analysis of famine, in contrast, does not diminish the importance of production but emphasizes that virtually all of the world's famines since World War II have been due to governmental mistakes or wars. That is, famines are caused by political action that inhibits production or political action to withhold food, not by farmer's failures to produce or, by implication, the lack of sufficient production technology. Low food production and subsequent availability can cause famine and starvation, but it is one of many possible causes. Food security and access to food are often more important than food supply. Undernourishment, starvation, and famine are influenced by the working of the entire economy—not just food production and agricultural activities as Avery (1994) claims.

Conway (1997, p. 33) agrees with Avery's view but differs in his view of what must be done. Conway argues that it will be possible for the world to provide enough food for everyone if (the important word) we (not just the United States, but all countries):

- Increase food production at a greater rate than in recent years;
- Do this in a sustainable manner, without significantly damaging the environment; and
- Ensure the food produced is accessible to all.

The latter is, of course, the hardest part. Conway's (p. 134) is a more nuanced analysis of the problems of feeding the third world than Avery's (1994). He establishes five priorities to achieve the goal of feeding all:

- Higher yield per hectare;
- Produced at less cost;
- With less environmental damage;
- Creating employment and income opportunities for the landless; coupled with
- Pricing, marketing, and distribution policies that ensure the poor gain.

Conway (p. 86) recognizes that the green revolution of the 1960s was dominated by the strongly held view that "a healthy, productive agriculture would necessarily benefit the environment. Good agronomy was good environmental management." Avery (1994, 1997) concedes this dominant view and asserts it is correct. Agricultural scientists have believed the same thing—good agricultural science is automatically good environmental management.

The technological optimists (the Cornucopians) who have read this far will tend to agree with the appraisals of agricultural technology offered by Avery and

Wildavsky (1979, 1997). The more cautious will tend to agree with Tenner (1996) and the New Zealand Parliamentary Commission (2001) that surprises are an inevitable concomitant of analytical science and caution, therefore, is obligatory. Many other sources of conflicting views can be found and could have been included in this chapter, but have not been because the list becomes long and greater clarity does not follow from additional sources. In addition to Avery and Wildavsky, the Cornucopians include Ridley (2010), Simon (1984, 1995, 1981), Khan et al. (1976), The Heritage Foundation, The Hudson Institute, the Competitive Enterprise Institute, and the Washington Legal Foundation. The cornucopian view is illustrated by a few words from Simon (1995):

> *Technology exists now to produce in virtually inexhaustible quantities just about all the products made by nature—foodstuffs, oil—and make them cheaper in most cases than the cost of gathering them in the wild, natural state (p. 25). We have on our hands now—actually in our libraries, the technology to feed, clothe and supply energy to an ever-growing population for the next 7 billion years (p. 26).*

The cornucopian view and its true believers have expanded human enterprise beyond long-term global carrying capacity (Rees, 2010).

Among the Jeremiads one finds Paul Ehrlich, Lester Brown, Wendell Berry, Wes Jackson, the Audubon Society, the Sierra Club, and Greenpeace International. Most people are left in a quandary by the plethora of confusing information from the polar views. If they choose to try to work their way out they may become convinced by the rhetoric from one side, or just stop thinking about the issues as their quandary deepens.

The remainder of this book will try to demonstrate that underlying each set of views on important societal and agricultural issues is an ethical position that is often unexamined and may even be unknown by its possessors, until it is identified and examined. Knowing the ethical foundation for a position on any issue is, I suggest, an important step toward discussion and resolution of any human dilemma. The issue to be resolved is not Wildavsky's (1997) question of whether or not accusations against technology are true but the more fundamental ethical question—What ought to be done? What is the right thing to do? The answer to the question about agricultural or any other technology will not be found in science, although science is essential to the quest. Science tells us what it is possible to do. When one knows what is possible, it is neither the data nor the conclusions that are most important. What is important, indeed decisive, are the reasons one chooses and can support with clear arguments that determine what ought to be done.

In his discussion of the promise and problems of technology, Peterson (1978) concludes with a parable to illustrate the importance of being able to make proper judgments in any human dilemma. To make the proper judgment one must be aware of and examine the reasons that support the judgment. If the reasons for an action are not made explicit, one risks making a decision based solely on economic, social, or political criteria and while it is appropriate to consider these criteria, they alone are not sufficient for making the best decision.

Peterson notes the parable was invented or translated by the English author, Somerset Maugham and was used by John O'Hara as the title of his first novel—*Appointment in Samarra*:

> *The chief steward of a wealthy merchant in ancient Basra went into the marketplace one day, and there he encountered Death, dressed as an old woman. On seeing the steward, Death suddenly drew back. In great terror, the steward returned immediately to the merchant's house and said, "Master, today I saw Death in the marketplace, and she made a menacing gesture to me. Please let me ride to Samarra, so that I can escape." The merchant replied, "By all means, take my fastest horse and ride."*
>
> *Later the same day, the merchant himself went into the marketplace and he, too, saw the old woman. He stopped her and said, "My servant told me that he met you here this morning and that you threatened him." And Death replied, "Oh, no—I did not threaten him. I was surprised to see him here in Basra, because tonight I have an appointment with him in Samarra."*

## References


Akamine, E.K. (1948). Plant growth regulators as selective herbicides. *Hawaii Agric. Exp. Station Circular* 26:1–43.

Asgrow. (1998). Survey, UK, PJB Publications Ltd. and World Crop Protection News 6/26/98 and 9/18/98.

Avery, D.T. (1994). *Saving the Planet With Pesticides and Plastic: The Environmental Triumph of High-Yield Farming*. Indianapolis, IN, The Hudson Institute.

Avery, D.T. (1997). Saving the planet with pesticides, biotechnology and European farm reform. *Brighton Crop Prot. Conf. Weeds* 1:1–18.

Bolley, H.L. (1908). Weeds and methods of eradication and weed control by means of chemical sprays. *N. Dak. Agric. Coll. Exp. Station Bull.* 80:511–574.

Carson, R. (1962). Silent Spring(1962). *Boston, MA*, Houghton Mifflin Co.

Clayton, M. (2005). Old culprit hits birds—maybe people. *Christ. Sci. Monitor* April:14, 15.

Committee on the Future Role of Pesticides in US Agriculture (2000). History and context. Pp. 17–32 *in* The Future Role of Pesticides in US Agriculture. Washington, DC, National Academy Press.

Colburn, T., D. Dumanoski, and J.P. Myers. (1996). *Our Stolen Future*. New York, Dutton Books.

Conway, G. (1997). *The Doubly Green Revolution: Food for All in the Twenty-First Century*. Ithaca, NY, Comstock Publishing Associates. A Division of Cornell University Press.

Crafts, A.S. and W.W. Robbins. (1962). *Weed Control*. New York, McGraw Hill.

Dibbell, J. (1996). Everything that could go wrong. *TIME Mag.* May:56.

Donaldson, D., T. Kiely, and A. Grube. (2002). *Pesticides Industry Sales and Usage: 1998 and 1999 Market Estimates*. Washington, DC, Biological and Economic Analysis Division, Office of Pesticide Programs, and US Environmental Protection Agency.

Economist. (2011). In place of safety nets. April 23, p. 18.

Egan, T. (2006). *The Worst Hard Time: The Untold Story of Those Who Survived the Great American Dust Bowl*. New York, Houghton Mifflin Co.

Goldman, L. (2004). *Childhood Pesticide Poisoning: Information for Advocacy and Action*. Châtelaine, United Nations environment programme.

Heap, I. (2003). International survey of herbicide resistant weeds. http://www.weedscience.com (accessed January 2005) and http://www.weedscience.org/In.asp (accessed December 2010).

Hamner, C.L. and H.B. Tukey. (1944a). The herbicidal action of 2,4-dichlorophenoxy acetic and 2,4,5-trichlorophenoxyacetic acid on bindweed. *Science* 100:154–155.

Hamner, C.L. and H.B. Tukey. (1944b). Selective herbicidal action of midsummer and fall applications of 2,4-dichlorophenoxyacetic acid. *Bot. Gaz.* 106:232–245.

Holling, C.S. (1998). Two cultures of ecology. *Conserv. Ecol. (online)* 2(2):4–5. (accessed November 2003).

Hopkins, W.L. (1994). *Global Herbicide Directory*. Indianapolis, IN, Ag Chem Information Services.

Isaacson, W. (2007). The last battle of Vietnam. *TIME Mag.* March.

Khan, H., W. Brown, and L. Martel. (1976). *The Next 200 Years—A Scenario for America and the World*. New York, W. Morrow and Co.

King, L.J. (1966). *Weeds of the World: Biology and Control*. New York, Interscience Publications, Inc.

Kirby, C. (1980). *The Hormone Weedkillers: A Short History of Their Discovery and Development*. Croyden, British Crop Protection Council Publications.

Kirschenmann, F. (2005). The death and rebirth of everything. *Leopold Lett.* 17(2):3.

Kirschenmann, F. (2010). Some things are priceless. *Leopold Lett.* 22(1):5.

Kroma, M.M. and C.B. Flora. (2003). Greening pesticides: a historical analysis of the social construction of farm chemical advertisements. *Agric. Human Values* 20:21–35.

Marth, P.C. and J.W. Mitchell. (1944). 2,4-dichlorophenoxyacetic acid as a differential herbicide. *Bot. Gaz.* 106:224–232.

Montgomery, D.R. (2007). *DIRT: The Erosion of Civilization*. Berkeley, CA, University of California Press.

New Zealand Parliamentary Commission for the Environment. (2001). Key Lessons From the History of Science and Technology: Knowns and Unknowns, Breakthroughs and Cautions. Wellington, NZ.

Peterson, G.E. (1967). The discovery and development of 2,4-D. *Agric. History* 41:243–253.

Peterson, R.W. (1978). *Technology: Its Promise and its Problems*. Boulder, CO, Colorado Association of University Press.

Pokorny, R. (1941). Some chlorophenoxyacetic acids. *J. Am. Chem. Soc.* 63:1768.

Rees, W. (2010). What's blocking sustainability? Human nature, cognition, and denial. *Sustainability: Science, Practice, and Policy* 6(2):13–25.

Reiss, M.J. (2001). Ethical considerations at the various stages in the development, production, and consumption of GM crops. *J. Agric. Environ. Ethics* 14:179–190.

Ridley, M. (2010). *The Rational Optimist: How Prosperity Evolves*. New York, Harper Collins Publishers.

Ryan, G.F. (1970). Resistance of common groundsel to simazine and atrazine. *Weed Sci.* 18:614–616.

Sen, A. (1981). *Poverty and Famines: An essay on Entitlement and Deprivation*. Oxford, Clarendon Press.

Shulevitz, J. (2011). The toxicity panic. *New Republic* April:11–15.

Simon, J. and H. Khan. (1984). *The Resourceful Earth—A Response to Global 2000*. New York, Basil Blackwell.

Simon, J. (1995). *The State of Humanity*. Cambridge, MA, Blackwell.

Simon, J. (1981). *The Ultimate Resource*. Princeton, NJ, Princeton University Press.

Slade, R.E., W.G. Templeman, and W.A. Sexton. (1945). Plant growth substances as selective weed killers. *Nature (Lond.)* 155:497–498.
Smith, A.E. and D.M. Secoy. (1975). Forerunners of pesticides in classical Greece and Rome. *J. Agric. Food Chem.* 23:1050–1055.
Smith, A.E. and D.M. Secoy. (1976). Early chemical control of weeds in Europe. *Weed Sci.* 24:594–597.
Tenner, E. (1996). *Why Things Bite Back: Technology and The Revenge of Unintended Consequences*. New York, A.A. Knopf, Inc.
Thompson, P.B. (2010). *The Agrarian Vision: Sustainability and Environmental Ethics*. Lexington, KY, University of Kentucky Press.
Troyer, J.R. (2001). In the beginning: the multiple discovery of the first hormone herbicides. *Weed Sci.* 49:290–297.
van den Bosch, R. (1989 [1978]). *The Pesticide Conspiracy*. Berkeley, CA, University of California Press.
Waggoner, P.E. (1994). How much land can ten billion people spare for nature? Council for Agricultural Science and Technology (CAST), Ames, IA. Task Force report no. 121.
Wildavsky, A. (1997). *But is it True? A Citizens Guide to Environmental Health and Safety Issues*. Cambridge, MA, Harvard University Press.
Wildavsky, A. (1979). No risk is the highest risk of all. *Am. Sci.* 67:32–37.
Zimdahl, R.L. (2010). *A History of Weed Science in the United States*. London, Elsevier, Inc.
Zimdahl, R.L. (1995). Introduction. Pp. 1–18 *in* A.E. Smith (ed.). *Handbook of Weed Management Systems*. New York, M. Dekker, Inc.
Zimmerman, P.W. and A.E. Hitchcock. (1942). Substituted phenoxy and benzoic acid growth substances and the relation of structure to physiological activity. Contrib. Boyce-Thompson Institute. Selective herbicidal action of midsummer and fall applications of 2,4-dichlorophenoxyacetic acid. *Bot. Gaz.* 106:232–245.

# 4 A Brief Introduction to Moral Philosophy and Ethical Theories

*Some there have been who have made a passage for themselves and their own opinions by pulling down and demolishing former ones; and yet all their stir has but little advanced the matter, since their aim has been not to extend philosophy and the arts in substance and value but only to … transfer the kingdom of opinion to themselves.*

***Sir Francis Bacon***

Discussion about agricultural ethics should not be regarded as something that is so scholarly that it risks being unaware of the realities of agriculture or the realities of the ethical life. That is, it should not be regarded as just an academic matter. Moral philosophy is not just an area of academic discourse that is fine in theory but not useful in practice. It is the opposite. The primary aim of moral philosophy is to ask and help us answer the question, how we ought to live.[1] Moral philosophy and its ethical theories are, at a minimum, "efforts to guide one's conduct by reason. Ethical theories and the values that undergird them support our daily life." An ethical theory that is useless in practice, that is, in life, is just that—useless. Ethical theories are not prescriptive. They support practice and supply a foundation, known or unknown, that is used to form reasons for action or inaction. The reasons help us justify why we do what we do and how our actions align with what we ought to do. Personal feelings lead toward actions. Mine, I am sure, are different than yours for many of life's dilemmas. That is why in our lives and in agriculture we must depend on reason, not just personal opinion, when deciding what ought to be done.

The following, briefly stated, complex moral dilemmas illustrate the importance of reason as opposed to personal opinion. These dilemmas are widely used in moral philosophy discussions.

The first moral dilemma:

*A runaway train is heading down the tracks toward five workmen who cannot be warned in time. You are standing near a switch that if activated will divert the train to a siding, but there is a single, unsuspecting workman there. Would you throw the switch, killing one to save five? Suppose the workman was on a bridge and you could push him onto the track where his body would stop the train. He is bigger than you are.*

[1] We are discussing no small matter, but how we ought to live (Socrates in Plato's Republic). The opening paragraph and much of this chapter has been significantly aided by Rachels (2003) and Rachels and Rachels (2010).

**Agriculture's Ethical Horizon. DOI: 10.1016/B978-0-12-416043-9.00004-0**

Would you throw the switch to save five people and kill one? Would you push the worker off the bridge? Why or why not?

The second moral dilemma:

> *Your lifetime dream of a luxurious cruise is shattered when the ship sinks. You are adrift with strangers in a sinking, overloaded lifeboat in the North Atlantic ocean. There is no possibility of a quick rescue. Even though all passengers have life jackets, the lifeboat is sure to sink and all will die in the frigid water. One person is awake and alert but seriously ill and will not survive several days exposure. All will be rescued, but no one knows when. Throwing the seriously ill person overboard will prevent the boat from sinking and save all.*

Will you do it? Why or why not? Many in the boat will object.

These are moral dilemmas because there is no obvious solution, no clear answer that all agree on. Moral debate and ethical criteria are required to think about and discuss why some solutions are appropriate and others are socially unacceptable and abhorrent. My experience when I have brought ethics into discussion of agricultural issues has been that my colleagues, while willing (even eager) to engage in the discussion, are not persuaded to continue the quest. Ethics seems to be regarded by agricultural scientists as something that is peripheral to the conduct of their science, rather than central. It is regarded as something that is purely academic. There is no reluctance to begin discussion of the ethics of agriculture, but there is a sense that it is unnecessary. Agriculture, after all, deals with producing food and fiber for humankind and, the general view is that the practice of agriculture is *a priori* ethical. Agricultural scientists, indeed most people, think they are, and actually are, fundamentally ethical in their daily lives. If moral energy must be expended, it most commonly goes toward defending claims against us, and "that includes protecting the state of our soul as purely private, purely our own business" (Blackburn, 2001, p. 4).

If Blackburn is correct that the majority of ethical reflection is purely personal and if agricultural practice is already ethically correct (agriculture feeds people), what is the problem? The problem is that this claim ignores several problems, including environmental problems, the practice of agriculture has created: soil erosion, pesticide pollution of soil and water, pesticide harm to human health, pesticide harm to non-target species, fertilizer (especially nitrogen) pollution of water, human displacement by labor-saving technology, and so forth (see Aiken, 1984). The claim that agriculture is ethically correct ignores the charge that modern technology that permits one person to do the work of a thousand also permits that person to wreak the environmental havoc of a thousand (Peterson, 1978). Environmental and other problems caused by agriculture have largely been externalized, thus transferring the effects and costs of agriculture to the general society. Agriculture's practitioners thereby avoid ethical challenges. The claim also ignores key ethical questions—What ought to be done? What is the right thing to do? The position ignores the core of morality: "When moral worth is at issue, what counts is not actions, which one sees, but those inner principles of action that one does not see" (Blackburn). A goal of moral philosophy is refinement of ethical theories that have been developed by careful thought and are consonant with inner moral principles and available scientific knowledge (Comstock, 1995). Agriculture has largely ignored the necessity of formulating an ethic, a moral foundation, which enables defense of its actions.

The apparent reluctance of agricultural scientists to raise and resolve ethical issues arises primarily from the unexamined assumption that their activity (their science and the practice of agriculture) is already ethically correct. Agricultural people have not needed an ethicist to point out that their ethical responsibility was production of healthy, nutritious, abundant food (Dundon, 2003). They know they have done this well. The reluctance also arises from at least three other sources. *First* is the fact that people engaged in agriculture (farmers, ranchers, and scientists) have not been educated to question agricultural practice. Agricultural education does not emphasize and often completely ignores the moral aspects of the problems agricultural practice creates. Consequently, students and those who practice agriculture do not have the linguistic or intellectual tools to discuss the ethics of agricultural practice and are not encouraged to do so. Agriculture education emphasizes study of how to employ technology to increase production and profit. It does not emphasize questioning the technology or its results. *Second*, there is a pervasive idea that ethics is largely, if not completely, subjective. Ethical decisions, it is claimed, are just a matter of opinion and opinions are strongly influenced by one's culture. There are no ethical facts available to support one opinion as opposed to another. Thus, there is no factual basis to question or defend an ethical position. The claim is that no one can prove their ethical view is the correct view. Ethical claims are not the kind of rational truth referred to in Chapter 1. They do not consist of rational statements defined mathematically, which are publicly verifiable, literal, definitive, and precise. Opinions are just that. They are personal views that lack objective truth. Ethical views, from this perspective, are just one opinion against another. *Finally*, McKibben (2003, p. 186, citing Rothman, 1998, p. 37) points out that when scientists do deal with a moral question, "even the language of moral inquiry quickly turns technical." Acting immorally is translated as "acting without knowing the consequences, taking risks." Playing God, an expression from the realm of personal knowledge, is vague but becomes understandable, in the scientific mind, when it is translated as acting without knowing the consequences (Rothman, 1998) (although one presumes, God always knows the consequences of an action). Reducing moral questions to consequentialist objections makes such questions manageable for scientists. The objection to the technology is reduced to the technical problem of how to reduce risk from the technology, the solution to which is a scientific, quantifiable matter (Sandler, 2004).

## Science and Emotion

Modern agriculture is based on science, which presents a defensible, objective tale of "what is" and "how things work." Agriculture's productive success has been achieved by the careful application of science and technology, largely without reference to what is regarded as the subjective, emotive story of what ought to be.

However, emotions actually produce quite reasonable human behavior, especially from the point of view of survival (Damasio, 1999, p. 54). Human emotions are frequently inseparable from the ideas of reward or punishment, pleasure or pain, approach or withdrawal, and of personal advantage and disadvantage. Inevitably emotions are inseparable from the idea of good and evil (p. 55). Nevertheless, neither philosophy nor science has trusted emotion, and science, especially, has dismissed emotion as unworthy of trust. Damasio (p. 38) reports that Darwin, William James, and Freud all wrote extensively on different aspects of emotion and "gave it a privileged place in scientific discourse." But emotion has not been trusted in scientific inquiry or in the laboratory. Emotions are too subjective, elusive, and vague (p. 39). Emotions are "at the opposite end from reason, easily the finest human ability. Reason, scientists assume, is presumed to be entirely independent from emotion" (p. 39). Thought about emotion places its origin in the brain where reason also occurs, but relegates it to a lower neural status. Emotion was not rational and even studying it was probably not rational (p. 39). Singer (2000, p. xix) tells us that "If emotion without reason is blind, then reason without emotion is impotent." Both are required in all fields of human endeavor, including, or perhaps one should say, especially, agriculture.

However, these twentieth century assumptions about the exclusively emotional basis of values and ethics and therefore what ethics is about are unwarranted and not defensible. Both science and ethics are highly value-laden and full of emotion. When we examine the "nature, origins, and methods of science the logical conclusion is that science is ineluctably involved in questions of value, is inescapably committed to standards of right and wrong, and unavoidably moves toward social aims" (Glass, 1965). While Glass recognizes that ethics in science rests on moral integrity, he does not deny that emotion is involved in determining what a scientist chooses to study. Scientists may become irritable, excited, demanding, and maybe even irrational when debating the importance of an experimental approach or the interpretation or value of results. Values are the things we humans hold most dear. They are matters that demand enthusiastic participation (science, skiing, music), or things toward which we feel most compassionate (children, the homeless, stray animals, a favorite charity, small businesses), or they are foundational to beliefs which cause us to behave in certain ways (scientific objectivity, religion, political affiliation) and accept some propositions wholeheartedly (no new taxes, smaller government) while rejecting others (foreign aid, universal health care). If we attempt to exclude emotion from judgment, value may consist only of those things which one holds to be most important (national defense, preservation of national parks, feeding the poor). Our values are how we define what is most important—they are how we put first things first. Be it ever so logical, the process is not without its emotional aspects (Shahn, 1957, p. 95). In fact, the more important a matter becomes in all fields of human endeavor, the more emotionally laden it becomes. Scientists learn that science does not make value judgments. They have a passion for objectivity, a value of the highest order. Important scientific matters should be decided objectively with the facts not with subjective, emotionally laden opinions. Science seeks to understand the world, not judge it (Pinker, 2008). If scientists use and rely only on objective criteria they risk failing to recognize where the objective intellect leaves off and emotion begins. I modestly claim that an important objective of this book is similar to Rawls' view of the task of philosophy: not to hold up a mirror for self-congratulation, but to recall us to what is best in us when we have failed or betrayed our ideals (see Larmore, 2008).

## Universal Values

Given the common assumption that values and ethical decisions are all subjective and culturally influenced, it is logical to posit that there cannot be any universal values shared by all humans that extend across cultures and time. Kidder (1994) however, proposed an "underlying moral presence shared by all humanity" that erases borders, races, and cultural traditions. These may be the values one might teach students or hope that children learn as accepted universal behavior; they are the right things to do and ways to behave. The elements of a universal code of ethics, universal values, identified by Kidder (1994) after interviews with people from 16 nations include:

| | |
|---|---|
| Love: | love of all people, compassion, helping one another |
| Truthfulness: | keeping promises, not lying or being deceitful |
| Fairness: | fair play, justice for all, the pursuit of equality |
| Freedom: | the desire for liberty, political democracy |
| Unity: | the need for community, solidarity, cooperation |
| Tolerance: | acceptance of thinking differently, listening to different views |
| Responsibility: | for actions, for the future, for others |
| Respect for life: | do not kill |

Josephson (1989) expresses the values above in slightly different words. Things that are right are the things that help people and society. In his view, they include compassion, honesty, fairness, and accountability, which he regards as absolute universal values. For Josephson, the essence of ethics is some level of caring. Blackburn (2001, p. 127) cites things that virtually all human beings care about for themselves: safety, security of possessions, satisfaction of basic human needs, and a basis for self-respect.

> *Every society that is recognizably human will need an institution of property (some distinction between "mine" and "yours"), some norm governing truth-telling, some conception of promise-giving, some standards restraining violence and killing. It will need some devices for regulating sexual expression, some sense of what is appropriate by way of treating strangers, or minorities, or children, or the aged, or the handicapped. It will need some sense of how to distribute resources, and how to treat those who have none. In other words across the whole spectrum of life, it will need some sense of what is expected and what is out of line.*
>
> *(Blackburn, 2001, pp. 22–23)*

Societies have other similarities that relate to their ethical foundation. Virtually every culture has its version of the Cinderella story and some version of the Golden Rule[2] is found in all religions (Van Eenwyk, 1997).

Many will immediately note that each of the sets of good values noted above is violated regularly by individuals and nations. Therefore, one must ask, how can they be universal? One can only agree that each is violated regularly but we must also recognize that humans in most societies lament and note violations as failures not achievements. Tharoor (1999) addresses the same issue and arrives at the same conclusion: there are universal human values. There is objective truth in ethics in that humans, independent of race, nation, or culture, can agree on some standards as being right. These are things that are right not just for me or you, or just for today, but for all people in all places always. Tharoor (1999) notes the philosophical objection that nothing can be universal because all rights and values are limited by culture. Because there is no universal culture, there can be no universal human rights. Of course, Tharoor is correct, there is no universal culture but the "number of philosophical common denominators between different cultures and political traditions make universalism anything but a distortion of reality" (Tharoor, 1999). Universality does not presuppose uniformity (Tharoor, 1999). Basic human rights, the things on which people across cultures agree, can be universalized and should be promoted even though they do not now exist in all nations. These include:

- The right to a normal course of life.
- Freedom from torture.
- The right not to be enslaved, physically assaulted, arbitrarily arrested, imprisoned, or executed.
- The expectation of tolerance of difference.

Again, many will note that these rights are regularly denied in many nations. Because this is true, they will ask, how can such rights be regarded as universal? What seems to be universal in this view is not what is good, but that some people are bad and some governments deny basic human rights.

Agricultural students are not taught to question the possible negative effects of agricultural technology, and they learn, as many people do, that humans, by nature, are greedy, aggressive, and compulsively and selfishly acquisitive. These are what Gould (1993) calls the "vexatious issue of human nature." We are natural beasts and nature is really red in tooth and claw. Humans express these natural instincts differently but they define who and what we are. He claims that events of great rarity make evolution, his analogous claim is that human events of great rarity make history; "human history is made by warfare, greed, lust for power, hatred, and xenophobia". But then he says, wait a minute, even if such events create our history, it does not follow that the

[2] *Christianity*—All things whatsoever ye would that men should do to you, do ye even unto them. *Confucius*—Never do to others what you would not like them to do to you. *Judaism*—Thou shalt love they neighbor as thyself. *Islam*—No one of you is a believer until he loves for his brother what he loves for himself. *Hinduism*—Men gifted with intelligence … should always treat others as they themselves wish to be treated. *Taoism*—Regard your neighbor's gain as your own gain, and regard your neighbor's loss as your own loss.

occurrence and nature of these events define human nature. Human nature is defined by what he calls "Ten Thousand Acts of Kindness." We do not go about our daily lives hitting or threatening to hit each other. We do not fight regularly. We make way for the elderly and the infirm. We hold doors for others (perhaps, if we are old, we hold doors open for women). We say "thank you" and "pardon me." We smile a lot, even at strangers. Gould asserts that we are basically nice folks and our real nature is not defined by the infrequent bad (albeit major) events that determine the fate of nations. We are incorrect in assuming that "the behavioral traits involved in history-making events must define the ordinary properties of human nature." Violent events make the news each day. Violence dominates what we watch on television. Kindness is fragile and easily overlooked because it is so common, but it is what dominates daily life. The universal values identified by Blackburn, Kidder, and Tharoor are what characterize human life. Most people ascribe to and practice them. Some of the most basic ethical issues and values are not just opinions that differ between cultures, races, genders, and time; indeed, they are universal.

A story I used in the agricultural ethics class will affirm, but I hope not over-emphasize the point. Early in the semester, I tell the class that only blonde women will be able to receive an A for the semester. The best men and brunettes or red-headed women may achieve is a B. I justify the decision because I am the Professor in charge of the class and it is my responsibility to set the standards by which student performance will be judged. Not surprisingly, over several years and among different students, the reaction has always been the same—That is not right! You can't do that! When pressed to explain why it is not right, the students always offer the same reasons. It is not fair! It is not right to award an A to blonde women! Being blonde and female has nothing to do with class performance! It is arbitrary! If I maintained my right to establish a blonde/female grading criterion, students would appeal to my Department Head and the Dean, and they would win the argument.

It is a simple case to illustrate that all ethical questions are not just matters of opinion. There are ethical standards that are not just subjective and, to the student's surprise, there is widespread agreement on many things. Hair color and gender are not appropriate grading criteria. Beating one's girlfriend or wife is not OK. Slavery is not right. All humans should be free of torture and the threat of bodily harm. Moreover, there is widespread agreement on the reasons why some things are right and others are not. Most ethical matters are not just up for grabs; they are not just your opinion versus mine.

## Ethics in Agriculture

Discussion of foundational ethical principles is essential if agricultural practice is to change in the sense that some of the problems agriculture practice causes are to be reduced or solved and if public challenges to agricultural practice are to diminish. Many professions have a view that they are better than, more ethical than most others (Josephson, 1989). Each profession tends to see the shortcomings of others and justify their moral laxity by saying that what is observed represents aberrant behavior, not the professional standard. For agriculture, ethical lapses often are identified by outsiders whom agricultural folk claim don't really understand the essentiality of agriculture and its production technology (Aiken, 1984). Outsiders, it is claimed, demand a higher standard than they practice. Besides, it is argued, some values must be sacrificed (environmental quality) to higher values (increasing production) if agriculture is to progress toward fulfilling its primary task: feeding the world. Agricultural people often argue that they are not the only ones guilty of ordering values. They claim *Tu quo que*: Thou also, or more simply, you do it, too. A claim, which, of course, does not absolve one of guilt.

As we explore what ought to be done rather than just what can be done, we will find surprising agreement about the standards by which one decides what ought to be done and why some things are the right things to do. Once one decides between the clearly unethical and the ethical thing to do, there will still be conflicts, there will still be choices. If we can agree that it is usually not ethically correct to lie, steal, cheat, or harm others, then we still have to decide what is the right thing to do to address agriculture's problems. We will face moral dilemmas that do not involve a choice between what is ethical and clearly not ethical but between two alternatives, neither of which is all bad. The result of choosing is often not clear when the choice has to be made. That is why we say we are caught on the horns of a dilemma (di = two, lemma = propositions). Such choices are common human experiences— To be or not to be, whether or not to eat the apple, Sophie's Choice, the lady or the tiger (Van Eenwyk, 1997, p. 31). Moral dilemmas are common in agriculture and an ethical foundation is needed to help all decide between two choices where each has strong supporting arguments. For example:

1. Should we increase agricultural production, to feed more people, regardless of the environmental harm the technology that creates the production causes?
2. Should we raise animals in confinement if it is harmful to the animals but reduces the cost of meat for consumers?
3. Should we mine water from deep aquifers to maintain irrigated farms in dryland areas even though the production system is not sustainable?
4. Should we change the US soybean production system to decrease soil erosion?
5. Should we decrease nitrogen fertilizer use in the Mississippi basin to reduce the effects on fishing and ecological stability in the Gulf of Mexico?
6. Should family farms be protected and preserved or allowed to die because they are economically inefficient, that is, they can't make sufficient profit?
7. Should we give more food aid to developing countries?
8. Should we accept or reject agricultural biotechnology?
9. Should we reduce herbicide and other pesticide use in American agriculture?

All items in this partial list are difficult dilemmas for agriculture and each has a moral dimension. It is time, indeed past time, for all involved in agriculture to think about and address these and similar questions. The next generation of agriculture's practitioners, scientists, and teachers should be equipped with the intellectual tools that are required to guide decisions about agriculture's existing and future ethical dilemmas (Chrispeels, 2004). Because each of the above questions is fundamentally an ethical question, we now turn to a brief description of a few of the most relevant ethical theories that may guide us as we try to address the moral questions that agricultural practice compels us to consider. Ethical theories do not and cannot provide final, definitive answers. They are guides. They serve to remind us of what we know is right.

## Contemporary Normative Ethics

*We are discussing no small matter, but how we ought to live.*

*Socrates, as reported by Plato in The Republic*

Moral philosophy is the branch of learning that deals with the nature of morality and the theories that are used to arrive at decisions about what one ought to do and why. Much has been written about moral philosophy and the theories that support ethical decisions. One of the best, brief explanations of moral theories is found in Rachels and Rachels (2010).

*If we want to discover the truth, we must try to let our feelings be guided as much as possible by the arguments that can be given for the opposing views. Morality is, first and foremost, a matter of consulting reason. The morally right thing to do, in any circumstance, is whatever there are the best reasons for doing.*

*Rachels and Rachels, p. 11*

I will draw heavily on Rachels and Rachels' work and briefly present only five moral theories: ethical egoism, social contract theory, virtue theory, deontological or Kantian ethics, and utilitarianism. These moral theories are largely unexamined within agriculture but, I submit, may be operative among those who practice agriculture. The chapter concludes with the one theory (utilitarianism) that seems to dominate resolution of moral dilemmas in agriculture. I will not make any attempt to say all that could or ought to be said about morality, moral philosophy, or moral theories. That is not the purpose of this book and others (see Blackburn, Comstock, Marks, Rachels and Rachels, and Singer) have discussed these things well.

A note of caution—the encounter with meaning in ethics often cannot be controlled by the seeker of meaning. The seeker of scientific truth (what Chapter 1 describes as rational truth) pursues something that can be defined mathematically, and is publicly verifiable, literal, definitive, and precise. In ethics, there are no moral facts that can be raised up and offered to others in the same way. Meaning in ethical discussions, as distinct from scientific truth, comes about as a result of the application of reason in a relationship of openness and trust (Van Eenwyk, 1997, p. 79). The relationship may seem to resemble a chaotic mixing in which those who seek meaning and what is experienced as they seek become so intertwined that a new symmetry, a new understanding, may come into being. For many, this may seem to move too close to the realm of personal truth or subjectivity, which, in science, is the least worthy of being called truth. But in moral philosophy and ethical decision making, the use of human reason to search for meaning is not just subjective. Invisible, beneath every moral decision, beneath every gut reaction or feeling about what is the right thing to do, there is a moral foundation—a moral theory. To say it is all just subjective is an artifice we use to avoid examining the reasons for ethical decisions (Marks, 2000, p. 13).

**Highlight 4.1**

Aiken (1998) proposed five proper goals for agriculture:

1. Profitable production
2. Sustainable production
3. Environmentally safe production
4. Satisfaction of human needs
5. Compatibility with a just social order

As agriculture's practitioners think about the ordering, comparative desirability, and the difficulty of achieving one or more of these goals, it is inevitable that they will ask to what extent current methods of agriculture lead us toward or away from a productive, sustainable, and environmentally sensitive agricultural system. It is clear that, except for large producers, agriculture is not especially profitable and is becoming less so, even for large producers. Our present system fails to achieve the first goal for many agricultural producers.

Many question if our present capital, energy, and chemically dependent, but highly productive agricultural system is sustainable. If agriculture is viewed purely as a production enterprise and its efficiency or achievement is measured solely in terms of output per acre, per person, or per hour of human labor, then it is efficient. If, however, efficiency is measured in terms of sustained production over an infinitely long time, then our present system may not be sustainable. The second of Aiken's criteria is not satisfied. Because agriculture is the single, largest human interaction with the environment, it is inevitably disruptive. Ecological stability is a goal to be constantly approached but possibly never reached. That means agriculture should not use production and profit as its only criteria of success. Agriculture should not be just an extractive industry; it must be restorative. Its required technology must complement, not harm nature. Farming systems that restore ecological systems are desirable. The proper use of science is not to conquer nature but to live in it. The "ultimate proper goal of farming is a not the perfection of crops, but the cultivation and perfection of human beings" (Fukuoka, 1978). Goal three is elusive but achievable.

Agriculture as we know it in the world's developed countries satisfies most human needs and most human wants (e.g., strawberries any month of the year). But do we live by bread alone? Are small, rural communities important, and, if so, are they important only to their residents or do they have value for all?

Is the agricultural landscape (those amber fields of grain) valuable? Should a society take pride in the fact that less than 2% (in 2005, in 2009 less than 1%) of its population feeds everyone else, which heralds the decline of rural life and low income for many who must leave the land? Should satisfaction of human nutritional needs and wants always take precedence over ecological needs? Do our food needs and wants support an unsustainable agricultural system? Whose responsibility is it to achieve agricultural sustainability? These are among the many moral questions those engaged in agriculture ought to be addressing.

Too many people in my town, in my state, in my country, and in our world are regularly homeless, hungry, and malnourished. That implies that agriculture is not compatible with a just social order and not directed toward achieving justice for all. Achieving a just social order is, by definition, a social, not a production goal. Agriculture has not typically been regarded by its practitioners as a social enterprise. But, all must eat to survive. To eat all must be able to grow their own food, harvest it from nature, steal it, buy it, or be given it. Most humans cannot grow their own or hunt and gather food. Most do not steal. Most buy their food and a few are given food. A just social order has not been an agricultural goal. It is a moral goal where agriculture and society intersect.

## Ethical Theories Relevant to Agriculture

Ethics, in one sense, is a branch of philosophy that studies and examines values relating to the standards of human conduct and moral judgment (Rollin, 1999). In this sense, ethics deals with the basic principles that determine the rightness or wrongness of certain actions and with the goodness or badness of the motives and ends of those actions. In another sense, ethics is a system of moral principles that govern and define how people should (ought to) behave in a particular group or culture (Rollin) (e.g., Christian ethics, medical ethics). In this sense, ethical behavior is comprised of the reasons that govern people's views of right and wrong, good and bad, fair and unfair (Rollin). Underlying and supporting moral behavior are ethical theories that describe reasons for choosing a particular action and what it is that makes it moral. Most people do not use moral reasoning. They use moral rationalization; "a post-hoc process in which ... they search for evidence to support … an initial intuitive reaction" (Haidt, 2007). That is, they conclude what should be done and work backward to a reasonable justification (Pinker, 2008). This does not mean that the conclusion (what ought to be done) is wrong. Intuitive reactions are often what one ought to do. The problem is that the justification for intuitive reactions is commonly subjective and therefore more difficult to defend than actions based on sound moral reasoning. For more details of the arguments that support each moral theory, see Blackburn (2001), Holmes (1993), Marks (2000), Rachels (2003), or Rachels and Rachels (2010).

## Ethical Egoism

The central idea of ethical egoism is that all will be right if all people do whatever will best promote their best interests. To be ethical, one must always act to promote one's own good. All acts ought to be self-interested acts.

It is a simple and appealing theory. What could be better than always doing what is good for oneself? Many farmers and ranchers would welcome a moral theory that permits them to do what is best for them and their operation without regard for anyone else or anything else. Ethical egoism does not say that one should always maximize personal good. That would be egoism and have no ethical component. Those who ascribe to ethical egoism are not necessarily more selfish or egotistical than others may be. If the ethical egoist is asked to cut the birthday cake, the largest piece may go to someone else because, in the long run, that serves the cake cutter's best interest. Therefore, this standard does not mandate that one avoid actions that help others, because the helper may have an ulterior motive of self-interest. Ethical egoism denies that frequent losses of money in Las Vegas or heavy drinking of alcohol, while momentarily enjoyable, are acceptable, because in the long run, they work against self-interest.

Ethical egoism does not require altruistic acts and may often dictate against them, but it does not rule them out, especially when such acts may be perceived to serve one's own interests, in the long run. As Rachels (2003) points out, ethical egoism makes most altruistic acts less than desirable because they tend to de-emphasize the importance of self-interest. Ethical egoism is compatible with commonsense morality in that it supports actions that contribute to promotion of one's best interests: do not lie, keep your promises, and do not harm others.

In spite of its appeal and in spite of the fact that many people act as ethical egoists, especially in economic decisions, the theory has serious faults that prevent its adoption as a universal standard. Most importantly, it fails to explain how to resolve conflicts of interest. For example, consider the case where two siblings have inherited a farm on the death of their parents. One considers it to be in her best interest to maintain the property as a farm and become a farmer. The other considers it to be in his best interest to sell the land to a developer and reap large profits. Both are acting in their own best interest and ethical egoism does not offer any way to resolve the conflict of interests (Holmes, 1993; Rachels and Rachels, 2010). As Holmes points out, because ethical egoism is a consequentalist position (what happens is what counts), and we know that we cannot always predict the results of an action, we may be wrong and an action may not be in our best interest.

It is also clear, as the farm example shows, that ethical egoism may be logically inconsistent. Both siblings are acting in their best interest and it is wrong to harm another by denying them the ability to act in their best interests. No moral theory can be both right and wrong at once. Finally, Rachels (2003) suggests that ethical egoism is unacceptably arbitrary because it divides the world in two—my interests and the interests of all others. The claim that my interests are always more important than the interests of others just won't hold. It works against common sense. Everyone is interested in eating and we all must eat. My interest in eating should not always trump the equal interest of someone who is poor but has the same need I do.

## Social Contract Theory

Under this theory, the right thing to do is to follow the rules that rational, self-interested people have agreed upon and established for their mutual benefit. Rachels (2003) provides a good explanation of the essence of social contract theory as a moral foundation for a society. The basis was first set forth by the British philosopher, Thomas Hobbes (1588–1679) who posited that all people want to live in a peaceful, cooperative society where all are governed by a set of moral rules that all have agreed to for their mutual benefit. It is an unwritten, but known, contract among those in a society. The idea is that people can freely reach an agreement about the principles that will govern all of them (Holmes, 1993). Because those who make it will be affected by the social contract, it follows that they should be the ones who create it. Rachels (2003) says the idea is feasible because people have an equality of need and there is a scarcity of goods that people want. In addition, there is an equality of human power that can be created by the contract. Some are smarter or richer but they cannot and should not prevail forever against others who act together. Finally, we cannot rely on the goodwill or altruism of others. All are primarily self-interested and while they may help others in need, it is more likely that they will not, so society needs to be ordered in such a way that the vital interests of all will be served.

Hobbes suggested that because self-interest governs human behavior, all will be doing what they can to get what they need. Scarcity means that many will be competing for the same goods and life will become intolerable in the face of relentless competition of all against all. To escape the inevitable societal collapse when governments collapse and although there are laws there is no one to enforce them, we create rules that govern behavior and relationships.

Rachels quotes an unnamed critic of social contract theory—the social contract "isn't worth the paper it's not written on." The problem is that the social contract is an historical fiction. There has never been one and no one has ever signed or been asked to sign one. There is an implicit contract we all participate in and members of any society become part of (presumed signers) by their action according to the rules of their society (Rachels). A more important problem with the theory is that it is easy to exclude the poor, the dispossessed, the homeless, and the uneducated. These members of a society may not know the rules intended to govern members' behavior and do not benefit from the social contract.

A modern view of social contract theory is that set forth by Rawls (1971). He proposes a thought experiment centered on the kind of society we want to live in. In this experiment, each person will be acting from what Rawls calls the original position. From this position, each is asked to select the principles that will create a proper society. The society will accomplish what Rawls regards as the essentials of a well-ordered society. It will be a social structure that will advance the good of its members, be regulated by a public conception of justice that the basic institutions of society satisfy, and that the people know they satisfy. Everyone will accept the same principles of justice, and society will be stable (Holmes, 1993). Those who choose the principles are assumed to be rational and self-interested as they select the principles that will govern relations among themselves. Because we are all self-interested, each of us will want to choose principles that favor our position in society, indeed, our position in the world, although Rawls seems to assume all who choose are Americans. Therefore, as decisions about the society are made, each person will operate behind a "veil of ignorance." The veil prevents each person from knowing what position they will occupy in the society. Because of the veil, no one can know, as the contract is being prepared, whether they will be rich or poor, educated or uneducated, male or female (Rawls does not mention gender), healthy or ill, capable or not so capable of achievement, or young or old. The veil makes it impossible to choose social policies (parts of the social contract) that favor one position (e.g., to each according to need) over another.

No Rawlsian society has ever been constructed and it is not at all clear that people of one culture would agree on the principles or that people from different cultures would arrive at the same principles. The problems of cultural relativism seem inevitable. However, even if no such society has been created, the thought experiment compels consideration of what the characteristics of such a society ought to be and how closely present social contracts come to justice for all. It may not be far off the mark to suggest that there is great disparity between the social contract in an ideal democracy and democracy as it exists.

Even though most citizens of most societies do not think often, or perhaps, not at all, about agriculture and its practitioners, it is reasonable to propose that agriculture has obligations and can expect rewards under all social contracts. The first expectation of agriculture's social contract is production. Those who practice agriculture are expected to provide food for those who do not. Failure to do so will violate the unwritten terms of the social contract. Secondly, because agriculture is the largest and most widespread human interaction with the environment, its practitioners are expected to care for the environment. Perhaps they do not do that well, but the expectation is real. Finally, those engaged in agriculture can expect a reasonable reward. Government subsidies (albeit primarily for a few major crops: wheat, corn, soybeans, cotton, and tobacco) are evidence of societal concern for agriculture and of the inadequate rewards the free-market system provides for production and environmental care.

## Virtue Ethics

Rachels (2003) defines a virtue as "a trait of character, manifested in habitual action, that is good for a person to have." Kidder (1994) and Josephson (1989) enumerated some of the habits and actions a virtuous person demonstrates. In general we like to believe that a person will do well by doing good, by acting virtuously, or, at least, by avoiding doing bad (Blackburn, 2001). Different roles may demand different virtues, a farmer, mechanic, or a teacher can each be virtuous but the specific virtues are different (Rachels, 2003). Traits that all virtuous people seem to share are generally recognized: compassion, dependability, fairness, honesty, loyalty, patience, and tolerance (see Section Universal Values). These shared traits and actions speak to our common humanity. They are the characteristics of a good person. It is assumed their actions will demonstrate virtue.

Most philosophers regard virtue theory as part of an overall ethical theory rather than as sufficient unto itself. Virtue theory tells one how to act and what characteristics are good but it does not lead one toward a complete conception of the reasons why particular actions are right. The essence of virtue theory is to ask what traits of character make one a good person? In agriculture, one can ask what traits of character make one a good farmer, rancher, or citizen? The extensive work of Wendell Berry draws heavily on a virtue perspective to describe the value of family farms and farmers. Much of his work reflects the thought of Thomas Jefferson, who in his *Notes on the State of Virginia* said: "Those who labor in the earth" have (if anyone has) been chosen to receive God's "peculiar deposit for substantial and genuine virtue." Those who cultivate the earth are the most valuable citizens. They are the most vigorous, the most independent, and the most virtuous because they are tied to their place by lasting bonds. Jefferson and Berry (1977) regard a farmer's self-interest as coincident with the public good. But family farms in the United States are failing and as they are lost, Berry suggests we lose virtues important to our society, that they embody and perpetuate, when we lose the family farm or ranch. We lose "much that we need as human beings: a daily personal relationship with nature; a social contract that works; a sense of connection with others; a sense of inhabiting a place for the long haul" (Knize, 1999). Berry suggests the family farm is failing because it belongs to a value system and a kind of life (a virtuous life) that are failing. As family farms fail, local schools, local businesses, the domestic arts of homemaking, and cultural and moral traditions, are also failing. The loss of family farms is coincident, in Berry's view, with the loss of the desirable societal virtues of thrift, generosity, and neighborliness.

Most will agree that the traits Berry describes as those of the generalist are virtuous. They include: caring for the environment, regard for the health of all creatures, an awareness of the importance of the carrying capacity of the farm or ranch and thus environmental carrying capacity, working well as opposed to working quickly, a regard for the natural order of things, an emphasis on serving a place (a community) rather than an institution or an organization, and the ability to think qualitatively (how good or how well) rather than simply quantitatively (how much). Most people would regard these traits of character as virtuous and desirable in any occupation. Those in agriculture who have them, one assumes, will do good things and achieve good results.

If a virtue is a trait of character manifested in habitual action (Rachels, 2003), are there specific agricultural actions, in addition to the generalist traits identified by Berry (1977) that are agriculturally virtuous? Clearly the answer is, Yes, but examples are required. A farmer or rancher who allows the soil to erode while gaining short-term profit from such neglect is not acting virtuously. The person is a bad farmer or rancher who may be temporarily doing well economically by practicing bad agriculture. An ethic of virtue includes or presupposes an ethic of right conduct (Holmes, 1993, p. 80). Sandler (2004) says that "the promotion of a technology must not be contrary to justice" (a virtue), as it would be if those who benefit from the technology are not those who bear its burdens. Pesticides benefit users and developers but the external costs (e.g., environmental pollution, harm to humans and other species), which are often quite high, are borne by all. Sandler (2004) also notes the case of farmers whose tax dollars are used to support agricultural research at public universities to develop genetically modified organisms that may harm the farmers are compelled to pay taxes to support the research. Compelling, virtue based, arguments raise objections to such actions. Sandler (2004) proposes there are no compelling aretaic (virtue based) reasons to justify promoting genetically modified agronomic crops (GMOs) (e.g., alfalfa, corn, cotton, soybeans, sugar beets). In the first edition of this book, a few of these crops (corn, cotton, soybean) made up nearly the entirety of genetically modified crops cultivated in the world. Given the widespread acceptance of GMOs, it is clear that Sandler's claim has not been persuasive among farmers. The basis for his claim is that promoting genetically modified crops is not virtuous because it has done little or nothing to relieve suffering of the poor, eliminate global inequality, or reduce the negative ecological effects of modern agriculture. This issue will be addressed in detail in Chapter 8.

## Deontological or Kantian Ethics

The German philosopher and devout Christian Immanuel Kant (1724–1804) described a demanding ethical standard that he called the categorical imperative: "Act only on that maxim through which you can at the same time will that it should become a universal law." The operative words under this ethical stance are right and wrong as applied to intent. Consequences are not unimportant but intention is paramount. Personal happiness achieved through ethical egoism is not an objective of Kantian ethics, in fact it is unimportant. Kantian ethical standards are much more demanding than other ethical positions. They say that independent of one's wishes, all people have a moral duty to act in certain ways. One ought not to lie, cheat, etc., period. When one's spouse is dressed and ready to go to a special event and enters the room and asks, "How do I look?" If one knows it is wrong to lie, then the truth must be told, regardless of the consequences. It is a demanding moral standard.

Creating universal moral laws becomes a problem when those laws create conflicts. One of the clearest examples is the well-known Dutch fishing boat story. The Captain of the Dutch fishing boat had picked up several Jews during World War II. With a group of Jews in the hold, the boat was proceeding to a safe harbor when a Nazi patrol boat appeared and stopped the fishing boat. The Nazi Captain asked who was aboard. One is not to lie, ever, and one also has a duty not to allow the murder of innocent people, which is exactly what the Dutch Captain knew would happen if he told the truth. The absolute rules of Kantian ethics do not help much with this kind of dilemma where two absolute rules—do not lie, do not cause harm to others, conflict.

However, many claim that there are some things that just should not be done, ever. I expect most humans and most societies would agree that no one should be another's slave (albeit slavery still exists). No person (and perhaps no society) should kill humans randomly. Even though it occurs, most people in most societies, if challenged, would agree that no one should be tortured. Rape is always wrong. It does not matter how much personal pleasure may be gained or good that may be accomplished by doing one of the above things, they are never right, never acceptable in civilized society. Humans have a duty to do what is right, because it is right. We do what is right not because it makes us happier, is convenient, or makes money, we do it because it is the right thing to do, regardless of the consequences, which are often nearly impossible to know in advance. We do right acts not because we achieve what we want but because we know that they are always right.

Can such absolute rules be applied to agriculture? Some suggest that agricultural practice should never cause environmental harm, that animals should never be treated cruelly, that water should not be mined to produce crops. The rightness or wrongness of these practices is not debated actively within the agricultural community, which seems to resist absolute moral standards, unless it is the imperative to produce more. The inevitable debate about the appropriateness of absolute moral standards in agriculture will benefit from Rachels (2003, pp. 127–129) analysis of Kant's basic idea—the requirement for absolute moral rules. It is generally agreed that many pesticides can harm humans, harm non-target species, and pollute the environment. None of these is a good thing and if the maxim is similar to the medical maxim of do no harm, many pesticides would be forbidden. However, it is possible to conceive of conditions where continued pesticide use would be permitted (e.g., to control mosquitos and prevent malaria infection). That is, even though bad consequences could follow, there might be circumstances where use would be acceptable. Rational human beings may violate an absolute rule against polluting the environment, if, and only if, they do so for reasons that we would be willing for all to follow. Sandler (2004) suggests that genetic crop modification can be acceptable if it is the only reasonable solution to prevent mass human starvation and the environmental risk is trivial. Rules need not be absolute if the reasons provided for violating the rule are good enough for us to be willing for all to accept the action. The agricultural community has not debated its rules for action and under what conditions it might be acceptable to violate or modify an absolute rule.

## Utilitarianism

Utilitarianism was developed by the British philosopher Jeremy Bentham (1748–1832) and elaborated by John Stuart Mill (1806–1873). It judges actions by their tendency to create the greatest balance of happiness or pleasure over pain or suffering for all affected by an act. The goal humans share no matter what they do, according to Bentham (1879), is to avoid pain and secure pleasure. That is what happiness is.

Happiness, pleasure, pain, and suffering are words that lack precision, but inevitably refer to something that is experienced or felt. They are, in Damasio's (1999) words, the feeling of what happens. Therefore, consciousness is assumed.

An important qualifier, within the utilitarian standard, is that each person counts as one but not more. That is, all affected by an act are given equal consideration. It is a laudable democratic principle, but immediately compels the question: Are all humans equal? The operative words in this ethic are good and bad as applied to consequences. The utilitarian claims that actions are to be judged right or wrong solely by the consequences that follow. In contrast, Kant emphasized the primacy of intent, not results. Bentham's "Principle of Utility" requires that when there is a choice between actions and social policies, one must choose that which has the best overall consequences for all who will be affected by the act. After an act has been done, the utilitarian standard requires that the consequences be assessed by the amount of happiness or pleasure that is created and its balance over pain and suffering. The goal is to increase the former and eliminate or decrease the latter. Happiness is the only good that matters because it is intrinsically good.

But are consequences really all that matter? Good consequences are often the result of bad intentions, although both may look the same. Moses Maimonides (1135–1204), the great Spanish Rabbi, philosopher, and physician, distinguished among seven levels of charity, which Singer (2000, p. 258) describes briefly. The lowest level is to give reluctantly. Second is to give cheerfully but not in proportion to the distress of those in need. Next is to give cheerfully and proportionately, but only when asked. Fourth is to give cheerfully and proportionately without being asked but to give directly to the needy person and potentially shame them. Fifth is to give so that the giver does not know to whom the gift has been given but the recipient knows who the giver is. In the sixth case one knows who the beneficiary is but remains unknown to them. Finally, one gives so one does not know who one gives to and the recipient does not know who the giver is. When considering these seven cases, one wants to say that consequences, which may be the same in each case, are not all that matter, intentions differ among the seven levels of charitable giving and intentions matter. We want to give more credit, more recognition to the person who gives in the latter way than to one who humbles the recipient or one who wants to gain public credit for a charitable act. The utilitarian standard does not allow us to distinguish between them.

In agriculture and in its sub-disciplines, the good consequence emphasized as a justification for action is production of abundant, safe food and fiber. It is undeniable that food and fiber production produce great happiness among all who are fed and clothed. The agricultural debate then should not be about this good utilitarian goal. It should be about the reasons that support the primacy of production as the best goal and about the techniques used to produce food and fiber. Production is used to justify environmental, human, and non-target species harm (Dundon, 2003). Environmental protection is not ignored, but it rarely rises to first place in the utilitarian calculus of what agriculture ought to do. Questions of justice for agricultural workers are not ignored, but do not have sufficient force to counter the argument of the need for ever greater production and profit. Family farms and those who work them, the communities, and local businesses they support (see Goldschmidt, 1998) can, inevitably must, be sacrificed to large, more economically efficient farming operations. Economic rationality rules and justice is not considered. Economic rationality has a role to play in agricultural decision making, but decision making should not stop with determination of the most efficient use of resources (Dundon, Madden, 1991). Efficient resource use does not consider, indeed it may often ignore, achieving the greatest level of happiness over suffering for all affected by an agricultural act. The utilitarian standard requires that all affected by an act count equally but many agricultural decisions that focus on total production or productive efficiency per person do not pause to consider who will be affected or what the human or environmental effects of an action will be.

A major part of the justification for agricultural pesticide use is productive efficiency. Hand weeding and hoeing are arduous, unpleasant tasks. Herbicides are easy to use, much more economically efficient, and crop yield usually increases. What happens to the labor displaced has not been considered by weed scientists. It is argued that people who move from arduous weeding to other tasks are undeniably better off, they are happier. The implicit assumption is that other work is available and those displaced will be able to find and qualify for it. The productive efficiency of pesticides trumps real harm to farm workers. They don't count much and just treatment is often denied so profit and production can be maintained or increased. In fact, pesticide use often denies the right not to be harmed (Dundon). A major critique of utilitarianism is that people's rights may be trampled because good results, defined as higher production, will be achieved.

Agriculture and those who practice it have, in my view, a clear utilitarian standard that is unexamined. Because food production, all agree, is undeniably good, the importance of other elements of utilitarianism—justice, who counts, rights—are ignored. Of course, it is impossible for anyone to consider everyone who could, in theory, be affected by an agricultural act. What is a farmer in the Missouri river drainage of eastern Montana to do when she learns that nitrogen fertilizer used to produce the lentils and sunflowers grown on her farm also is, in theory, harming fish, shrimp, and fishermen in Louisiana who depend on the waters of the Gulf of Mexico for their livelihood? It may be that the utilitarian standard is too demanding. Must that Montana farmer consider the children of the fisherman in Louisiana? They will never meet. The farmer may have her own children who are primary. Should one (can one?) be required to sacrifice one's own happiness for the presumed happiness of distant, unknown people? In addition, it is impossible for us to consider all as equal. Certainly my children and my spouse are more important to me than yours. My neighbors are more important than distant people. I am charged to be my brother's keeper, but utilitarianism does not help me know who my brothers are.

Rachels (2003) and Rachels and Rachels (2010) note the objections to utilitarianism, but do not relieve us easily of the obligations. It may be true that justice and human rights may be denied and personal obligations may be distorted or abandoned, if we adhere too closely to the demands of utilitarianism. However, it is a clear moral standard that provides a good guide for choosing the rules that govern how we practice agriculture. It is not just individual acts (Act Utilitarianism) that matter but the set of rules (Rule Utilitarianism) that is optimal from a utilitarian point of view (Rachels, 2003, p. 113) that are most important. Individual acts (use of nitrogen fertilizer in Montana) will be judged right or wrong according to their acceptability under the optimal utilitarian rules. The debate to determine what the best rules are has barely begun within the agricultural community.

## Applying Ethics in Agriculture and Agricultural Science

This chapter began by suggesting that discussion of ethics in agriculture should not be regarded as something that is so scholarly that it risks being unaware of the realities of agriculture or the realities of the ethical life. That is, ethics should not be regarded as just an academic matter. The preceding, albeit brief, discussion of ethical theories is, in my view, very relevant to agricultural practice. The ethical theories and the values that undergird them, even when unknown and unexamined, are present in agricultural practice. The implicit utilitarianism of agriculture and agricultural science, even though it is often unknown to those who espouse it, is the common justification for action: to provide the greatest good for the greatest number of people. It provides the reasons why agriculturalists do what they do. Explorations of the reasons for action and inaction are essential to address agriculture's ethical dilemmas. However, it is highly likely that the implicit, unexamined utilitarian ethic that pervades agriculture is an ethical bias and prevents agriculture's practitioners from dealing openly and honestly with ethical challenges. If one is already sure that agriculture and agricultural technology are ethically correct because production, the highest value, benefits the greatest number, then there is no necessity to explore ethical challenges to this position. This book posits that agriculture lacks a strong ethical foundation that is necessary to sustain progress toward the desirable goal of sustainability.

In agricultural practice one often encounters what James (2003) identifies as Type I and Type II ethical problems.[3] Type I problems are important because of difficulty in deciding what ethical norm should apply. Different individuals have different and conflicting values or ethical perspectives. The utilitarian goal of achieving the greatest good for the greatest number via more production may not justify all actions. For example, weed scientists applaud weed science technology because it is labor saving. Weeding, especially hand hoeing, is an arduous task that no one likes to do for long or impose on another. Weed control techniques that save labor are regarded as good but if the development of labor-saving technology results in loss of employment for developing country people that may be bad. Some will claim that it may be better to have a job weeding than no job at all. But, one must ask if weed scientists should work to reduce the arduous labor involved in weeding crops

[3] The terms are similar but James' (2000) use of type I and II ethical problems is not identical to the use of these error terms in statistics. In statistics, a type I error occurs when the null hypothesis (without effect, consequence, or significance) is rejected when it is true and a type II error occurs when the null hypothesis is accepted when it is false.

even if that leads to unemployment of people with few or no options for employment? Labor-saving technology is good for large farms and for profit, but it may be harmful to laborers who lose jobs and to small farmers who are driven out of business because they cannot afford the new technology. What ethical standard, what moral theory should be used to decide? The same arguments apply to development of labor-saving agricultural machines.

There is a tendency in agriculture to rely on technology to solve the problems it creates. A new, better pesticide will solve the problems the old one created. A more efficient machine (usually larger and more expensive) will solve more problems. New cultivars, resistant to a disease or an insect, will be better until they too fail. Sandler (2004) claims that the advent of biotechnology is "another attempt to meet our material needs by domination and manipulation" of nature. He points out that it is doubtful that most scientists, politicians, and business people who advocate biotechnology think of themselves as promoting a detrimental practice of domination (see Federoff and Brown, 2004, for an opposing view of biotechnology, see Chapter 8). Nor are they motivated by a desire to do so, but that, in Sandler's view is "exactly what they are attempting to do." Biotechnology is "within the tradition of manipulating and dominating the agricultural environment." It fails to exhibit the requirements of virtue because it illustrates hubris not humility. However, from a utilitarian perspective it may be good because many people will benefit. Arguments on both sides of these issues are reasonable and the result is what James (2003) calls a Type I dilemma (because of the difficulty of deciding what ethical norm should apply). Such ethical dilemmas are largely ignored within the agricultural community.

Type II dilemmas are also common in agriculture. These occur when there is a general social consensus on what ought to be done but there is significant incentive to violate the social consensus, that is, to violate the social contract. There is consensus that those engaged in agriculture should not do things that lead to irrevocable environmental or human harm. Widespread use of pesticides and nitrogen fertilizer may lead to both. Pesticides used for crop protection in the world's developed countries have been used improperly in developing countries and manufacturers have often been fully aware of improper use, do not approve it, but may be unable to stop it short of withholding the product from the market, which certainly will diminish their happiness. Other pesticides have been tested and used in developing countries but have not been marketed in developed countries because of environmental, human, or non-target harm. It is not unreasonable to suggest that this is equivalent to using developing country agriculture and people as test cases for pesticide development in full knowledge of the potential for harm. Such testing could never have been done in the United States because more stringent health and environmental regulations prohibit it. These illustrate Type II problems where one knows what ought to be done but proceeds to act against the consensus because the potential for profit encourages, or loose environmental regulations permit, the action. Some US farmers may use a pesticide when they know it is not permitted (illegal), but they also know it works well.

The latter case is illustrated well by the example of the effective carbamate insecticide/nematicide, aldicarb (trade name Temik). It was approved for use in several crops including cotton, beans, oranges, peanuts, pecans, sorghum, soybean, sugar beet, sugar cane, and sweet potato. It has high acute mammalian toxicity ($LD_{50}$ = 0.93 mg/kg) and has some soil stability plus water solubility that has led to its being found in groundwater after use over several years. For many years, it was regarded as a safe, environmentally favorable, specifically targeted, and especially effective insecticide. It was soil applied and incorporated as a granular formulation.

In 1985, aldicarb appeared as an illegal residue in watermelons in California and Oregon (Green et al., 1987; Marshall, 1985; MMWR, 1986). It was not registered for use on watermelons. It could have been applied illegally and therefore would illustrate a Type II problem. It could have been applied by someone other than the grower as a criminal act, or it could have resulted from soil residues after legal use on a preceding crop.

The presence of aldicarb was discovered and a food alert was broadcast by the California Department of Health Services and the Oregon Department of Agriculture on the Fourth of July 4 weekend. The alerts suggested that anyone with cholinergic symptoms that appeared within 2 h of eating a watermelon should see a physician immediately. The symptoms included twitching, uncontrolled muscle activity, nervousness, and anxiety. Such symptoms occur because aldicarb interferes with the action of acetylcholine, which is important in transmission of nerve impulses. In California, 1350 cases were reported from all regions and 493 were classified as probable by the State Health Department (MMWR, 1986), Oregon reported 206 cases with 27 probable (Green et al., 1987). There was a huge public expense and massive public alarm created by a few watermelon growers who may have done what they knew they should not do to protect a valuable crop, which then lost much of its value because of what they did. Eventually all watermelons in the California distribution chain were destroyed and Oregon recalled all watermelons.

The action of a Wyoming rancher I met several years ago illustrates a similar although less consequential Type II error. He had a supply of, and regularly used, the highly effective but unapproved (for the particular use) broadleaved herbicide, Tordon. He knew it was illegal but, in his words, "It sure works to kill the larkspurs (a poisonous weed) and if I don't kill 'em, they will kill some of my cattle."

Thus, we observe conflicts that arise from the difficulty of deciding what ethical norm should apply (a Type I error) or the equally vexatious problem of knowing what the general social consensus is but proceeding to violate it because there is significant incentive to do so (a Type II error) (James, 2003). Type I problems can be solved when reasoned discussion occurs. Type II problems occur when people do what they know they should not do because of other benefits. That is people give higher value to production, efficacy, or the potential for economic gain than they ought to. Agricultural examples can be found in Conroy et al. 1996, Murray, 1994, and Wright, 1990.

Steinbeck and Ricketts (1941) accurately characterized the ethical dilemma that one finds in agricultural and other fields of human endeavor.

> *There is strange duality in the human which makes for an ethical paradox. We have definitions of good qualities and of bad; not changing things, but generally considered good and bad throughout the ages and throughout the species. Of the good, we*

*think always of wisdom, tolerance, kindliness, generosity, humility; and the qualities of cruelty, greed, self-interest, graspingness, and rapacity are universally considered undesirable. And yet in our structure of society, the so-called and considered good qualities are invariable concomitants of failure, while the bad ones are the cornerstones of success. A man … while he will love the abstract good qualities and detest the abstract bad, will nevertheless envy and admire the person who though possessing the bad qualities has succeeded economically and socially, and will hold in contempt that person whose good qualities have caused failure. When such a … man thinks of Jesus or St. Augustine or Socrates he regards them with love because they are the symbols of the good he admires, and he hates the symbols of the bad. But actually he would rather be successful than good.*

*Steinbeck and Ricketts, p. 96*

## Multiple Strategies Utilitarianism

Rachels and Rachels (2010, pp. 177–180) conclude by asking, What would a satisfactory moral theory be like? They suggest Multiple Strategies Utilitarianism. It does not focus exclusively on motives, consequences, intentions, acts, or rules as the theories outlined above do. It is utilitarian because the ultimate goal is to maximize the general welfare. It recognizes two critical things: diverse strategies may be used to accomplish the goal, and ethics must be grounded in action rooted in a desired future rather than in the status quo (Gardner, 2006, p.154).

If the multiple strategies approach is acceptable, it will still be necessary to think about and choose the appropriate criteria to be used to

1. judge the value of a technology or practice, and
2. to determine if all components of agriculture are maximizing the general welfare.

I suggest the following 11 *value criteria* should be included. All may not be applicable to every ethical decision.

1. *Equity*. Is it fair to all users and those who will be affected? For example, herbicides can increase yields, may reduce soil tillage for weed control, and generally benefit users; but pollution of soil and water, injury to non-target species and humans are costs borne by all. Herbicides reduce the need for farm labor without concern for those displaced.
2. *Food security*. This includes elements of human rights although whether or not food is a human right has not been resolved. Security includes nutritional adequacy, availability, and access. Availability is not just a production problem.
3. *Environmental soundness*. One must assume that, in the best of all possible worlds, no technology or practice would cause or contribute to environmental degradation. It would improve the environment and enhance sustainability. Unfortunately, there are several agricultural examples of the opposite effect: unwise use of pesticides, irrigation that causes soil salinization, tillage that causes soil erosion, water mining for irrigation, fertilizer and pesticides that affect non-target species, and chemical and energy demands of unsustainable crop monoculture systems.

4. *Profitability*. Profitability for users of technology is a reasonable expectation. Although this is generally true it is often achieved because many real costs are externalized (costs not paid by the user).
5. *Safety and risk*. Harm and the potential for harm to users of technology and to others must be considered. Measuring risk is an empirical scientific activity, but judging safety is a normative political activity. Risk is not just a technical issue; distribution over time and whether the risk is voluntary or involuntary are important.
6. *Quality of life and human dignity*. Some technologies (herbicides) eliminate drudgery (weeding) and improve human life. Some create routine drudgery (hoeing, assembly line work, a clerk who types all day). Pesticides decrease the need for human labor, but they may create a conflict among profit, equity, and environmental soundness.
7. *Aesthetics*. Ideally agriculture should make the world appear better and be better.
8. *Human and animal health*. Clearly, technologies that improve human, animal, and environmental health are more desirable than those that do not. No one advocates technology that worsens health.
9. *Consent*. A major tenet of democracy is that people should not be subjected to harm or the risk of harm without their consent.
10. *Sustainability*. The common agricultural view is that production and profit must be sustained. Others argue that maintaining the environment on which agriculture and life depend should be a higher value. It is not uncommon to claim that maintaining rural agriculturally based communities is or ought to be a high value (see Goldschmidt, 1998; Berry, 1984). Their persuasive arguments suggest that protecting human health, species, and species diversity should also be highly valued.
11. *Institutional roles*. What is the proper role of agricultural institutions such as Land-Grant Colleges of Agriculture, the US Department of Agriculture, and large private corporations?

Because agriculture is the essential human activity, debating the importance and relevance of these value criteria is a critical first step toward determining if agriculture and its technology achieve the ultimate goal of maximizing the general welfare. If a technology or practice is judged acceptable under all eleven criteria, decisions are easy. Resolving conflicts among criteria requires good scientific and ethical decision making.

The primary emphasis, the ethical justification, of agricultural research has been to increase productivity and profit: worthy goals. Agricultural scientists and farmers also have personal and professional goals. Similar to most people, they want to help create a world that is just, peaceful, generally prosperous, democratic, free of prejudice of all kinds, and humane. These worthy goals are not agriculture's exclusive responsibility. Achieving them is a responsibility shared with all segments of society. But, as mentioned in Chapter 2, agriculture has specific responsibilities, including those listed in Table 4.1:

1. Achieving sustainable production practices
2. Decreasing pollution
3. Eliminating soil erosion
4. Eliminating harm to other plant and animal species
5. Ending habitat destruction (see Green et al., 2004)
6. Ending the detrimental effect of agriculture on other species
7. Ending water pollution and mining water for irrigation

**Table 4.1** Comparison of Value Criteria Suggested for Agriculture, Agriculture's Specific Responsibilities, and Scientific Societies' Ethical Guidelines

| | **Suggested Specific Responsibilities of Agriculture** | **Shared Ethical Guidelines of Scientific Societies** |
|---|---|---|
| **Suggested Value Criteria** | **Goals (from list below) in common** | **Goals (from list below) in common** |
| Equity | | |
| Environmental soundness | 2, 3, 5, 7 | 5 |
| Profitability | | |
| Safety and risk | | |
| Quality of life and human dignity | | |
| Aesthetics | | |
| Human and animal health | 4, 6 | 6 |
| Consent | | 7, 8 |
| Sustainability | 1 | |
| Institutional roles | | |
| Professional standards | | All |

The code of ethics of many agricultural[4] and other scientific societies share general ethical guidelines. The primary emphasis is professional ethics, as illustrated by the following list of guidelines common to most scientific societies.

1. Data must not be fabricated. Uphold the highest standards of honesty.
2. Experimental results in opposition to a particular hypothesis must be included.
3. Plagiarism is not permitted.
4. Disclose conflicts of interest.
5. Field research must minimize environmental harm.
6. Animal research must minimize pain and suffering.
7. The life, health, privacy, and dignity of human subjects must be protected.
8. Human subjects must be volunteers informed about the research.

I have suggested criteria appropriate for judging the value of a technology or practice and thus determine if agriculture maximizes the general welfare. Table 4.1 compares the criteria with agriculture's specific responsibilities, and scientific societies' ethical standards. The comparison shows the failure of professional agricultural societies to deal with their moral responsibilities beyond their specific responsibilities and scientific ethical standards. Within the societies it is argued that professional societies are just that. They have no larger moral responsibility. Members conduct scientific work (they do science) and believe it is the responsibility of others, indeed their privilege, to evaluate the contribution of the scientific work to the general welfare. This

[4]Many scientific societies have a statement of ethical principles on the societies' websites. Given the author's experience and career path, it is worthy of note that no weed science society has a statement of ethical guidelines.

is not a valid or supportable position. A contrasting view is that although individual scientists must bear ultimate responsibility for their work, there is clearly an opportunity for scientific societies to influence the ethical climate in which they operate; to deal with what ought to be done and the resolution of agriculture's present and future dilemmas. Indeed, the willingness of any profession to engage in critical self-examination of its goals is a mark of maturity. The foregoing is not to denigrate the necessity of statements of professional expectations which define the expected character and proper scientific conduct of a society's members.

Members of professional agricultural societies are distinguished as individuals and collectively as a group with shared goals, beliefs about the value of those goals and the appropriate means of achieving them. This information should be available to the general public so it can be understood and evaluated. Scientific societies are visible, stable, enduring institutions that are custodians of a discipline's core values, which are usually unexamined and not written. Societies are gatekeepers, whose oversight of the trust relationship between individual members and the general public is critical to the advancement of science. It is time for professional societies to expand their vision and include as part of their mission a statement of ethical criteria they believe will maximize the general welfare.

## References

Aiken, W. (1984). Ethical issues in agriculture. Pp. 247–288 *in* T. Regan (ed.). *Earthbound: New Introductory Essays in Environmental Ethics*. Philadelphia, PA, Temple University Press.

Aiken, W. (1998). The goals of agriculture and weed science. *Weed Sci.* 46:640–641.

Bentham, J. (1879). *An Introduction to the Principles of Morals and Legislation, Oxford.* my source, New York, 1948. New York, Hafner Publishing Co.

Berry, W. (1977). *The Unsettling of America: Culture and Agriculture*. New York, Avon Books. Pp. 7–8.

Berry, W. (1984). A defense of the family farm. Pp. 162–178 in Home Economics. San Francisco, CA, North Point Press.

Blackburn, S. (2001). *Being Good: A Short Introduction to Ethics*. New York, Oxford University Press.

Chrispeels, M.J. (2004). Preface. Agricultural Ethics in a Changing World. Lancaster, PA, American Society of Plant Physiology.

Comstock, G.L. (1995). Do agriculturalists need a new, an ecocentric, ethic? *J. Agric. Hum. Values* Winter:2–16.

Conroy, M.E., D.L. Murray, and P.M. Rosset. (1996). *A Cautionary Tale: Failed US Development Policy in Central America*. Boulder, CO, L. Rienner Pub.

Damasio, A. (1999). *The Feeling of What Happens: Body and Emotion in the Making of Consciousness. A Harvest Book.* New York, Harcourt Inc.

Dundon, S.J. (2003). Agricultural ethics and multifunctionality are unavoidable. *Plant Physiol.* 133:427–437. Pp. 7–18 *in Agricultural Ethics in a Changing World*. Lancaster, PA, American Society of Plant Physiology.

Federoff, N. and N.M. Brown. (2004). *Mendel in the Kitchen: A Scientist's View of Genetically Modified Foods*. Washington, DC, J. Henry Press.

Fukuoka, M. (1978). *The One–Straw Revolution.* New York, Bantam Books. P. 103.
Gardner, G.T. (2006). *Inspiring Progress—Religion's Contributions to Sustainable Development.* New York, W.W. Norton & Co.
Glass, B. (1965). The ethical basis of science. *Science* 150:1254–1261. Also Pp. 43–55 *in* R.E. Bulger, E. Heitman, and S.J. Reiser. (1993). *The Ethical Dimensions of the Biological Sciences.* Cambridge, UK, Cambridge University Press. Goldschmidt, W. (1998). Conclusion: The Urbanization of Rural America. Pp. 183–198 *in* K.M. Thu and E.P. Durrenberger (eds.). *Pigs, Profits and Rural Communities.* Albany, NY, State University of New York Press.
Goldschmidt, W. (1998). Conclusion: The urbanization of rural America. Pp. 183–198 *in* K.M. Thu and E.P. Durrenberger, (eds.), Pigs, Profits and Rural Communities. State Univ. of New York Press, Albany, NY.
Gould, S.J. (1993). Ten thousand acts of human kindness. Pp. 275–283 *in Eight Little Piggies: Reflections in Natural History.* New York, W.W. Norton & Co.
Green, M.A., M.A. Heumann, H.M. Wehr, L.R. Foster, L.P. Williams, J.A. Polder, C.L. Morgan, S.L. Wafner, L.A. Wanke, and J.M. Witt. (1987). An outbreak of watermelon-borne pesticide toxicity. *Am. J. Pub. Health* 77:1431–1434.
Haidt, J. (2007). The new synthesis in moral psychology. *Science* 316:998–1002.
Holmes, R.L. (1993). *Basic Moral Philosophy.* Belmont, CA, Wadsworth Pub. Co.
James, H.S. (2003). On finding solutions to ethical problems in agriculture. *J Agric. Environ. Ethics* 16:439–457.
Josephson, M. (1989). Interview on pages 14–27 of Moyers, B.A. World of Ideas: Conversations with Thoughtful Men and Women About American Life Today and the Ideas Shaping Our Future.
Kidder, R.M. (1994). Universal human values: finding an ethical common ground. *The Futurist* July/August:8–13.
Knize, P. (1999). Winning the war for the West. *Atlantic Monthly* July:54–62.
Larmore, C. (2008). Behind the veil. *The New Republic.* February:43–47.
Madden, P. (1991). Values, economics and agricultural research. Pp. 285–298 *in* C. Blatz (ed.). *Ethics and Agriculture.* Moscow, ID, University of Idaho Press.
Marks, J. (2000). *Moral Moments: Very Short Essays on Ethics.* New York, University Press of America.
Marshall, E. (1985). The rise and decline of Temik. *Science* 229:1369–1371.
McKibben, B. (2003). *Enough: Staying Human in an Engineered Age.* New York, Times Books, H. Holt and Co.
MMWR (Morbidity and Mortality Weekly Report). (1986). Epidemiologic notes and reports aldicarb food poisoning from contaminated melons—California. April 25 35(16):254–258.
Murray, D.L. (1994). *Cultivating Crisis: The Human Costs of Pesticides in Latin America.* Austin, TX, University of Texas Press.
Peterson, R. (1978). *Technology: Its Promise and Its Problems.* Boulder, CO, Colorado Assoc. University Press.
Pinker, S. (2008). The moral instinct. *The New York Times.* http://www.nytimes.com/2008/01/13/magazine/13Psychology-t.html?scp=1&sq=%22the+moral+instinct. (accessed September 2009).
Rachels, J. (2003). *The Elements of Moral Philosophy,* 4th Ed. New York, McGraw-Hill, Inc.
Rachels, J. and S. Rachels. (2010). *The Elements of Moral Philosophy,* 6th Ed. New York, McGraw-Hill, Inc.
Rawls, J. (1971). A Theory of Justice. Cambridge MA. Harvard Univ. Press.

Rollin, B.E. (1999). Part I—Theory. Pp. 3–32 *in Veterinary Medical Ethics: Theory and Cases*. Ames, IA, Iowa State University Press.
Rothman, B.K. (1998). *Genetic Maps and Human Imagination*. New York.
Sandler, R. (2004). An aretaic objection to agricultural biotechnology. *J. Agric. Environ. Ethics* 17:301–317.
Shahn, B. (1957). *The Shape of Content*. Cambridge, MA, Harvard University Press.
Singer, P. (2000). Introduction. Pp. xiii–xx *in Writings on an Ethical Life*. New York, Harper Collins Pub., Inc.
Steinbeck, J. and E.F. Ricketts. (1941). *Sea of Cortez; a leisurely journal of travel and research, with a scientific appendix comprising materials for a source book on the marine animals of the Panamic faunal province*. New York, Viking Press.
Tharoor, S. (1999). Are human rights universal? *World Policy Journal* Winter 1999/2000:1–6.
Van Eenwyk, J.R. (1997). *Archetypes & Strange Attractors: The Chaotic World of Symbols*. Toronto, Canada, Inner City Books.
Wright, A. (1990). *The Death of Ramon Gonzales: The Modern Agricultural Dilemma*. Austin, TX, University of Texas Press.

# 5 Moral Confidence in Agriculture[1]

Ethics is about what individuals, groups, and societies ought to do. Ethical analysis involves thought about and analysis of what one does and attempts to provide reasons, based on moral theory, to show why one's actions are the right things to do. It must probe stated and unstated goals and the values presupposed by them. A good analysis will try to develop a rational way of deciding how an individual ought to live and how a profession ought to proceed toward realization of its goals (Singer, 1994). Such an analysis might lead one toward a set of "rules, principles, or ways of thinking that guide, or claim authority to guide, the actions of a particular group" (Singer). It will make explicit what is valued and may lead to action on those values.

This chapter argues that those engaged in agriculture, whether practitioners, research scientists, extension agents, technology developers, or technology suppliers share an unexamined moral confidence about the goodness of their activity. The chapter also argues that the basis of that moral confidence is not obvious to those who have it or to those not involved in but who may be curious about agriculture and its technology. In fact, the moral confidence that pervades agricultural practice is potentially harmful because it is unexamined by most of its practitioners. Regular debate about ethics and the applicability of moral theories to agriculture have not been part of intellectual discourse within agriculture.

Perhaps the best that can be sought in the moral realm is not absolute certainty but the best available option, which many defenders of agricultural practice appear to believe is what they now have. We live in a world of moral ambiguity, but it is also a world where some moral values seem to have been accepted by all cultures (Josephson, 1989; Kidder, 1994; Kroeber and Kluckhohn, 1952). Although there is evil in the world, it is not unreasonable to argue that no culture tolerates indiscriminate, harmful lying, stealing, or violence; incest is universally forbidden; no mature culture values human suffering as a desirable goal; all cultures have death rituals; protecting children is universally regarded as good in civilized societies; and all societies despise cowardice and applaud bravery.

Even though there is moral certainty on these matters among societies, that same degree of moral certainty may not be desirable or achievable for the world's major environmental interaction: agriculture. Absolute moral certainty and the consequent lack of moral debate stifle discussion and moral progress. Debate about what agriculture ought to do, rather than complete agreement on what agriculture must do or be, is preferred. Debate about morality should reveal the foundational theories and

[1] This is a modified version of a paper which appeared in the *American Journal of Alternative Agriculture* 17:44–53. 2002. Reprinted with permission.

**Agriculture's Ethical Horizon. DOI: 10.1016/B978-0-12-416043-9.00005-2**

values and thus provide a guide for action. Moral theories are the often invisible foundation on which our judgments rest. Exploration of foundational theories (perhaps one of those described in Chapter 4) will expose them to debate and discussion. When one understands the foundation of moral judgments, that ought to lead to more confident judgments, but foundational values are not the answers to problems. They are ways to assist construction of the personal, social, and cultural world we inhabit. Exploration of the moral confidence posited for agriculture will not reveal a single guiding principle, the use of which will solve all moral dilemmas in agriculture. It will reveal several principles that are used in the morally ambiguous, pluralistic world in which agriculture is practiced. Nagel (2010) points out that there are many value questions for which there is no obvious answer and about which people disagree. In our morally ambiguous world, it is only after evaluating all moral arguments that one can arrive at an open commitment to a position, which leads to a view and, one hopes, proposals to address a particular moral dilemma. Although it is a definite position, one should remain open to change as new arguments become persuasive. A position which is agreed upon by everyone may not describe what one ought to do or be. What everyone agrees on may lack a clear and ethical foundation that supports the claimed truth. It is important to note that some scientific facts (the square root of 9 is 3) are just true, as are some moral claims (one should not beat children to compel certain behaviors). Nothing else makes them true. But it is equally important to recognize that a value judgment's truth is not dependent on a physical fact (Nagel).

## The Benefits and Costs of Modern Agriculture

Hugh Sidey, a contributing Editor of TIME magazine, delivered the 1998 Henry A. Wallace lecture to the Wallace Institute for Alternative Agriculture. Sidey (1998) quoted Dumas Malone, a biographer of Thomas Jefferson, who said, "The greatness of this country was rooted in the fact that a single farmer could produce an abundance of food the likes of which the world had never seen or imagined, and so free the energies of countless others to do other things. So much of recorded history is about the struggle of individuals and families to feed themselves. That changed dramatically in this country." Sidey contends that the story of the productivity and success of American agriculture is the greatest story never told. Few, if any, other segments of the American scientific-technological enterprise have amassed such an impressive record of predictive, explanatory, and manipulative success over many years. American agriculture has been a productive marvel and is envied by many other societies where hunger rather than abundance dominates. Examination of the yield records for nine US major crops[2] during the twentieth century shows that yield increases have varied from two to sevenfold (Warren,

[2] Corn, cotton, peanut, potato, rice, sorghum, soybean, tomato, wheat.

1998). No yields decreased. Scientific advances that led to these steady yield increases include development of higher yielding cultivars; synthetic fertilizers; improved insect, weed, and disease control; better soil management; and mechanization. Warren suggested the rate of yield increase does not appear to be slowing. Avery (1997) points out that without the yield increases that have occurred since 1960, the world would now require an additional 10–12 million square miles (roughly the land area of the US, the European Union countries, and Brazil combined) for agriculture to achieve present levels of food production. Avery claims that modern high-yield agriculture is not one of the world's problems, but rather the solution to providing sufficient food for all, sufficient land for wildlife, and protecting the environment. Degregori (2001, p. vii) sees the dilemma of agriculture and its modern technology as "one of the great paradoxes of the twentieth century." He argues that the "century was characterized by economic and technological gains of unprecedented rapidity as shown by all economic indicators." The noneconomic indicators (life expectancy, human health, and the increase in per capita food supply) "were just as spectacular" and fed an increasing population that all experts believed could not be fed. The paradox is that the agricultural and other technology that allowed much of this to happen, have been "under external attack for almost the entire century." Degregori (p. viii) believes the attack has been unwarranted, is not supported by the evidence, and represents a pervasive anti-technology bias. In his view, much of the difficulty of discussing the issues surrounding modern agricultural technology is "that public discourse is being driven by emotional language" (p. 125). He is correct, but his accusation fails to see the relevance of such language as pointed out in Chapter 1. Degregori agrees with Avery and Sidey in his claim that "the anti-technology, anti-rational views have gained such a stronghold in many areas of academia and other elite groups" that they will have adverse consequences for many of the world's vulnerable people. The view is that technology is not the problem, it is the solution to a host of problems.

American agricultural producers and research scientists are proud of their achievements. Our food production system, including growing, distributing, processing, and preparation, are now all part of a large, vertically integrated commercial system (Blatz, 1995). The family farm as an independent, self-supporting entity, and a cultural icon is dying. The agricultural legacy that came from family farms and is the heart of the Jeffersonian agrarian tradition that few Americans now experience but most value, no longer serves as an immediate, experiential source from which citizens derive social values or moral sustenance.

The abundant production that all involved in agriculture value highly is, in Berry's (1977) view, illusory. Berry argues that the abundance is illusory because "it does not safeguard its producers, and in American agriculture it is now virtually the accepted rule that abundance will destroy its producers" and the land base. The evidence supports Berry's prediction. Stauber et al. (1995) provide census data that verifies farm population decline from 1940 to 1990 in Iowa (72%), Minnesota (77%), Montana (74%), and North Dakota (82%). The data, for the same period, show that as farm population declined farmland remained nearly constant because farm size increased in Iowa (88%), Minnesota (89%), Montana (121%), and North Dakota (123%). A 1998 US Department of Agriculture report found that 300,000 small family farms went out of business between 1979 and 1998 while the share of agricultural dollars received by farmers dropped from 21% in 1910 to 5% in 1990. In 2001, the number of farms and ranches in the US declined by 0.7% to 2.16 million compared to 2.17 million in 2000. This was the second-largest decline in farm numbers since the 1.4% drop in 1991. From 2003 to 2006 the number of farms continued to decline from 2.12 to 2.11 million (0.17%).[3] But from 2006 to 2007, the number of farms increased 0.16%. The reason provided by the National Agricultural Statistics Service (NASS) of USDA is:

> *Extensive list building efforts and the augmentation of the area frame sample allowed NASS to capture more of the small farms with less than $10,000 in value of agricultural sales. Additionally, 2007 was a year of relatively high commodity prices. As the value of farm commodities increase, more very small operations are able to meet the $1,000 value of farm sales threshold to qualify as a farm in the census.*

More small farms were counted but total land in farms declined by 15.3 million acres (0.16%) and average farm size decreased from 440 to 418 acres from 2003 to 2007. In the early 2000s, the number of farms declined in 23 states, remained the same in 22 states, and increased in CA, CO, OK, TN, and TX (Alt. Agric News, 2002). This trend continued from 2003 to 2006, farm size decreased in 34 states, remained the same in 11, and increased only in Alaska, Montana, South Carolina, South Dakota, and Utah.

Economic sales data indicate the concentration of agricultural production on large farms. There may be more farms, but small ones don't produce much. In 2007, 42% of farms sold less than $5,000 annually, whereas 10% had sales greater than $250,000 and 5.5% were above $500,000. Farms that sold less than $5,000 had 6.7% of farmland and those that sold more than $250,000 had 47%. Those in the low group had 87 acres or less, while those in the high group had more than 1,500 acres. Farms that sold more the $1 million had at least 3,000 acres. The farm subsidy program created during the depression (1930s) to help small farms now distributes $12.4 billion a year in government payments primarily to large-scale commercial farms. Between 1995 and 2002, 10% of the largest farms (mostly corporate agribusinesses) collected two-thirds of all crop subsidies (Standaert, 2003). In 2006, about 1% of all farmers participating in commodity programs received over $100,000 in payments. Over 18% of all payments were received by farms with average gross income over $1.35 million (ERS, 2008).

[3]The primary source for the data in this and the next paragraph is the website of Agricultural Statistics Board of the National Agricultural Statistics Service (NASS) of the USDA. Accessed January 2011.

### Highlight 5.1

The rise of corporate agriculture and the consequent domination of the economies of farm communities has changed the face of the United States and destroyed many rural communities. Today, fewer than 1% of Americans are farmers, many on small or medium-sized farms. Most of corporate agriculture's profits and the subsidies distributed by the US government for crop production go to farms with more than $500,000 in annual sales, which account for over 45% of agricultural production in the United States. Small and medium-sized farms have difficulty competing with corporate agriculture. The farmer has only a few options to survive:

1. seek outside employment,
2. borrow money and thereby accumulate debt, which they may never be able to pay back,
3. lease the farm to a large farm, or
4. sell.

Leasing allows the farmer to continue to live on the land. Selling is convenient and profitable for those near cities who can sell to someone who wants to 'develop' the land and grow houses or condos instead of corn.

Corporate agriculture has concentrated agricultural business. Large, vertically integrated corporations control everything from seed to selling, piglet to pork or heifer to hamburger. Few actually farm. Monsanto sells 87% of the cotton seed in the United States. Five beef packing companies (IBP, ConAgra, Cargill, Farmland National Beef, and Packerland Packing Co.) control 79% of the beef market. Four firms process more than 50% of all poultry. Corporate influence extends to the US Congress, where it is one of the most generous and potent lobbying groups.

Williams, T. 2010. The corruption of American Agriculture. Washington, DC, Americans for Democratic Action Education Fund.

Therefore, one is led inexorably to the conclusion that Berry (1977) is right. The US agricultural system favors large farms. Small farms are just that: small, and therefore can be and are neglected. Those who farmed the land, created communities, and supported small businesses are slowly disappearing and few notice. Farms are not the only businesses in decline and where the number of independent owners has decreased. Blockbuster rented one-third of all VCR movies in 2005 (and filed for bankruptcy in early 2011). The advent of DVD movies combined with use of the Internet took advantage of the public's desire to rent, not own, entertainment. In the 1990s, 11,000 independent pharmacies closed and more than 40% of independent bookstores failed. In 2005, five supermarket firms (Albertsons, Ahold, Kroger, Safeway, and WalMart) surged from controlling less than one-quarter of all grocery sales to 42% (Harkinson, 2004); as much as 85% in some states. In 2007, three supermarket chains (in order of sales volume: Walmart, Kroger, and Safeway) alone sold 42% of the sales of 50 supermarket chains in the US and 74% of the top 10. Walmart and Sam's Club had 26% of the sales of the top 10 firms. In addition, 100 chains handle 40% of restaurant spending, and "Walmart controlled a third of the market for products ranging from dog food to diapers" (Harkinson). The US has at least 1,800 micro-breweries that control just 5% of the beer market. The four largest breweries (Anheuser Busch—InBev, SAB-Miller, Heineken, and Carlsberg) control a bit more than half of the global beer market. That dominance is small compared to Coca-Cola and PepsiCo, which have three-quarters of the soft drink market. PepsiCo is the largest food-and-beverage company in the US and second largest in the world after Nestlé. This information, albeit limited, indicates but does not prove that some businesses are increasingly controlled by large corporations, which is also true for agriculture. In 2005, four firms controlled 60% of US grain trade, the four largest meat packers controlled 70% of the US beef supply, and five energy companies dominate the US petroleum supply.

The effects of these trends in farming and related community businesses were foreseen by Goldschmidt in 1947 and verified in his 1998 study. He claims that "large-scale, labor intensive, technologically innovative production … made industrial farming possible but not necessary; social policies were needed for that." These policies that made the decline of family farms inevitable were devised in a system where "social relationships were money-based and social standing was money-driven." The singular goal of the US agricultural system was "to gain wealth, without the least concern for the welfare of those whose lives" were being destroyed. There was also no concern, or, at best, minimal concern, for the effects of the money-driven system on the environment on which agriculture and life are dependent. The monetary rewards of the agricultural system are handsome for the survivors: profit. The social rewards of belonging to a caring community, the spiritual satisfaction of serving a larger public purpose, and the communities themselves and the businesses they support have been sacrificed to the bottom line (Goldschmidt). These losses are the social costs of technological improvements in agriculture. Rolston (1975) does not specifically mention agriculture but his thought encourages consideration of the other costs of agricultural technology and, of greater or equal, concern what agricultural practice and the implicit moral position does to the "flora, fauna, species, ecosystems, and landscapes." It is consideration of these that, in Rolston's view, "reveals the character of that society." These costs, these harms, are neither necessary nor desirable. Goldschmidt argues the losses are not trivial and should not be thought of as simply the inevitable price of progress. He claims that changes in agriculture shown in the data above, have made a big, and largely unnoticed, difference to our nation. His work supports the inevitability of the decline of other small businesses as small farms disappear.

The highly productive agriculture that Sidey (1998) applauds is a business similar to any other business where producers seek high production at low costs. Each strives to adopt new technology rapidly to stay ahead of other producers and gain a competitive edge that leads to greater profit. In a very real sense, the farmers' purpose is to keep farming—to survive. Their purpose is not to work out ways for all to survive and thrive, but to gain as much of the market and profit as possible so others cannot (Blatz, 1995). If a neighboring farmer's autonomy or survival is threatened by this system, it is not viewed as a systemic problem but as the neighbor's failure to adapt and survive. US agriculture has become industrialized not only in terms of its size and methods of operation but also in the values its practitioners espouse. The guiding purpose of each farmer is to produce as much as possible at the lowest cost of capital and labor to generate maximum profit (Blatz). Much rhetoric is heard about the necessity and obligation of American farmers to feed the world through our production, which Keeney (2003) considers a goal that has failed US agriculture. The failure is a combination of technologies that have pushed farmers off the land, created few new markets, and an environment that suffers from diminishing soil fertility plus nutrient and pesticide runoff (Keeney).

We also hear that keeping food costs low for consumers is a requirement of our agricultural system in spite of the fact that this goal leads to the moral wrongs that Berry (1977) deplores: the destruction of producers who care for each other and destruction of the land. It is reasonable to claim as Berry, Blatz, and Jackson (1980) have that the highly productive, capital, energy, and chemically intensive American agricultural system is environmentally unsustainable at present levels of production. It is also reasonable to claim that US food costs are not low. Low cost is an illusion, because we actually pay for our food three times. First when we buy food, second when we pay federal taxes that support massive farm subsidies, and finally when we pay for the environmental clean-up and health care costs that arise from agricultural pollution (Pretty, 2003).

At this point, many readers will assume that this chapter will plead for a return to small-scale farming and for support of the possibly false claim of the moral virtue of family farmers and farming communities. I, a Luddite, want to abandon the great achievements of scientific agriculture. On the contrary, the quantitative claims of US agricultural abundance are true. Sidey (1998) is correct. But blind acceptance of that may lead societies to assume that agricultural abundance is assured. No society should assume its agricultural abundance is assured. The system that produces food should not be regarded as similar to a factory, which with the right inputs can manufacture abundance at will (Blatz, 1995). Food is essential to life, but it comes from the land, not from money (Berry, 1999). Therefore the land that produces food is essential. If the foundational values of the food production system do not place protection of the land, the source of agricultural abundance and of food as an essential obligation, and regard food as just another industrial commodity that can be purchased by those with money, then the ethics of the system ought to be a subject of societal concern.

It is common knowledge that the poor are hungry, even in the richest countries. It is a problem not amendable to an easy technical solution or social algorithm that will lead to its solution everywhere or quickly. In face of the undeniable success of agriculture in the developed world, its practitioners should at least ask the qualitative question about whether hunger is a problem of insufficient production, inequitable distribution, or irresponsible consumption by the rich (Thompson, 1989). Many, certainly not all, agriculturalists often see more production as the solution because they consider hunger to be primarily a problem of insufficient production (Avery, 1997).

The other possibilities, inequitable distribution of abundance and over consumption (see Halweil and Mastney, 2003) are not considered or are readily dismissed. Sen (1981, pp. 162–166) concluded that starvation is characteristic of some people not having enough food. It is not accompanied by there not being enough food. Low production and subsequent availability can cause famine and starvation, but it is only one of many possible causes.

As agriculture's productive capabilities have been enhanced by science-based technological discoveries, it is not surprising that the pursuit of production has conflicted with other societal values (Thompson, 1989). Any technology has effects in addition to those intended. For example, air in urban areas is polluted partially because of automotive exhaust: an unplanned effect. Similarly, agricultural technologies have undesired and often unanticipated effects (Thompson). As mentioned in Chapter 1, nitrates from fertilizer increase production *and* pollute water. New cropping techniques may expose soil to excessive erosion. Monocultural agriculture leads to loss of species diversity. From 1903 to 1983 US agriculture lost 80.6% of tomato varieties, 92.8% of lettuce varieties, 90.8% of field corn and 96.1% of sweet corn varieties, 86.2% of apple varieties, and only four potato varieties are grown in the US (Kimbrell, 2002, pp. 7181). Many technologies save labor but do not consider what happens to the people whose jobs are lost. Pesticides increase production by limiting pest damage, but they pollute soil, air, and water and harm nontarget species. New agricultural technologies have reduced the risk of production failure for producers but increased the risk of harm to other species, including humans. Agricultural technologies are significant contributors to the pervasive environmental and human health problems that identify modern society (Gerrard, 2000). Advances in medicine and environmental science have made all of us more cognizant of risks that may have been unknown in the past (Thompson). Agricultural technology has always exposed people to risk. In the past, most of the risk was borne by the user of the technology. Now many risks of agricultural technology are known, and it is known that they are borne by users and others. Technology developers, sellers, regulators, and users, in their moral confidence have not secured or even considered the importance of securing the public's consent to use production technology that exposes people to involuntary risk (Thompson). For example, with the present US population ($308.7 \times 10^6$ in December 2010) and the fact that 1.24 billion pounds of pesticides were used in the US in 2000 (5.7 billion were used in the world), one can only conclude that four pounds of pesticide are being applied somewhere by someone for every American. Annually, we voluntarily consume five pounds of butter per capita and eight pounds of coffee, neither of which may be good for perfect health. But to think that each of us may be involuntarily exposed to four pounds of pesticide bothers a lot of people. It is no longer sufficient to claim that all is well as long as those who use pesticides do so in accordance with label directions. It is insufficient to claim that there is not a problem with pesticides, people just think there is. That is not a moral argument. At best, it is an incorrect empirical claim. Thompson says:

> *Agricultural producers and those who support them with technology may have been seduced into thinking that, so long as they increased food availability, they were*

*exempt from the constant process of negotiating and renegotiating the moral bargain that is at the foundations of the modern democratic state. Democratic societies will not entrust their water, their diets, or their natural resources blindly into the hands of farmers, agribusiness firms, and agricultural scientists. Agricultural producers must participate in the dialog that leads to social learning and social consensus about risks, and they must be willing to contribute the time and resources needed to understand the positions of their fellow citizens, and to make articulate statements of their own position.*

For most nonagricultural segments of society, these are not new demands. For agriculture they are. Agriculture has been so confident of its narrow pursuit of increased production that its practitioners have frequently failed to listen to and understand the position of others (e.g., environmental groups, modern agrarians, organic practitioners). The claims of those who object to modern agricultural technology are rejected as unworthy of serious thought. That represents a risky narrowing of viewpoints (Gerrard, 2000). As pointed out by Feyerabend (1987, p. 34), who cites Mill, views one may have reason to reject may still be true, and rejection of those views without careful consideration implies infallibility on the part of those who reject. It also fails to recognize that even though the bulk of a view may be wrong, some useful truth may be present. No one view ever has all the truth. Alternative views help hone and sharpen one's view and defenses against objections. Third, and this appears to be common, a refusal to address alternative views, to debate, leads to one's view becoming just a prejudice. Finally, only by considering contrasting opinions can one understand the true meaning of one's view of any issue. Without contrast, one's view becomes "a mere formal confession" (Feyerabend).

Agriculturalists have not taken the time to articulate any value position other than the value of production and have not articulated the reasons that value ought to retain its primacy. They have not considered other views of issues and their views often become just formal confessions of an unexamined view they assume others will support because it is so logical to them. Unintended consequences of agricultural technology are never just scientific or production questions. They always include a moral dimension that demands thought about fundamental values and the ethical foundation of those values.

**Highlight 5.2**

The dominant social goal for agriculture is to produce abundant food and fiber and to assure producers a profit for their work. The modern, developed country agricultural production system has succeeded well with the production goal but less well with the goal of profit for the producer. Agricultural science and the agricultural establishment (producers, manufacturers, marketers, and so on) have not paid much attention to agriculture's effects on social and environmental goals. These things are often regarded as beyond agriculture's purview.

More production is nearly always viewed as better and for the US economy as a whole. Increasing the supply of things seems to be the primary goal of national economic policy.

The United States, with less than 4% of the world's population, uses nearly 30% of the earth's resources to maintain our wonderful lifestyle. We are the world's great consumers. More than 60% of Americans are overweight and that is estimated to cost more than 12% of annual health care expenses. This is all part of what Durning (1992) calls "the conundrum of consumption." It is:

*Limiting the consumer life-style to those who have already attained it is not politically possible, morally defensible, or ecologically sufficient. And extending that life-style to all would simply hasten the ruin of the biosphere. On the other hand, reducing the consumption levels of the consumer society, and tempering material aspiration elsewhere, though morally acceptable, is a quixotic proposal. It bucks the trend of centuries. Yet it may be the only option.*

Durning adds the corollary conundrum that "this historical epoch of titanic consumption appears to have failed to make the consumer class any happier." US citizens living in the 1990s were, "on average, four-and-a-half times richer than their great-grandparents were at the turn of the century, but they are not four-and-a-half times happier." The Worldwatch Institute reported that the number of Americans who described themselves as very happy declined from 35% in 1957 to 30% in 2001 (Worldwatch Magazine, March/April 2001). But, when one Googles 'happy Americans' in 2011, the percentage is reported to be as high as 84%. Happiness is personally defined and is an elusive concept; one to which we all aspire.

Thus, it seems reasonable to ask the agricultural community a moral question. Should the endless quest for greater production make and keep producers or US citizens happy. Should agricultural practices maximize happiness in the United States, among farmers and ranchers? Under what conditions should increasing production be reduced in importance when compared to maintenance of small farms and rural communities? Should the agricultural community consider ranking preserving environmental quality ahead of production?

## Goals for Agriculture

Production of abundant food and fiber must remain a goal of agriculture. If agriculture does not produce, it fails to fulfill an essential goal of interest to all members of any society. When one asks what agriculture's goals ought to be, reliable production of an adequate supply of safe food must remain near the top of any list. However, because we live in a culturally pluralistic world, we are compelled to ask what other goals ought to be considered by agriculture and when and why one or more of these may take precedence over production. Without describing all possible or desirable goals for agriculture it is possible, for purposes of discussion, to divide goals into two groups: social and environmental.

## Social Goals for Agriculture

Proper social goals for agriculture were dealt with by Aiken (1984). He, as Day (1978) did, ranked profitable production as the primary goal of agriculture.[4] Nearly all agricultural practitioners would agree. Aiken then suggested sustainable production, environmentally safe production, meeting human needs, and contributing to a just social order as additional goals which may often be of greater moral importance than profitable production. This array diverges from the dominant agricultural view. Few agricultural voices speak of achieving a just social order. There is no apparent objection to achieving a just social order but it is not an agricultural goal. Thompson's (1986) analysis of Aiken's article ascribes the ranking to a straightforward principle for ordering of duties or rights. "The principle would hold that one right or duty is more basic than another whenever the moral benefits associated with the second can be extended to all people only after the first has been protected or fulfilled." Thus, "a right to life is more basic than a right to an education." One presumes that education is desired only by the living and that they must be fed, so the agricultural goal of production of sufficient food is not unreasonable.

Agriculturalists begin to diverge from others when discussion of sustainability occurs. Many in agriculture see sustainability as achievable by modification of the present system and do not countenance abandoning the system that has been so successful. Achieving sustainability is regarded as a scientific matter. However, because agriculture is the largest and most widespread human interaction with the environment, achieving sustainability will have social and ecological effects. One must acknowledge that while agriculture may be the most widespread, it is not the most destructive interaction. Highways, cities, and dams, and so forth are. Sustainability can be achieved best when the discussion includes consumers and producers (Thompson, 1986) and considers environmental effects. Thus, achieving a sustainable agricultural system is a societal not just an agricultural responsibility, which should be greeted with pleasure by those in agriculture.

The agricultural market as part of the greater market of all goods and services distributes agricultural supplies to producers and produce to consumers. Markets are very powerful mechanisms, but they often are not just. If they were just, then America would not have hungry people. Those who emphasize the primacy of production must recognize the connection between what is produced, the market that distributes it, and a just social system.

Thompson (1986, 2010) acknowledges the persistence of the agrarian ideal in the American mind. As family farms and rural communities disappear, the virtues they instilled in past generations (help your neighbor, be kind to animals, help those in need, respect the family, respect your elders, and so on) are still valued by our society. Churches and schools try to teach these virtues although schools are frequently chastised for teaching values. I suggest that most people still want children to learn what they regard as traditional values and mourn their loss. We cannot figure

[4]A difference is that Aiken (1984) followed with four other goals and compared them. Day (1978) suggested production was the only proper goal for all of agriculture for all time.

out how or where to teach what used to be obtained by osmosis from the culture. Thompson (1986) suggests that one way to encourage these values is to have them "prominently displayed in the social purpose of an economically central and vital activity such as agriculture." To accomplish this, those in agriculture are going to have to abandon the singular pursuit of production as their only goal and incorporate the conscious pursuit of social goals as part of the practice of agriculture. This necessitates debating what the right goals are. It also assigns a large task to an already small and decreasing, dispersed minority of the US population. Who will prominently display such values? Who will speak for agriculture?

**Highlight 5.3**

Some regard using food as a weapon to gain political alliances or punish political opponents as a means justified by the desirable ends it serves. However, a counter-argument to this position is that it is morally unacceptable to mistreat anyone or to cause humans to suffer, especially if they are poor and powerless and can be manipulated by those in power. If they are just wars using food as a weapon is morally reprehensible because the weapon cannot be effectively directed. When a country attempts to use food as a weapon it fails. The weapon, directed toward those in power, affects the powerless.

Rich nations of the world and their citizens regularly face the question of whether or not to give food aid to countries in need and, if so, how to give it and how much to give. It is reasonable to claim that an individual cannot do much to change what happens in other, far-away parts of the world. Because so many (but not all) of the hungry are in distant places, people can argue that they do not know anyone there and they are not sure if what they are able to give will actually help those in need. Some also worry that what is given may be intentionally misdirected and help the wrong people. Thus, it is reasonable to conclude that what one ought to do, what it is prudent, is to do nothing.

All Americans ought to struggle with the question of food aid. Singer (1996) notes that the "result of everyone doing what he really ought to do cannot be worse than the result of everyone doing less than he ought to do, although the result of everyone doing what he reasonably believes he ought to do could be." After making this claim, Singer offers a moral challenge:

*If it is in our power to prevent something very bad from happening, without thereby sacrificing anything else morally significant, we ought, morally, to do it.*

This only requires us to prevent what is clearly bad. We do not have to work to create something good. And one only has to prevent something bad from happening, if one does not have to sacrifice anything of comparable moral significance. For example, if one is all dressed up and walking by the lagoon on my campus when suddenly a child cries from the middle of the lagoon where the water is deep, one must help. Dirtying, perhaps ruining, one's best clothing, simply doesn't matter. One's clothing is not of comparable moral significance. Many people have, but Singer's moral challenge does not require anyone to sacrifice their own life to save another. It is of comparable moral significance.

The principle does not make any distinction between cases where someone is the only person who can prevent something very bad from happening and where several can act. It is true that when someone is in physical trouble and nearby if it is someone with whom we have had personal contact we are more likely to help. But, and this point is crucial, it does not claim that we ought to help someone who is near or dear rather than one who is far away and unknown. A needy person who is near does not have any greater claim on our ability to help than one who is far away that we will never meet or know. Mother Teresa offers a reason why we do not act:

*If I look at the mass, I will never act. If I look at one, I will.*

Most Americans accept that, in some way, we are our brother's keeper. It is highly likely that we will never meet the brothers that we are, somehow, obligated to keep. Nevertheless, our moral obligation demands that we figure out how to catch them when they fall.

## Environmental Goals for Agriculture

Environmental goals for agriculture should not be, and perhaps cannot be, divorced from social goals. Sustainability is regarded by those in agriculture as primarily a production and secondarily an environmental goal. I will call those in this group advocates of what is called conventional or industrial agriculture. Others, advocates of alternative agriculture, see sustainability as a social goal. The view depends on what one wants to sustain. For the first group, to sustain usually means to protect the productive resource (soil, water, gene pools) while maintaining and increasing production. This is consistent with the popular myth (the operative paradigm) that continued growth is the only way to economic health and human well-being (Kirschenmann, 2007). Intellectually honest economists and all ecologists know that continued growth of anything (knowledge is an exception) is impossible in a limited ecosystem, which the earth is. Those in the second group agree that protecting the productive resource is important, but rank it below sustaining environmental quality, family farms, small communities, rural life, small businesses that serve agriculture, small communities, and a way of life that is disappearing. They affirm the importance of place and the values derived from commitment to meaningful relationships in a community (DeLind, 2006). The sum is a sense of belonging to and having responsibility for something larger than one's self. It is not so much a theological as it is an ecological claim. This is not merely a semantic debate. It goes to the heart of what agriculture ought to do; what values it ought to espouse. Leopold (1947) addresses the essential issue regarding land when he suggests that the last great moral divide is between those who regard land as property that they own and can do with as they please and those who see land as part of the community to which they belong. Leopold based his land ethic on the absolute human need for community. We humans want to belong. In Leopold's view, we need to protect the land so there will be a community left to which we can belong. For Leopold, protecting the land meant ensuring its capacity for self-renewal. It is what ought to be done. Assuring self-renewal is a moral responsibility. Agriculture has a major responsibility because it has the potential to care for or harm so much land. This is a different view from protecting land only as a productive resource. Land, as Leopold claimed, is the basis of life. It is the necessary resource for a productive and profitable agriculture (Thompson, 1988). Without the land, the soil, there will be no agriculture, so land must be regarded as something more than other productive resources (e.g., fertilizer, machines, irrigation water, pesticides, or seed). Its importance to agriculture should not be thought of only in monetary terms. The ecological imperative of interconnectedness must be stressed more than a simple production ethic allows. To harm or destroy the land is not just a matter of profit and loss because such actions destroy that which is essential to life, and that certainly raises a moral question.

The challenges and problems of defining and ranking social and environmental goals for agriculture are that they involve values. Which values do and which should serve to define the questions raised. It is generally not recognized in agricultural science that values are not peripheral to the science and technology but are foundational (Capra, 1996). Scientists know they are responsible for the scientific integrity of their work and for its intellectual contribution. They do not as readily assume responsibility for the moral aspects of their work.

The biotechnology bandwagon, which has captured university, state, and federal research systems is a good example. Just as the pesticide bandwagon did from the 1950s through the 1970s (Cate and Hinkle, 1993), the biotechnology bandwagon is now rolling. In the earlier era, massive resources turned toward learning how to use pesticides to make agriculture more productive. Now the same institutions are turning from the talents of agronomists, entomologists, pathologists, weed scientists, chemists, and toxicologists to the talents of molecular biologists and biochemists to create new agricultural resources. In the first era, pest management problems were solved through chemistry, and now genetic engineering (genomics) is becoming dominant (Cate and Hinkle, 1993). However, agricultural systems are not products to be discovered, wrapped in a shiny package, and sold. They are continually evolving toward changing goals and objectives. Effective management of any pest is fundamentally an exercise in applied ecology (Cate and Hinkle), which is not achieved by one more product from the genetic engineer or chemist. Neither group has the education and often not the inclination to address the ecological, social, and moral questions that arise. The true social and environmental costs of any technology are rarely measured by the developer. They are not part of most profit and loss calculations, and many such costs are not even known, although all are paid by someone. Pesticide developers have been alerted to ecological, social, and moral questions by many authors (Carson, 1962; Murray, 1994; Perkins, 1982; Van Den Bosch,

1978; Wright, 1990; and others). The continuing but diminishing controversy about genetically modified organisms has a familiar ring.

Anyone can dismiss any criticism of technology by saying, "Well, it's not true for me." This makes our personal beliefs, our assumptions, absolutely secure, and then there is no reason for beginning the difficult task of examining them (Melchert, 1995). How any idea strikes us, especially one that is critical of our activity or profession, is not a reliable guide to the validity of the comment or to how we ought to respond. Our first reaction, our intuition, may be mistaken. It is best to know and consider the arguments that support the idea or criticism. In science, the data or theory that best explains the observations usually wins. In ethics, the best reasons win. It is wise to avoid the temptation to ignore good reasons that disagree with our assumptions. We often define what we are for by knowing what we are against (Thompson, 2010). Thus we find the truth we want; not by inventing it, but by allowing the emergence of only the part of the whole truth we want to hear (Barrow, 1995). When agriculturalists think of the future of agriculture, it is important that they recognize that their vision of the future (their truth) affects the decisions they make (Harman, 1976) and how they practice or recommend agriculture be practiced. The research and teaching we do now involves a view of a future we expect, desire, or fear (Harman, 1976). As our view of the future changes, the decisions we make each day change. Our view moves from what is, to how we would like things to be, to what we perceive to be good. We then move quickly to a description of what we ought to do to achieve the good we desire. It is in this transition that we depart from the domain of science and enter the domain of evaluation, from which an ethic can be, and frequently is, developed (Rolston, 1975). If the good we desire is inspiring, it will impel each of us and others to action. If our view is uninspiring or wrong, there may be no common image of what is worth striving for. Then a discipline or a profession will decline because it lacks an inspiring vision and adequate motivation.

Most of my colleagues in Land Grant University Colleges of Agriculture are confident that their research and teaching are morally correct. They defend their objective approach to science and their objectivity in defending agriculture and agricultural research against emotional attacks from people who don't understand either. I often hear, "People should not complain about our agricultural system with their mouth full" a conceit that dismisses the complaint without hearing it. The scientist's frequent appeal to the value of objectivity in science is evidence of a lack of awareness of the inevitable subjectivity of science. Neils Bohr, a pioneer in quantum mechanics, suggested the scientifically heretical idea that the mental decisions of the investigator influenced the outcome of an experiment. His point was that the observer somehow created the reality observed (Shlain, 1998). This does not mean that there is no such thing as objectivity or that all of science has to be discarded. It does mean that science has "to admit its nemesis—subjectivity—into its calculations." No science, including agricultural science, is immune to the nemesis of subjectivity. For example, a study done of the faculty of the college of agriculture at Washington State University (Beus and Dunlap, 1992) showed faculty members slightly more in favor of conventional[5] (industrial) agriculture than farmers statewide were. They were slightly less conventional than proponents of conventional agriculture and far more conventional than known proponents of alternative agriculture. Beus and Dunlap also found that women, younger faculty, and faculty not raised on farms were somewhat more likely to endorse the alternative agriculture paradigm than their counterparts. These faculty groups have made a transition from the scientific to the ethical realm, perhaps without knowing they have changed. They moved quickly and easily from observation of what is, to knowing that it is good and ought to continue (Rolston, 1975). The supporters of the status quo of conventional agricultural research direction and practice are morally confident of the rightness of their unexamined position.

An additional example is provided by a series of articles in the *High Country News*. Jones (1994) accuses land grant universities of having "elevated efficiency and scale of production above all other values." Their efforts have nearly eliminated the original land grant constituency of small family farms because rural people and the environment have not been central concerns. Jones suggests this is because the faculty and administration are tightly connected with traditional rural interests and faculty have lacked the freedom and encouragement to implement new ideas. She laments the lack of creative ideas flowing from land grant universities about such rural problems as land use planning in rural areas, grazing, or logging. Land grant college faculty, in Jones' view, have handed over to others the intellectual oversight of major agricultural issues. As apologists for exploitive industries, land grant universities "usually value the economic interests they serve above the public interest" (Wuerthner, 1994). Jones concluded that the institutions have been challenged and "in their own, ponderous way, they are responding to that challenge." As the Washington State study (Beus and Dunlap) shows, younger faculty are aware of agriculture's problems and are inclined, as the university system permits, to deal with them.

**Highlight 5.4**

A major environmental pollution problem has been created by confined animal feeding operations (CAFOs). CAFOs house a few thousand to several thousand animals in confinement on a small acreage. They are an economically efficient way to produce pork or chicken with a minimal requirement for human labor. CAFOs are attractive to farmers because it allows them to specialize production operations as they attempt to maintain income. However, 400 hogs can generate more than 12,000 gal of manure each day and disposal is a problem.

[5] My use of conventional means the presently dominant industrially dependent, capital, chemical, and energy dependent agricultural production system.

Neighbors complain of olfactory pollution and the liquid manure lagoons built to store the waste have become part of the problem in addition to the fact that they may rupture and pollute nearby water.

Scientists with the USDA's Agricultural Research Service have developed a combination of technologies that separate solid from liquid waste, recover organic matter, remove ammonia from wastewater, and transform phosphorus removed from wastewater into calcium phosphate, a marketable fertilizer. The system removes more than 97% of total suspended solids, 95% of total phosphorus, 99% of ammonia, 98% of copper, 99% of zinc, 99% of biochemical oxygen, and more than 97% of odor-causing components of hog manure. This modern, efficient system will significantly reduce the undesirable olfactory effects of CAFOs and permit further expansion of confinement rearing of animals.

However, publicity about this desirable scientific advance does not even mention any concern about whether confinement rearing of animals causes animals to suffer. Efficiency and profit are all that matter and both will be improved as science eliminates the undesirable environmental effects of CAFOs. Whether humans have a right to treat animals in ways that cause them to suffer becomes an unasked and irrelevant question (see Chapter 10).

## Expanding Agriculture's Moral Scope

To suggest expanding agriculture's moral scope is not to suggest that agriculture and its practitioners lack moral standards or that all past achievements and values must be abandoned. It also does not suggest that this book is about to reveal a new, correct set of moral standards for agriculture that will facilitate analysis of problems and ease the path to a solution. That is, proper moral standards will reveal what ought to be done and provide reasons for the action. Moral behavior is central to our humanity but its origins are obscure (Greene, 2005). It is probably not innate. Therefore, expanding agriculture's moral scope includes asking where moral values come from, and what are or ought to be the source of moral values for agriculture. The standard must involve addressing the debate between advocates of conventional/industrial and alternative agricultural systems. Pelletier et al. (2000) identify three viewpoints which should be considered (Table 5.1). The enduring question is, Whose values should count in developing a firm ethical foundation for agriculture? The following chapters will address but not answer the question. The answer should come from continuing discussion and analysis of competing views.

**Table 5.1** Three Views of Value Issues Relative to Food and Agriculture

| Viewpoint | Characteristics |
|---|---|
| Social justice | Primary concern is hunger. They recognize the importance of agriculture to communities but are concerned first and foremost with addressing social inequities. |
| Pragmatist | Values contribution of agriculture to local communities but are not concerned about and deny the salience of environmental or social externalities. Generally are unsympathetic to social consequences of agricultural practice. [See Beus and Dunlap (1992), conventional agriculturalists.] |
| Visionary | Values contributions to communities and are highly concerned about environmental and social externalities, but only mildly concerned about social inequities. (See Beus and Dunlap, alternative agriculturalists.) |

## The Utilitarian Standard

In fact, as suggested in Chapter 4, agricultural research and agricultural policy have had an identifiable utilitarian ethical standard since inception (Thompson et al., 1991a). The clear emphasis on increasing production and reducing production costs to increase profit identifies the utilitarian ethical standard to provide the greatest good for the greatest number that has been implicit in agriculture. This standard, accepted but largely unexamined within agriculture, has assumed that increasing production and reducing cost will optimize the social benefit of agriculture. There has been almost no dialog within agriculture about the correctness of the standard for all agricultural issues. One result has been that many scientists, ignorant of their own social context and all results of their technology, have, without questioning, accepted the loss of small farmers and rural communities as part of the necessary cost of achieving the greater goal of maintaining a cheap food supply (Stout and Thompson, 1991). The utilitarian standard is evaluated in terms of the outcome(s), for those affected by an act, and agriculturalists use the outcomes that are easily observed to evaluate what they do. They measure total production, crop yields and profit, and the latter, according to the USDA as reported in the Nov. 28, 1999 *New York Times*, has fallen 38% since 1997. They conclude that their efforts are ethically correct because good results have and, it is assumed, will continue to follow increased production. The morality of an act according to the utilitarian standard lies in its outcomes, its consequences. It does not focus on intentions as other moral standards do (see Chapter 4). The basic utilitarian principle, by itself, does not tell us whether peace is better than war, truth better than lying, or hoes better than herbicides. The utilitarian decides what is right by looking at the consequences (Singer 1981, p. 64). In agriculture, the cry for justice by the poor or the pleas of those concerned about loss of environmental quality is overwhelmed by the desirable consequences of increased production. The cries and pleas are often regarded as simply pusillanimous appeals that will not produce as much happiness as increased production.

Utilitarian thinking allows individual research scientists to believe that each research program is ethically correct because of its perceived good effects. Utilitarian standards are able to assign moral responsibility to any agricultural program or research area without considering the entire system (Thompson et al., 1991b). Agricultural scientists often see agriculture as a system of separable goods (e.g., seeds, bags of fertilizer, liters of pesticide, and so on) rather than as an integrated system. This view is a result of our reductionist heritage. Agriculturalists have learned to deal with highly divisible technologies (separable goods) rather than with nondivisible parts of the system (e.g., tractors, dams, irrigation systems) that demand a more holistic perspective.

None of the foregoing should be interpreted as an attack on the personal moral standards of individual scientists. I assume that anyone who has a research or teaching position knows the ethical norms (Holt, 1997). But as Ruttan (1991) said, "agricultural scientists have been reluctant revolutionaries." They have wanted to change agricultural practice and results but have neglected the revolutionary effects of their efforts on society. They have believed that their work could be reduced to their little pieces of agriculture and then added to the system without changing the whole system. Increasing production was the goal, and it was believed that could be accomplished without creating or at least without out being bothered by the creation of other, revolutionary effects (Ruttan). The ethical standards Holt assumes operated only within the immediate confines of the scientific enterprise; all else was external and could, therefore, be neglected. Therefore, agricultural research has been and is done in a morally narrow and ambiguous context with little recognition of either.

## The Relevance of the Western Agricultural Model

The post-World War II shift in the developed world to intensive farming systems with modern chemical and energy technology led to major increases in plant and animal production. These systems maximized production through specialization, increased scale of production units, minimized labor requirements, and maximized use of technological inputs. They allowed Western nations to fulfill more adequately than any societies have before what Ponting (1991) calls "the most important task in all human history"—finding ways to extract from the ecosystem enough resources to maintain life. The concomitant problem, in Ponting's view, is that human societies have had difficulty balancing their "various demands against the ability of ecosystems to withstand the resulting pressures." Countries that employ intensive agricultural systems have met the needs and many of the wants of their citizens, a high value; but, in the view of many, they have made excessive, unsustainable demands on the ecosystem, which was less valued. We in the West use this story of success (Sidey, 1998) in meeting human needs to support our belief in the universal relevance and applicability of intensive agriculture.

Huntington (1996) proposes that "this belief is expressed both descriptively and normatively." Descriptively, agriculturalists believe that all societies want to adopt Western agricultural techniques to increase production. They also believe that others are willing to accept Western institutions and practices to achieve these good ends. If they seem "not to have that desire and to be committed to their own traditional values, they are victims of a 'false consciousness,'" for example, failure to adopt genetically modified seed. Normatively the Western agriculturalist proposes that all societies ought to adopt our methods, institutions, and the associated values, because they embody "the highest most enlightened, most liberal, most rational, most modern, and most civilized thinking of humankind." Huntington says belief in the universality of Western values and culture suffers from three problems: "it is false, it is immoral, and it is dangerous." In looking at other cultures through our Western lenses and with our Western assumptions about what is good, we make the further error of assuming we are learning about what other people's conceptions of the world are rather than observing how the world really is. Only we understand how the world really is and therefore, we assume we know how agriculture ought to be practiced. Thus, part of expanding agriculture's moral scope will be to give up some of our hubris about the moral correctness and values of our culture and its agriculture.

## Bottom-line Thinking

Goldschmidt's (1998) work claims that the singular goal of our agricultural system was "to gain wealth, without the least concern for the welfare of those whose lives" were being destroyed. There was little thought about the effects of the money-driven system on the environment. The social rewards of belonging to a caring community, the spiritual satisfaction of serving a larger public purpose, and the communities themselves and the businesses that they need and support were sacrificed to the bottom line (Goldschmidt). Convenience, ready availability, and low cost are not the only things that matter. There are other things that are important such as the presence of local businesses, friendliness, service, the essentiality of business to any community's survival, and local employment opportunities. If large-scale industrial farming eliminates or harms these things, then we should think again about the importance of the bottom line. The losses Goldschmidt speaks of are the social costs of agriculture's technological changes. In agriculture, bottom-line thinking is the norm and may be part of the hubris we must reconsider if we are serious about our communities, and our agriculture.

## Sustainability

As hubris and bottom-line thinking are reconsidered, there will inevitably be conflicting interests that arise from opposing world views, incompatible analyses based on different views of the nature of the problem, rising material expectations, and different views of sustainability (Allen, 1993). Just as we know good parents and apple pie are intrinsically good, everyone seems to favor sustainability, but there are many different views of what ought to be sustained. Expanding agriculture's moral scope requires that we give up the common two-track agricultural defense used when issues ranging from pesticides to sustainability and loss of small farms to animal treatment are raised. The first track has been to deny that the suggested problem exists, for example, pesticides don't harm people or wildlife, people who use them incorrectly do, and the loss of small farms is unfortunate but it is an economic not an agricultural matter. The second defense has been to explain, calmly, but forcefully, that the reforms advocated (e.g., reductions in pesticide use, maintaining small farms, humane animal treatment,) will increase the need for farm labor that is not available, make food too expensive, and diminish or eliminate the favorable balance of trade the US enjoys from its agricultural surplus. The argument claims that the public will not tolerate higher food costs to save a few small, inefficient farms. The reforms would surely diminish or eliminate the trade surplus, and neither is politically acceptable.

Expanding agriculture's moral scope demands considering challenging views of agricultural practice. For example, Ludwig et al. (1993) posit a "remarkable consistency in the history of resource exploitation: resources are inevitably over exploited, often to the point of collapse or extinction." The view is shared by Ponting (1991), Berry (1977), and Jackson (1980) and other commentators on modern agriculture. Ludwig et al. suggest that wealth or the prospect of wealth generates social and political power that is used to promote unlimited exploitation of a resource. Exploitation of land and farm communities are good examples. Scientific understanding is hampered by lack of controls, and the natural variability that is expressed on a large scale. Therefore, detection of effects is very difficult. The complexity of large agricultural systems encourages a reductionist approach to study and management that precludes observation of large-scale effects. In the view of Ludwig et al. the long-term outcome is a heavily subsidized industry that over-harvests the resource. That seems to be a perfect description of modern agricultural practice. Ludwig et al. suggest that sustainable exploitation is always preceded by over exploitation. If they are correct, then agricultural sustainability will not be achieved by adjustments to the present system. Experience over 3,000 years, as documented by Ponting (1991) and cited by Ludwig et al., suggests that good scientific understanding of exploitation, its causes, and the appropriate prophylactic measures are not sufficient to prevent destruction of the vital resource. It is a sobering commentary that must be considered and addressed by the agricultural community and the larger society in the quest for sustainability.

Kirschenmann and Youngberg (1997) understand the complexity of agriculture's striving for sustainability. They acknowledge that those involved in the discussion must consider biotechnology, continuing agricultural industrialization, public concern about food quality and safety, pesticide use and abuse, rural community deterioration, worker welfare, international trade, global competitiveness, farmland preservation, wildlife and habitat protection, public funding of agricultural research and extension programs, natural resource conservation, and how to identify and somehow include what have been external agricultural costs. None of these is a simple issue. Kirschenmann and Youngberg believe agriculture will develop along two paths. The first will be "larger, highly specialized, vertically integrated farms" with efficient input management and precision farming techniques. The second will be "small, intensively- managed, diversified operations" with low input and an emphasis on sustainability. Many of the latter will emphasize organic production methods (see den Hond et al., 1999). The two paths are not necessarily heading in opposite directions and should not be viewed as conflicting agricultural futures. Kirschenmann and Youngberg forecast that biotechnology will be the primary force that shapes tomorrow's agriculture, and that water availability and quality will be the major environmental issues. There will be increased reliance on export markets to sustain grower profitability while the primary concern of consumers will be food safety. There will be a resurgence of rural communities near cities but a continued, slow demise of isolated rural communities. Development of sustainable agriculture will be emphasized, but there will be a decreasing level of public support for agricultural research in publicly supported agricultural institutions.

## Conclusion

Agricultural scientists share with other scientists a passion for the truth. We hold the Socratic belief that the search for the truth is the very best way of life (Melchert, 1995, p. 452). We err when we engage in distorting the truth by hiding it from ourselves or pretending that it is other than what we know it is. We tend toward that mistake when we think that what has no price has no value, that what cannot be sold is not real, and that the only way to make something actual, to determine its value, is to place it on the market (Merton, 1960). Bottom-line thinking cannot answer all questions. Small-scale farmers may be necessary to care for the land, our most precious resource; small communities may be important to our national and personal character. Neither is susceptible to pure economic analysis. Merton also deplores in strong, environmentally oriented language "constructing a world outside the world, against the world, a world of mechanical fictions which contemn nature and seek only to use it up, thus preventing it from renewing itself and man." Strong environmental criticism is not foreign to agriculture, but it is not valued as a means to seek the truth about agriculture.

As agriculture searches for an ethical standard while being compelled to deal with criticism of its apparent bottom-line standard, and the environmental problems that have followed modern agricultural practices, one must ask, as Holt (1997) has, "how do we find our way?" Holt's answer comes in the context of his concern that "if there are no revealed truths, immutable principles, or even practical, utilitarian generalities, such as the importance of honesty and integrity, that apply to science, how can we determine what constitutes ethical behavior?" When there are no points of reference, no moral certainty, in what he sees as "a sea of uncertainty," how can one make ethical judgments? His answer epitomizes bottom-line thinking and is apparently utilitarian. He argues that in the global marketplace of which agriculture is now a part, "the question of whether it is ethical to conduct research on certain subjects, including products or services that might do harm to the environment or make food unsafe (e.g., chemicals) is moot. The need for effective pest control will be so great that no options can be precluded before research reveals the potentials." He goes on to suggest that research on technology that might cause dislocations (a locution that avoids the unpleasant connotation of suffering) of small farmers is also moot. Market oriented economies and diminished government interference in agriculture will "assure that less efficient and productive practitioners will be dislocated." Holt's concluding message for agricultural scientists is quoted in full below.

> *I think there is an important message for agronomists and other agricultural researchers and research administrators in this analysis of driving forces. It says that as you plan and implement research on the environment, natural resources, safety, social issues, or other themes, keep it within the context of quality, productivity, and efficiency. In order for any technology or information generated in that research to find fruition in practice, it will have to contribute to quality, productivity, or efficiency or at least not detract from them.*

Holt believes that the "broadly focused ethical debates of the past" are also moot. His view is consistent with Day's (1978) view that agriculture has only one proper goal: profitable production. This view ignores the necessity of balancing the often conflicting goals of profitable production and environmental quality or the loss of small farms and the rise of corporate agriculture. Agriculture's continuing problems with and concerns about pollution from pesticides and nitrates, animal welfare, soil erosion, loss of genetic diversity, genetic modification of plants and animals, and food safety are all swept under the carpet of profitable production that makes all other arguments moot. To be moot is first to be open to discussion or debate, to be debatable, and second to be so hypothetical as to be meaningless. It is the latter definition that Holt uses, and I fear it is the view shared by most in agriculture. It is a view that says "we are doing the most ethical thing of all, feeding the world and when you are hungry you will recognize the wisdom and rightness of our actions and methods." The view dismisses the cries for justice, environmental quality, or food safety as pusillanimous appeals that are moot.

In Holt's view, there is no compelling moral argument for saving family farms because of the economic realities of the global market within which agriculture must operate. Berry (1977) and others who argue for preservation of family farms, on the basis of a Jeffersonian agrarian appeal about the virtue of the small farm and how it was and remains the basis of our civilization obviously, from Holt's perspective, just don't understand the modern world.

I conclude that while agricultural scientists are ethical in the conduct of their science (they don't cheat, don't fake the data, give proper credit, and so on) and in their personal lives (they earn their wages, take care of family, respect others, are responsible for their actions, and so forth), they do not extend ethics into their work at a deeper level. Agricultural scientists are the reluctant revolutionaries that Ruttan (1991) identified, but they are also realists. Realists run agricultural research and the world; idealists like Berry and Jackson do not. Idealists attend academic conferences and write thoughtful articles and books from the sidelines (Kaplan, 1999). The action is elsewhere. The rule is, produce profitably or perish in the real agricultural world. Realism rules, and philosophical and ethical correctness are no more necessary for useful work in science than theological correctness is in religion (Rorty, 1999).

I find that true, but I want more. I want my agricultural colleagues to accept the burden of beginning the difficult task of doing a discriminating cultural and moral analysis of agriculture and its results. We must strive for a careful analysis of what it is about our agriculture and our society that thwarts or limits our aspirations and needs modification. That analysis must include the US Department of Agriculture, agricultural colleges, scientific societies, and the many commercial organizations, which serve and profit from agriculture. Those features that are beneficial must be nourished and strengthened and those that are not must be changed. To fit ourselves for this task, we must be sufficiently confident to study ourselves and our institutions and dedicated to the task of modifying both. It takes a combination of intellectual rigor and management skill to do this well. Management skill is required but difficult to achieve. Intellectual rigor annoys people because it interferes with the pleasure they derive from allowing their wishes to be the father of their thoughts (Will, 1999). Most people don't want their assumptions challenged, they want to use them.

To preserve what is best about modern agriculture and identify and oppose the abuses that modern technology has wrought on our land, our people and other creatures, and finally to begin to correct them will require many lifetimes of work (Berry, 1999). Agriculturalists must try to see agriculture in its many guises—productive, scientific, environmental, economic, social, political, and moral. It is no longer sufficient to justify all agricultural activities on the basis of increased production. Other criteria, many with a clear ethical foundation, must be included. Citizens of the world's developed countries live in a post-industrial, information age society. They do not, and no one ever will, live in a post-agricultural society. Societies have an agricultural foundation within their borders or in other countries. Those in agriculture must strive to ensure all that the foundation is secure.

## References


Aiken, W. (1984). The goals of agriculture. Pp. 29–54 *in* R. Haynes, R. Lanier (eds.). *Agriculture, Change and Human Values—A Multidisciplinary Conference*. Gainesville, FL, University of Florida.

Allen, P. (ed.). *Food for the Future: Conditions and Contradictions of Sustainability*. New York, NY, J. Wiley and Sons.

Alternative Agriculture News. (2002). Number of US Farms and Ranches Declines in 200120(4):2, New Report Finds. See (http://usda.mannlib.cornell.edu/reports/nassr/other/zfl-bb/fmno0202.txt).

Avery, D. (1997). Saving the planet with pesticides and biotechnology and European farm reform. Pp. 3–18 *in* British Crop Prot. Conf.—Weeds.

Barrow, J.D. (1995). *The Artful Universe*. Boston, Back Bay Books. P. 246.

Berry, W. (1977). The agricultural crisis as a crisis of culture. Pp. 39–48 i*n The Unsettling of America: Culture and Agriculture*. New York, Avon Books.

Berry, W. (1999). In distrust of movements. *The Land Report* 65(Fall 1999):3–7. Salina, KS, The Land Institute.
Beus, D.E. and R.E. Dunlap. (1992). The alternative–conventional agriculture debate: where do agricultural faculty stand? *Rural Sociology* 57:363–380.
Blatz, C.V. (1995). Communities and agriculture: Constructing an ethic for the provision of food and fiber. Pp. 207–240 *in* Decision making and agriculture: The role of ethics. Int. Conf. on Agricultural Ethics. Rural Research Centre, Nova Scotia Agricultural College, Truro.
Capra, F. (1996). *The Web of Life: A new scientific understanding of living systems*. New York, Anchor Books.
Carson, R. (1962). *Silent Spring*, 25th Anniversary Edition. Boston, MA, Houghton Mifflin Co.
Cate, J.R. and M.K. Hinkle. (1993). *Integrated pest management*. Special Report. National Audubon Society.
Day, B.E. (1978). The morality of agronomy. Pp.19-28 *in* J.W. Pendleton, (ed.). *Agronomy in Today's Society*. Madison, WI, American Soc. Agron. Special Pub. No. 33.
Degregori, T.R. (2001). *Agriculture and Modern Technology*. Ames, IA, Iowa State University Press.
DeLind, L.B. (2006). From the Editor. *Agriculture and Human Values* 23: 269–270.
den Hond, F., P. Groenewegen, and W.T. Vorley. (1999). Globalization of pesticide technology and meeting the needs of low–input sustainable agriculture. *American J. Alt. Agric* 14:50–58.
Durning, A.T. (1992). *How Much is Enough? The Consumer Society and the Fate of the Earth*. New York, W. W. Norton & Co.
ERS (Economic Research Service of US Dept. of Agriculture). (2008). Farms receiving government payments chapter of the ERS Briefing Room on Farm Income and Costs. http://www.ers.gov/Briefing/farmincome/govtpaybyfarmtype.htm/
Feyerabend, P. (1987). *Farewell to Reason*. London, UK, Verso.
Gerrard, S. (2000). Environmental risk management. Pp. 435–468. *in Environmental Science for Environmental Management*. London, UK, Prentice Hall Co.
Goldschmidt, W. (1947, 1978). *As You Sow: Three Studies in the Social Consequences of Agribusiness*. Montclair, NJ, Allanheld, Osmun & Co.
Goldschmidt, W. (1998). Conclusion: The urbanization of rural America. Pp. 183–198 *in* K.M. Thu, E.P. Durrenberger (eds.). *Pigs, Profits, and Rural Communities*. Albany, NY, State University of New York Press.
Greene, J. (2005). Cognitive neuroscience and the structure of the moral mind. Pp. 338–352 *in* P. Carruthers, S. Laurence, S. Stich (eds.). *The Innate Mind: Structure and Contents*. New York, Oxford University Press.
Halweil, B. and L. Mastney. (2003). *State of the World—2004*. New York, W. W. Norton & Co.Worldwatch Inst.
Harkinson, J. (2004). Profits of place: A different vision of success emerges along main street. *Orion* Jan/Feb:59–63.
Harman, W.W. (1976). *An Incomplete Guide to the Future*. New York, W. W. Norton & Co. P. 1.
Holt, D. (1997). Practical ethics in agronomic research. *Advances in Agronomy* 60:149–190.
Huntington, S.P. (1996). *The Clash of Civilizations and the Remaking of the World Order*. New York, Touchstone Books. P. 360.
Jackson, W. (1980). *New Roots for Agriculture*. Lincoln, NE, University of Nebraska Press. San Francisco, CA. Friends of the Earth Press.
Jones, L. (1994). Land grants under the microscope. *High Country News*. May 1. Pp. 8–9.

Josephson, M. (1989). Interview on pages 14–27 of Moyers, B. *A World of Ideas: Conversations with Thoughtful Men and Women About American Life Today and the Ideas Shaping Our Future.*

Kaplan, R.D. (1999). Kissinger, Metternich, and Realism. *Atlantic Monthly*. June:73, 74, 76–78, 80–82.

Keeney, D. (2003). Feed the world: A failed policy. *Leopold Letter* 15(4):7.

Kidder, R.M. (1994). Universal human values: Finding an ethical common ground. *The Futurist* July/August:8–13.

Kimbrell, A. (ed.). (2002). *Fatal Harvest: The Tragedy of Industrial Agriculture*. Covelo, CA, Island Press.

Kirschenmann, F. (2007). How long will we continue to fiddle while Rome burns? *Leopold Letter* 19(2):5.

Kirschenmann. F. and G. Youngberg. (1997). Letter from the President and the Executive Director. Annual Report, H. A. Wallace Institute for Alternative Agriculture. Greenbelt, MD. P. 2.

Kroeber, A.L. and C. Kluckhohn. (1952). Values and Relativity. *In Culture, a critical review of concepts and definitions*. Papers of the Peabody Museum. Harvard University 47 (1):174–179.

Leopold, A. (1947). *A Sand County Almanac*, 1966 edition. New York, Ballantine Books. P. 238.

Ludwig, D., R. Hilborn, and C. Walters. (1993). Uncertainty, resource exploitation, and conservation: Lessons from history. *Science* 260(17):36.

Melchert, N. (1995). *The Great Conversation: A Historical Introduction to Philosophy*, 2nd Ed. Mountain View, CA, Mayfield Publishing Co. Pp. 63, 452.

Merton, T. (1960). Rain and the Rhinoceros. Pp. 9–23 *in Raids on the Unspeakable*. New York, New Directions.

Murray, D. (1994). *Cultivating Crisis: The Human Cost of Pesticides in Latin America*. Austin, TX, University of Texas Press.

Nagel, T. (2010). The facts fetish. *The New Republic*. Nov. 11. Pp. 30–33.

Pelletier, D.L.V., C. Kraak, McCullem, and U. Uusitalo. (2000). *Agriculture and Human Values* 17:75–93.

Perkins, J. (1982). *Insects, Experts, and the Insecticide Crisis*. New York, Plenum Press.

Ponting, C. (1991). *A Green History of the Earth: The Environment and the Collapse of Great Civilizations*. New York, Penguin Books. P. 17.

Pretty, J. (2003). Ecologist challenges traditional notions about agriculture. *Leopold Letter* 15(4):10.

Rolston, H. (1975). Is there an ecological ethic? *Ethics* 85:93–109.

Rorty, R. (1999). Phony Science Wars: A review of Hacking, I. 1999. The Social Construction of What? *Atlantic Monthly*. November:120–122.

Ruttan, V. (1991). Moral responsibility in agricultural research. Pp. 107–123 *in* P.B. Thompson, B.A. Stout. (eds.). *Beyond the Large Farm: Ethics and Research Goals for Agriculture*. Boulder, CO, Westview Press.

Sen, A. (1981). *Poverty and Famines: An Essay on Entitlement and Deprivation*. Oxford, UK, Clarendon Press.

Shlain, L. (1998). *The Alphabet Versus the Goddess: The Conflict Between Word and Image*. New York, The Penguin Group. P. 395.

Sidey, H. (1998). *The Greatest Story Never Told: The food miracle in America*. H. A. Wallace annual lecture. H. A. Wallace Inst. for Alternative Agric. Greenbelt, MD.

Singer, P. (1994). *Ed. Ethics*. Oxford, UK, Oxford University Press. Pp. 3–13.

Singer, P. (1996). Famine, Affluence, and Morality. Pp. 26-38 *in* W. Aiken and H. LaFollette (eds.). *World Hunger and Morality*. 2nd Ed. Upper Saddle River, NJ, Prentice Hall.

Singer, P. (1981). The expanding circle: Ethics, evolution, and moral progress. Farrar, Straus and Giroux, New York. Pp. 208.

Standaert, M. (2003). World farmers struggle with globalization issues. Newsdesk.org wysiwyg://79http://www.enn.com/news/2003–12–31s_11631.asp. (accessed, January 2004).

Stauber, K.N., C. Hassebrook, E.A.R. Bird, G.L. Bultena, E.O. Hoiberg, H. MacCormack, and D. Menanteau–Horta. (1995). The promise of sustainable agriculture. Pp. 3–15 *in* E.A.R. Bird, G.L. Bultena, J.C. Gardner (eds.). *Planting the Future: Developing an Agriculture that Sustains Land and Community*. Ames, IA, Iowa State Univ. Press.

Stout, B.A. and P.B. Thompson. (1991). Beyond the large farm. Pp. 265–279 *in* P.B. Thompson, B.A Stout. (eds.). *Beyond the Large Farm: Ethics and Research Goals for Agriculture*. Boulder, CO, Westview Press.

Thompson, P.B. (1986). The social goals of agriculture. *J. Agric. and Human Values* 3:32–42.

Thompson, P.B. (1988). Ethical dilemmas in agriculture: the need for recognition and resolution. *J. Agric. and Human Values* 5(4):4–15.

Thompson, P.B. (1989). Values and food production. *J. Agric. Ethics* 2:209–223.

Thompson, P.B. (2010). *The Agrarian Vision: Sustainability and Environmental Ethics*. Lexington, KY, The University Press of Kentucky.

Thompson, P.B., G.A. Varner, and D.A. Tolman. (1991a). Environmental goals in agricultural science. Pp. 217–236 *in* P.B. Thompson, B.A Stout. (eds.). *Beyond the Large Farm: Ethics and Research Goals for Agriculture*. Boulder, CO, Westview Press.

Thompson, P.B., G.L. Ellis, and B.A. Stout. (1991b). Values in the agricultural laboratory. Pp. 3–31 *in* P.B. Thompson, B.A Stout. (eds.). *Beyond the Large Farm: Ethics and Research Goals for Agriculture*. Boulder, CO, Westview Press.

Van Den Bosch, R. (1978). *The Pesticide Conspiracy*. Berkeley, CA, University of California Press.

Warren, G.F. (1998). Spectacular increases in crop yields in the United States in the twentieth century. *Weed Technol* 12:752–760.

Will, G.F. (1999). The Last Word. *Newsweek*. May 24:84.

Wright, A. (1990). *The Death of Ramon González: The modern agricultural dilemma*. Austin, TX, University of Texas Press.

Wuerthner, G. (1994). Get Lost. *High Country News*. May 1, 1995:21.

# 6 The Relevance of Ethics to Agriculture and Weed Science

*Agriculture experts and agribusinessmen are free to believe that their system works because they have accepted a convention which makes external and therefore irrelevant, all evidence that it does not work. External questions are not asked or not heard, much less answered.*

***Berry, 1977***

*Like it or not, ready or not, the age of agricultural ethics has arrived.*

***Ferré, 1994***

After describing a few relevant moral theories (Chapter 4) and suggesting that those in agriculture possess abundant, but perhaps inappropriate, moral confidence (Chapter 5), it is time to take up the task suggested toward the end of Chapter 3. That task is to demonstrate that underlying views on agricultural issues there is an ethical foundation that determines agriculture's ethical horizon. Knowing that foundation for any position is an important step toward addressing any of agriculture's ethical dilemmas. The question is, can one show how ethics and ethical theory can be applied to weed science, an important but small segment of agriculture? If such a connection can be made, it should be easy to extrapolate the connection to agriculture's other subdisciplines. Earlier it was proposed that scientific objectivity should include thought about what value judgments are made, might be made, and perhaps ought to be made by scientists in any discipline. That claim included the assumption that an ethical foundation underlies views on important agricultural issues and that the ethical position is usually unexamined and may even be unknown. In this chapter, I ask if such claims have any validity when applied to one of agriculture's subdisciplines: weed science.

I selected weed science as a test case because I know it best and because it represents much that is good and bad about agriculture. Weed science and, by implication, all agricultural disciplines, is deficient[1] because its implicit values are unexamined and its operative values are purely instrumental (i.e., they are purely means to an end). Agricultural scientists tend to view the values of groups (e.g., farmers, environmentalists) or populations (e.g., Americans) as aids or obstacles to improving the productivity and profitability of agriculture: the primary values. Other values are subordinate to the primary values—production and profit, which are uncritically assumed to be good. This chapter suggests that the primacy of these values should be debated, because production and profitability are but instrumental goals. They are the means to

[1] Much of this chapter is a revision of Zimdahl, R.L. (1998). Ethics in weed science. *Weed Sci.* 46: 636–639. Reprinted with permission.

Agriculture's Ethical Horizon. DOI: 10.1016/B978-0-12-416043-9.00006-4

**Highlight 6.1**

In his small book, *Too Many People*, Grant (2000, p. 7) presents a challenging calculation. Using the World Bank's data (*World Development Report*, 2002, 2010), he asks, "How many people could live at a decent level at present rates of economic activity?" His calculation begins with the average per capita GNP of the World Bank's high income countries (44 in 2000, 53 in 2006, and 69 in 2010) which he uses as "a crude surrogate for a good standard of living." He then assumes that the current world gross income GNP is environmentally sustainable, which is not the same as assuming it will not grow. A simple division follows:

For 2000 data (World Bank, 2002)

$$\frac{\text{World GNP in US \$}}{\text{Average GNP of 53 UN High-Income Countries}} \quad \frac{\$\ 31{,}171 \times 10^9}{\$\ 24{,}781 \text{ per capita}} = 1.26 \times 10^9 \text{ people}$$

For 2008 data (World Bank, 2010)

$$\frac{\text{World GNP in US\$}}{\text{Average GNP of 66 UN High-Income Countries}} \quad \frac{\$57.637.5 \times 10^9}{\$39{,}345} = 1.46 \times 10^9 \text{ people}$$

The logical conclusion is that the earth can support between 1.26 and 1.46 billion people at "a decent level." The earth had 6,901170,358 people on February 20, 2011 (http://www.census.gov/main/www/popclock). Grant assumes a "decent level" is the standard of living in the world's industrial countries. Following Grant's logical calculation, one may conclude: 1. Grant is wrong; 2. The poor will become poorer; or 3. The rich may remain rich, but will feel more threatened.

The UN/FAO estimates the world now has as many as 1 billion undernourished people, many of whom are children. The number of undernourished declined in the first half of the 1900s but is rising again. And, while the earth remains productive, it is heavily burdened (Clayton, 2005).

- Each year an area of the United States about 1 km wide stretching from New York to San Francisco is converted to nonagricultural uses. The rate of rural land lost to development in the 1990s was about 2.2 million acres per year. If the rate of loss continues about 110 million acres of land will no longer be agricultural by 2050. A major reason is that agricultural land is worth 5–10 times more for urban or industrial development than as crop land. From 1982 to 2007, each of the contiguous 48 states lost agricultural land. About 1% of US land is lost to agriculture each year. About 23 million acres of agricultural land, an area the size of Indiana, were lost in the 25 years.[1] Should this continue? Why?

- Coral reefs support thousands of species and cover more than 278,000 $km^2$ of the earth's surface. About 20% of the earth's coral reefs have been effectively destroyed with no immediate prospects for recovery. About 24% of the world's reefs are in imminent risk of collapse.[2] Should this continue? Why?
- Farmers' increased use of nitrogen fertilizer since 1985 has polluted waterways and coastal ecosystems. The number of ocean dead zones increased from 146 in 2004 to 405 in 2008.[3] About 35% of coastal mangrove swamps that serve as biological filters have been bulldozed. Ninety percent of global mangroves are in developing countries and are critically endangered and facing extinction in 26 countries.
- Ocean fisheries are nearly all over-fished, with stocks down 90–99% from pre-industrial levels. The stability and function of oceans, that support 90% of the earth's livable habitat, absorb carbon, and supply 70% of the oxygen we require has been compromised (Kirschenmann, 2007).

Ecology teaches that all systems are inter-related and inter-dependent. No one is sure if a butterfly's flight in central China will cause a storm in Iowa (or vice-versa), but the metaphor illustrates the potential interrelation of all things. During the next 50 years, the final period of major agricultural expansion, demand for food by an expanding population, and especially by the citizens of the UN's high income countries, will be a major driver of global climate change (Tilman et al., 2001). If past trends continue, Tilman et al. predict that another billion hectares (2.47 billion acres) will be converted to agriculture by 2050. Humans already release as much nitrogen and phosphorus to natural ecosystems as all other sources combined, and releases will increase. Pesticide use will increase, with all of its attendant problems. Tilman et al. project what is likely to happen, and Grant (2000) asks if all can be fed at a decent level. Few in agriculture are asking the moral question: Should any of this be allowed to happen? Is this what we, in agriculture, ought to be part of?

We may need what Conway (1997) calls a Doubly Green Revolution. But achieving it must include the moral questions: What is right for humans, for the earth, and for its other creatures. Agriculture is the single largest human interaction with the environment. Its conduct demands it be so. A doubly green revolution must prevent continuing environmental damage, as outlined above, *and* try to feed all. It is a huge scientific and ethical challenge, perhaps the largest ever.

There is no question that the agricultural system of developed nations is enormously productive. More of nearly all crops is produced each year with fewer producers and food costs less in developed countries. That accurate claim is one that any manufacturer would be proud to make. But those enormously productive technologies frequently have undesirable ecological effects. The doubly green revolution needs new technology that may not yet be readily available. Biotechnology (see Chapter 8) will help, but not quite yet. Organic agriculture holds promise but with present technology and production levels, it cannot feed the world of today no less the 50% increase many forecast by 2050.

We need to produce. We need to think about our lifestyle and that of others. We need to explore whether agriculture must stay industrially based or if it can move to an environmentally sensitive system(s) where biology and ecology rather than chemistry become the foundation and its ethical dimensions are not neglected.

[1] www.msu.edu/user/dunnjef1/rd491/landuse. Accessed March 9, 2011. www.farmland.org/news/pressreleases/American-farmland-trust-press-release-farm (accessed March 11, 2011).
[2] www.globalissues.org/article/173/coral-reefs. Accessed March 11. 2011.
[3] www.takepart.com/news/2010/09/14/ocean-dead-zones-spreading-in-us-waters. Accessed March 9, 2011.

meet human needs. As operative values are discussed, we should consider broadening the concept of productivity and efficiency to include basic environmental, resource, ecological, health, social, and political processes and the costs and benefits of agricultural technology (Dahlberg, 1982). The ultimate formulation of a sustaining ethic for weed science will not be as difficult as all of these criteria imply, if the ethic can be based on a value all might accept, such as meeting human needs (Burkhardt, 1986). Innovation in agriculture can be good, but perhaps it will be best when what agricultural scientists declare good is also judged good by a concerned populace that finds agricultural practice and innovations compatible with their view of what constitutes the good life, the good society, satisfaction of human needs, and a good environment.

As stated before, agriculture is the most significant and most widespread human–environment interaction. Therefore, public participation in decisions about the effects of agricultural technology, although uncommon, is inevitable. Public debate about values and the governing ethic should precede actions that we presume will lead to meeting human needs. Industrial weed scientists who design strategies to create change in weed control techniques, and university and public sector scientists who study and test components of these strategies must be willing to submit their assumptions about what is good and the reasons for these assumptions to public scrutiny.

It is not my intent to describe all aspects of an ethic for weed science. That is, this chapter will not describe what ought to be done to solve ethical dilemmas in weed science and then give the supporting reasons for action. I do not know what the precise characteristics of the discipline's ethical foundation should be, and it would be arrogant to suggest I do. I am certain the characteristics of acceptable values and the supporting ethical theory must be considered. As weed scientists plan their course, they must ask what is good, who benefits, who is harmed, and what is not considered; what are the externalities. I know I cannot and should not attempt to answer such questions for all weed scientists. I suggest exploration, a chance to hear and consider other views. Thus, the purpose of this chapter is to explore the elements of an acceptable ethic for weed science.

Science is a descriptive, explanatory enterprise, whereas ethics is normative (that which establishes norms or standards) and prescriptive (that which prescribes what ought to be done). Ethical debate elicits reasons for what we do and ought to do. In contrast to science, ethics does not rely on an established body of factual knowledge or presuppose and rely on a set of fundamental laws. Much of science proceeds via the scientific method aided by occasional bursts of inspiration, but ethical analysis does not follow the same predetermined method.

It is my perception that ethics and exploration of values are regarded with suspicion by many because it is assumed that bringing up the subject of ethics or values implies the need for not just change, but reform. To many, it implies that something is wrong and that those who are ethical (or who can at least use the jargon) are going to prescribe what is right, what ought to be done. There is a deeper reason for suspicion of ethics. Agricultural scientists think of themselves as having an abundance, even an excess, of concern for agriculture and society (Dundon, 1986). Agricultural scientists (including weed scientists) know their work is founded on the most ethical behavior of all: feeding the world. How can that be questioned? One suspects that those who raise ethical questions about agriculture have never understood its purpose. It is also clear that weed scientists share with other agricultural scientists a deep commitment to the value of "persons, institutions, and ways of serving both" (Dundon). Questioning the nature or depth of that commitment by outsiders is looked upon as meddling with deep-seated, essential values. Some meddlers may be regarded as kooks, whereas colleagues may be regarded as traitors. And the meddler who questions current practice and fails to put in place a new practice that the agricultural scientist can add to the field's competency is subverting the fundamental responsibility and competency of the profession, while bringing harm to its future (Dundon).

Value analysis is inherently a subversive undertaking that is likely to have the short-term effect of converting certainties into problems. Achieving value consensus is a meaningful, useful ideal when it enables scientists to proceed beyond analysis of values to a statement of collective preference for particular values (Bressler, 1978). It would be good to include a statement of value preference in any agricultural society's mission statement. The American Society of Agronomy and the Crop Science Society of America have published statements of professional and scientific ethics since 1992. The Soil Science Society of America adopted a statement in 1999.[2] The Weed Science Society of America does not have a similar statement.

Scientists learn that the scientist *qua* scientist cannot (should not) make value judgments. That is confirmed by statements in texts used in science courses. Value judgments are, at worst, the purview of politicians and, at best, of society in general (Rollin, 1996). Rollin (1995) suggests that when any discipline finds itself in a position where there is no value consensus, and clarification and discussion are required, "the most shrill and dramatic articulations and discussions of such issues will tend to seize the center stage." Rollin calls this Gresham's law of ethics, which argues that in the absence of informed expertise to counter and moderate shrill, dominating distortions, the distortions "will dominate the social mind and drive the legitimate ethical concerns out of awareness." I think this is the situation in which weed science, and, by implication, all of agriculture finds itself. Agricultural scientists have allowed the shrill voices, a few internal, mostly external, to dominate public perception of who

[2] www.agronomy.org/ethics, www.crops.org/ethics, www.soils.org/ethics. Accessed March 2004 and August, 2011. There were no significant changes in the identical statements of each society.

we are, what we do, and why we do it. Weed scientists have heard, and I suspect many believe, that the best offense is a good defense. But weed science has not had an offense; it has had only a defense against external distortions of its technology. A good offense must begin with an understanding of the ethical foundation, followed by discussion of the implicit values of the discipline and a statement of consensus on its values.

Day (1978) laid out one view of the best ethical foundation for all of agriculture in his presidential address to the American Society of Agronomy. He proposed that the basic and only morally defensible foundation for agronomy was to produce food and fiber. I have suggested that production is currently the dominant ethic for weed science and all of agricultural science (Zimdahl, 1998). Those who subscribe to this ethic usually recognize that its results have not been universally applauded, but they often fail to see that the lack of applause has been for good reasons. These reasons include valid objections from five perspectives (Danbom, 1997).

The *first* includes socially conscious individuals and organizations who suggest that production has been achieved at the expense of the environment and our natural resource base.

*Second* are modern agrarians (e.g., Wendell Berry) who note the loss of family and community values and the rise of selfishness and materialism that often accompany increased production.

*Third* are those who warn of the unsustainability of modern production practices.

*Fourth*, others warn that our abundance has been accompanied by deterioration, or at least a perception of deterioration, in food quality and healthfulness. *Finally*, there are those who argue that abundance has been obtained without regard for effects on small farmers, tenant farmers, and the rural and urban poor who have paid a disproportionate share of the price of our bounty.

Advocates of these views tend to agree that the high productivity of modern agriculture has been achieved by extensive use of technologies that often contribute to depletion of nonrenewable resources (e.g., soil and water). They suggest that the undeniable success of modern agriculture has been achieved by combining mechanization, genetic manipulation, fertilization, and pesticide chemicals and is dependent on a massive unsustainable fuel subsidy (Pais, 1982).

What is right about the production ethic, and there is much right about it, is that it allows less than one million Americans to feed the rest of the population of the United States. One US farmer in 1990 fed about 130 people in the United States and abroad, whereas in 1960, one farmer fed only 26 people. The 2009 data are even more startling. Farmers are 0.05% of all employed Americans (751,000). Therefore, one US farmer feeds 410 Americans (total 2009 population = 308 million) and an unknown number of others who rely on US exports. US farmers produced 34% of the world's soybeans in 1990, 36% in 2009, and 34% of the world's corn[3] in 1991 and 49.5% in 2009. Weed scientists take pride in and correctly claim credit for

[3] 1991 Data are from The Agricultural Council of America, 11020 King Street, Suite 205, Overland Park, KS 66210, USA. 2009 data were obtained from the Internet.

their contributions to this productivity. Their work has played a prominent role in increasing the world's food supply and decreasing the labor of weeding crops. They correctly claim they have made significant contributions to increasing local, national, and international food supplies. What is wrong with the assumed primacy of Day's (1978) production ethic is that it centered on the interests and rights of some humans while ignoring the rights and good of other living things (Comstock, 1995). It has created a system that ignores Liberty Hyde Bailey's (1915) admonition that "a good part of agriculture is to learn how to adapt one's work to nature."

Ethics were not part of what I was taught during my formal education, and consideration of the ethical aspects of my work has not been a regular part of my collegial or professional discourse. If my experience is typical, and I think it is, then a dispassionate observer might conclude that weed and other agricultural scientists have not thought about the ethics of their science or the particular set of values that have driven its development. The point is reinforced by review of the 2722 pages of the *Encyclopedia of Agricultural Science* (Arntzen and Ritter, 1994). There are 53 entries related to herbicides, eight related to herbicide resistance and one small column on environmental and toxicological effects. There is not a single entry in the four volumes on ethics or ethical issues.

The point was affirmed for me several years ago when I met with several weed science colleagues during scientific meetings to begin exploration of some value questions. At the beginning of the interview, I asked several background questions (Where were you born? What did your parents do? Where did you obtain your degrees?). About half way through the 1-h interview, I asked, "What values have driven your weed science career?" The first response to the question was a blank stare. To illustrate the question's intent, I offered a brief commentary about the importance of honesty in science. The answer subsequently received, in every conversation, was nearly identical: honesty in scientific work and never manipulating the data. This was followed by the value of earning one's pay, working hard, caring for family, caring for children, being a good neighbor, being a good citizen, being responsible for actions, and being trustworthy. Not everyone interviewed gave exactly the same answers or used precisely the same words, but each explained the values that had driven their career as a list of the characteristics that define a good person. No one questioned the value of agricultural production or its primacy in agricultural endeavors. No one questioned why greater production always was a higher value than environmental protection. No one questioned the value of increasing use of herbicides versus possible harm to nontarget species or humans. Everyone was and wanted to be ethically correct in their personal and professional lives but did not transfer any degree of ethical concern to their work except to agree that one must follow the scientific rules.

I have argued (Zimdahl, 1998) that weed scientists define a good agriculture as one that optimizes yield and maximizes profit. I knew I worked in an agricultural system that was driven by government programs that had the unintended effect of bestowing the greatest benefits on the largest farmers and encouraged oversupply of commodities and low prices to farmers. It was also a system that always kept supermarkets full of an abundant variety of food. Agriculture's problems were confounded by a production system that encouraged widespread use of agricultural chemicals and intensive mechanization that, when combined with the effects of government programs, discouraged diversification and crop rotation while encouraging monoculture. Unintended but real effects included farm consolidation, loss of rural communities and businesses, and much of what our society values about rural America (Danbom, 1997).[4] Our agricultural production system had led to agriculture being viewed as a source of problems by the public rather than as a public good. Not all of agriculture's effects are positive and each involves ethical questions. I am compelled to acknowledge and deal with the observation that much of what is involved in modern American agricultural production causes ecological deterioration, loss of biological diversity, loss of rural communities, and may impair human health. These things never appear in conventional financial balance sheets, and we therefore gain the false impression of untroubled and unending success (Ehrlich and Ehrlich, 1990). The dominant tendency in American agriculture is to judge the success of any technology or the whole enterprise solely with economic criteria. If it is not profitable, it must not be valuable. Agriculture's producers and scientists have been lulled into the false productionist belief, according to Thompson (1995). The proper goal is to produce as much as possible, regardless of the cost. The view is that measuring agriculture's success in terms of production of food and fiber is both a necessary and a sufficient criterion for evaluating agriculture's ethics (p. 48). We may know but tend to forget that "ecological criteria of sustainability, like ethical criteria of justice, are not served by markets" (Daly, 1996, p. 22) and their requisite economic criteria for success.

These observations describe a view of what American agriculture is. The related, but more important question is what weed science, an important part of American agriculture, ought to be. If we fail to understand the importance of beginning the discussion and striving for agreement on what we ought to do and be, we will only add, as Freudenberger (1994) suggests, momentum to the pace of human and natural resource depletion.

This chapter began with the premises that agricultural and weed science lack a well-constructed, carefully articulated ethic and that values are treated only instrumentally. It concludes by suggesting some components of an acceptable ethic for agriculture and weed science. Rachels and Rachels (2010, pp. 177–180) concluded with the same question: What would a satisfactory moral theory be like? Their suggestion, Multiple Strategies Utilitarianism (see Chapter 4), does not focus solely on motives, consequences, intentions, acts, or rules. The goal is to maximize general welfare by using diverse strategies that are often required to accomplish the goal. I suggest 11 criteria that could be considered among the diverse strategies used to accomplish the goal of maximizing general welfare. These could be used to judge the value of a technology and determine if all components of agriculture are maximizing the general welfare.

[4] For a good explanation of the things valued, see Thompson (1997). Agrarian values: Their future in US agriculture. Pages 17–30 in W. Lockeretz (ed.) *Visions of American Agriculture*. Iowa State University Press. Ames, IA and Thompson, P.B. (2010). *Agrarian Vision: sustainability and environmental ethics*. The University Press of Kentucky.

The multiple strategies utilitarian approach which has been embraced, albeit unknowingly, is a good beginning. The basic ethical words for a utilitarian are *good* and *bad*. The problem, outlined above, with the common agricultural version of utilitarianism is its singular emphasis on production as the only, or at least the best, way to achieve the greatest good for the greatest number. Multiple strategies, as the name implies, will combine a utilitarian base with some of the best elements of a rights-based ethic, where the basic words are *right* and *wrong*. One criterion for judging whether something is right or wrong is to determine the effect of the action on others. An act's correctness, rights theorists hold, is also to be judged by its compatibility with justice and the rights of all affected by the act. For me, the weed scientist's concern must be expanded to all humans, other species, and the environment. For the utilitarian, good consequences make an action right, and bad ones make it wrong. Increasing food production is good, but it may lose its primacy when the environment, the rights of others and the need for justice for all are considered. Production growth may be reaching the point where its environmental and social costs are increasing faster than its economic and human benefits (Daly, 1996, p. 151).

I suggest agricultural and weed scientists examine their science in the context of its applications and the values they affect. We could begin by struggling with the options for future directions that Kirschenmann (1993) offers. The *first* option is to stay the present course without knowing how long it will be possible. *Second*, weed scientists could emphasize a search for technological fixes such as another herbicide when resistance appears. These and similar agricultural technological fixes will adjust or modify, but probably not solve, problems that are now so apparent. *Finally*, he suggests we try to consider how science will contribute to regenerative (i.e., sustainable) food systems that will include technology but are not dependent on a series of technological fixes. Kirschenmann selects the third option and thinks it means agricultural science must become part of a caring culture. For him, that is a culture governed by making explicit that healthy prospects for future generations and a healthy landscape are paramount values. They can be part of the ethical foundation upon which we stand.

Scientists learn subtly, almost by osmosis, through their education and experience, that the scientific process tends to disregard the subjective role of humans in the construction of reality (Conviser, 1982). Scientists are often not fully cognizant of the political economy that shapes production of knowledge. Science, we learned, doesn't make value judgments! Berry (1977) offered two views of the construction of agriculture's reality: exploiter versus nurturer. Berry asked his readers to consider which of the two views best described the values agriculture ought to advance. I suggest Berry's characterization of the exploiter versus the nurturer poses questions that agricultural scientists should consider as they debate appropriate ethical norms. I have learned and taught that most simple, dichotomous divisions, if not wrong, are suspect, but are often useful for framing discussion. Berry chose eight ways to compare the two views (Table 6.1). My experience in weed science and among agricultural colleagues has affirmed that although the title may be vigorously denied, the results of our research and how we approach what we do affirms that we have been exploiters, not nurturers. We have been specialists whose major (often sole) interest

**Table 6.1** Two World Views (Berry, 1977)

| | View of Term | |
|---|---|---|
| **Comparative** | **Exploiter** | **Nurturer** |
| Focus | Specialist | Generalist |
| Standard | Efficiency | Care |
| Goal | Money—profit | Health |
| Concern | Productive capacity | Carrying capacity |
| Work | Little | Well |
| Competence | Organization | Order |
| Serves | Institution | Place |
| Thinking | Quantitative | Qualitative |

has been ensuring efficient, profitable production. Generalists who emphasize environmental health and carrying capacity can be found, but they are rare. I do not suggest that Berry's comparison is the only way to discuss agriculture's future. It is a certainty that many colleagues will vigorously disagree with how I have characterized the way they practice their science. However, considering Berry's contrasts will help focus the discussion and lead toward development of that which Ferré (1994) and others say we must have: an ethical foundation for agricultural science.

I close by asking the reader to consider the value of these simple words to the discussion:

> *The ultimate goal of farming is not the perfection of crops, but the cultivation and perfection of human beings.*
>
> *Fukuoka, 1978*

I find this to be a place to begin, a place that may help us learn, without risk of embarrassment, how to ask about what we need to know.

## References

Arntzen, C.J. and E.M. Ritter. (1994). *Encyclopedia of Agricultural Science* Four Volumes. San Diego, CA, Academic Press.

Bailey, L.H. (1915). *The Holy Earth.* Lebanon, PA, Sowers. P. 9.

Berry, W. (1977). *The Unsettling of America Culture and Agriculture.* New York, Avon Books. Pp. 7–8.

Bressler, M. (1978). The academic ethic and value consensus. Pp. 37–48 *in The Search for a Value Consensus. A Rockefeller Foundation Conference.* New York, Rockefeller Foundation.

Burkhardt, J. (1986). The value measure in public agricultural research. Pp. 28–38 *in* L. Busch, W.B. Lacy (eds.). *The Agricultural Scientific Enterprise: A System in Transition.* Boulder, CO, Westview Press. Westview Special Studies in Agricultural Science and Policy.

Clayton, M. (2005). A productive, but taxed, Earth. *Christian Science Monitor* (March):11–12.

Conway, G. (1997). *The Doubly Green Revolution: Food for All in the 21st Century.* Ithaca, NY, Comstock Publishing Associates, a division of Cornell University Press.

Comstock, G. (1995). Do agriculturalists need a new, an ecocentric ethic? 1994 Presidential address to the Agriculture, Food, and Human Values Society. *J. Agric Hum. Values* 12:2–16.

Conviser, R. (1982). Appropriate agriculture. Pp. 436–452 *in* R. Haynes, R. Lanier (eds.). *Agriculture, Change and Human Values—A Multidisciplinary Conference.* Gainesville, FL, University of Florida.

Dahlberg, K.A. (1982). Global aspects of agriculture and human values. Pp. 87–112 *in* R. Haynes, R. Lanier (eds.). *Agriculture, Change and Human Values—A Multidisciplinary Conference.* Gainesville, FL, University of Florida.

Daly, H. (1996). *Beyond Growth: An Economics of Sustainable Development.* Boston, MA, Beacon Press.

Danbom, D. (1997). Past visions of American agriculture. Pp. 3–30 *in* W. Lockeretz (ed.). *Visions of American Agriculture.* Ames, IA, Iowa State University Press.

Day, B.E. (1978). The morality of agronomy. Pp. 19–28. *in* J. W. Pendleton (ed.). *Agronomy in Today's Society.* Madison, WI: American Society of Agronomy Special Publ. 33.

Dundon, S. (1986). The moral factor in innovative research. Pp. 39–51 *in* L. Busch and W.B. Lacy (eds.). *The Agricultural Scientific Enterprise: A System in Transition.* Westview Special Studies in Agriculture Science and Policy. Boulder, Co, Westview Press.

Ehrlich, P.R. and A.H. Ehrlich. (1990). *The Population Explosion. A Touchstone Book.* New York, Simon and Schuster.

Ferré, F. (1994). No hiding place: the inescapability of agricultural ethics. Pp. 11–17 *in Agricultural Ethics: Issues for the 21st Century.* Madison, WI, American Society of Agronomy Special Publ. 57.

Freudenberger, C.D. (1994). What is good agriculture? Pp. 43–53 *in Agricultural Ethics: Issues for the 21st Century.* Madison, WI: American Society of Agronomy Special Publ. 57.

Fukuoka, M. (1978). *The One-Straw Revolution.* New York, Bantam Books. P. 103.

Grant, L. (2000). *Too Many People: The Case for Reversing Growth.* Santa Ana, CA, Seven Locks Press.

Kirschenmann, F. (2007). Managing with less part II: reinventing the human. *Leopold Letter* 19(1):5.

Kirschenmann, F. (1993). Rediscovering American agriculture. *Word World* 13:294–303.

Pais, J.D. (1982). Ethical dimensions of agricultural research. Pp. 869–893 *in* R. Haynes, R. Lanier (eds.). *Agriculture, Change and Human Values—A Multidisciplinary Conference.* Gainesville, FL, University of Florida.

Rachels, J. and S. Rachels. (2010). *The Elements of Moral Philosophy*, 6th Ed. New York, McGraw-Hill, Inc.

Rollin, B. (1995). *The Frankenstein Syndrome: Ethical and Social Issues in the Genetic Engineering of Animals.* New York, Cambridge University Press.

Rollin, B. (1996). Bad ethics, good ethics and the genetic engineering of animals in agriculture. *J. Anim. Sci.* 74:535–541.

Thompson, P.B. (1995). *The Spirit of the Soil.* New York, Routledge.

Tilman, D., J. Fargione, B. Wolff, C. D'Antonio, A. Dobson, R. Howarth, D. Schindler, W.H. Schlesinger, D. Simberloff, and D. Swackhammer. (2001). Forecasting agriculturally driven global environmental change. *Science* 292:281–284.

World Bank. (2002). *World Development Report.* New York, Oxford University Press.

World Bank. (2010). *World Development Report-2002—Development and Climate Change.* New York, Oxford University Press.

Zimdahl, R.L. (1998). Rethinking agricultural research roles. *J. Agric. Hum. Values* 15:77–84.

# 7 Agricultural Sustainability

*We may utilize the gifts of nature just as we choose, but in her books the debits are always equal to the credits.*

*There is no balance in either column.*

**M.K. Ghandi, 1961**
***In Search of the Supreme* (v.2, p.116)**

*Live as if you are going to die tomorrow, but farm as if you are going to live forever.*

**Nineteenth century English motto**
**Cited in Mepham (1998)**

One does not have to review much of the current writing on agriculture to discover that achieving sustainability[1] has obtained the generally revered status of motherhood (see Chapter 1), with one important difference. Nearly everyone is in favor of motherhood and there is little debate about its nature. Sustainability is similar in that everyone seems to favor its achievement. The difference is that in spite of the nearly universal adulation of agricultural sustainability, there is little agreement on its nature, on what is to be sustained, or on how it is to be accomplished. Pretty (1995) noted at least 80 definitions. Wackernagel and Rees (1996, p. 36) confirm this with their ecological claim that "conflicting interests, opposing world views, incompatible analyses, rising material expectations, and fear of change, have led to a disorienting array of interpretations of sustainability and how to achieve it." The US Department of Agriculture Sustainable Agriculture Research and Education (SARE)[2] program has three primary goals, the Three Pillars of Sustainability:

1. Profit over the long term
2. Stewardship of our nation's land, air, and water by
   - Protecting and improving soil quality;
   - Reducing dependence on nonrenewable resources, such as fuel and synthetic fertilizers and pesticides; and
   - Minimizing adverse effects on safety, wildlife, water quality, and other environmental resources.
3. Quality of life for farmers, ranchers, and in stable, prosperous farm families and communities

[1]The adjective "sustainability" is a fairly new word. It is not in my 1971 Oxford English Dictionary, nor in any of my other six dictionaries. It does appear as a derivative of sustainable, without a definition, in my compact edition of the Oxford English Dictionary.

[2]The goals can be found in an undated brochure, *What is Sustainable Agriculture*, produced by the SARE program. See http://www.sare.org (accessed March 14, 2011).

**Agriculture's Ethical Horizon. DOI: 10.1016/B978-0-12-416043-9.00007-6**

Douglass (1984) described three primary uses of sustainability as it was used in the literature.

1. *Sustainability of production*: Long-term food sufficiency, either domestic or worldwide. The practice of agriculture is the way the world is fed and economic cost-benefit analysis of agricultural science and technology is how one determines the best way to practice agriculture to sustain it and thus prevent massive starvation.
2. *Sustainability as stewardship*: A primary concern is ecological balance and environmental quality. The proper quest is to create an agricultural system that preserves and conserves renewable resources while not polluting the environment or disrupting ecological balance on which life depends. Achieving sustainability is not a production question. It is a matter of knowing the ecological consequences of any production system and then minimizing the negative consequences. In Westra's (1998, p. 175) view, in regard to the earth's natural systems, sustainability means maintaining "optimum, undiminished capacity for their time and location for sustained evolutionary development" (see Daly and Cobb, 1989).
3. *Sustainability as community*: Sustainability is achieved by creating a set of agricultural practices, which encourage certain virtues that undergird the vitality of local communities. It is these practices that are to be preserved or reinstated. This view is similar to the second one, but emphasizes maintaining the "social organization and culture of rural life."

**Highlight 7.1**

Buttell (2003) claims that "The essence of food and fiber production is that on one hand, the key production resources (seeds, tubers, soil, manures, and rain water) are renewable, thus potentially enabling agriculture to be a highly sustainable activity." But agriculture has actual and potential characteristics of an extractive industry, similar to mining, and accordingly has the potential to be, and has become unsustainable. Any enterprise which depends on environmental extraction is not sustainable. "In addition, food and fiber production include long-term nonenvironmental costs (e.g., impacts on workers, communities, regions, and consumers) to a greater or lesser degree" (Buttell).

No one is overtly against achieving a sustainable agriculture. But the concept continues to elicit cautious and some negative reactions from the agricultural community (Schaller, 1993). Agribusiness companies that create and sell products (fertilizer, pesticides, seeds, machines) to farmers and ranchers are concerned that business will decline if there is a rapid shift to low-input agricultural production systems. Many agricultural scientists and producers know that the present system is highly productive of food and fiber and fear that the quest for sustainability will reduce yield and profit. In Schaller's view, moving away from the present system will be a step backward to the era of low production and surging pest problems. Many scientists and producers also think they have been environmentally and socially responsible and resent what they think is unjust criticism. The production system is not perfect, but it is steadily improving, and those who create improvements should be given credit, not criticized. Schaller argues that while concerns about the effects of a new system are justified, they often arise because the new system demands a "fundamentally different way of thinking." When one has learned a system, whether it is food production or education, one masters the intricacies of the system and learning the requirements of a new system is resisted. Learning may involve acknowledging that the old system is part of the problem not the solution. The old system and way of thinking must be, at least, partially abandoned as one develops or learns entirely new way of conceiving the problem and it solutions.

The primary uses above are commonly known as the food sufficiency, stewardship, and alternative agriculture definitions. The requirements to achieve each differ. Prominent questions include:

- Is achieving food sufficiency possible if population continues to grow?
- How long can modern agriculture's production techniques be sustained?
- What can be sustained for how long?
- What must be done to achieve sustainability (Burkhardt, 1989)?
- Who benefits, at what cost, and in what place is it to be achieved (Pretty, 1995)?

Pretty claims that some technologies are not sustainable and must be modified; therefore, new ones appear. What must be sustained is "the process of innovation itself." The "old" technology tried, but it was clear to users that it was not sustainable. It is common to assume that newer technologies will be. Pretty seems to conclude that there will be a continuing need for new technologies to achieve sustainability. But this merely begs the question (Davison, 2001, p. ix).[3] As the definitions above emphasize, it is first necessary to decide what ought to be sustained and why. Is it production, stewardship, or community? Davison, speaking more generally, implicitly challenges the assumption that for agriculture to accomplish its accepted moral obligation to feed the world, it is necessary to persist with the technological skills that "presently define our practices." This assumption, this argument persists, in spite of evidence that "our practices, while very productive, cause some problems and do not address or solve others." Examples of each follow (p. 2).

- Problems agriculture contributes to or causes:
  - Loss of biological and cultural diversity
  - Contamination and loss of nonrenewable natural resources
  - Technological risks to humans, other species, and ecological health
  - Effects on global ecological life-support systems
- Related problems not addressed:
  - Overconsumption of resources by citizens of developed nations
  - Social, psychological, and physiological effects of overconsumption
  - Persistent destitution of a least 2 billion people
  - Widening economic gap within and between countries

[3] To beg the question is to assume the truth of the point raised in the question.

Proponents of the present agricultural system in the world's developed countries will almost immediately advert—these are real problems, but it is not my or agriculture's role to solve or address them. That is not what we do. Each of these problems can be framed as a moral issue, or a question of what we ought to do. It is important to recognize that achieving sustainability of any agricultural system is a societal not just an agricultural responsibility, which should be greeted with pleasure by those in agriculture. They do not have to do it alone. Agriculture's practitioners claim that they may not have caused any of the problems above, and if they did it was not intentional. It is equally reasonable to suggest that recognition of the agricultural enterprises involvement in the problem(s) is in order followed by affirming a willingness to work together to find solutions. It is what we ought to do. Involvement will require thought about the dominant view that other creatures and the natural world have no inherent rights, unless we choose to grant them rights. Commercial rights to profit, even to point of exploitation of the natural world, prevail over the rights of natural systems to survive. Our ethical traditions include views on human rights [e.g., suicide, genocide, homicide, abortion (still hotly debated)], but do not include harm to the earth, what Berry (1999) calls geocide. "Disengagement from such exclusive commitments to human exploitation requires an ethical stance and a courage of execution seldom found in contemporary human societies" (Berry). Progress toward achieving a sustainable agriculture system requires a reinterpretation of the basic idea, the foundation of our ethical stance.

## The Present Agricultural Situation: The Example of Weed Management

Whenever possible, today's farmers and ranchers take advantage of economies of scale to produce more and earn more profit. Getting big has been good from the economic point of view. Larger units tend to operate more efficiently with less cost per unit of production than small ones. Costs can be lower because larger size allows one to take advantage of volume discount buying and the efficiency of large planting and harvesting machines. Modern technology can increase productivity of labor and decrease machine operating cost per operating unit and per acre (Edwards, 1980). Edwards claims that most modern technology (e.g., tractors, harvesters, pesticides, new cultivars) has encouraged growth in farm size. Commodity pricing favors the large farmers, as do most government price support and subsidy programs. To survive and make a reasonable living, farmers are forced to become large, because the profit margin on a unit of product is low and declining with time.

In the 1950s, President Eisenhower's Secretary of Agriculture (1953–1961) Ezra Taft Benson, said to farmers, "Get big or get out." Twenty years later, Earl Butz, former Dean of Agriculture at Purdue University and Secretary of Agriculture under Presidents Nixon and Ford, told farmers to "Adapt or die" and "Get big or get out." He meant that farmers must adapt to the economics of agribusiness. Butz urged planting commodity crops (e.g., corn) "from fencerow to fencerow." The policy shifts begun during Benson's tenure and strengthened during Butz's tenure coincided with the rise of major agribusiness corporations, and the declining financial stability of the small family farm. These policies, Berry (1981) claims, always implicitly included the ruin of small farmers and farmland. Berry claims that making things bigger and more centralized makes them both more vulnerable and more dangerous to everything else. From 1910 to 1990, the share of agricultural dollars received by farmers dropped from 21% to 5% (Standaert, 2003). More telling, US census data[4] for nearly all states support the fact that farms are becoming larger and the number of farmers is declining, a trend that increased during Benson's and Butz's tenures.

After more than 100 years of agricultural research by land-grant colleges and agribusiness companies, yields of nearly all crops are high in the world's developed countries, the need for human labor is low, and input costs are high. Global per capita calorie availability rose by almost one-third from the 1930s to the late 1980s. Per capita food supplies rose by 40% in Africa, Asia, and Latin America (Eberstadt, 1995, p. 8). Modern, developed country agriculture is a chemical, energy, and capital dependent system that produces consistent high yields using required technology that creates persistent environmental problems, most of which are externalized. This leads to the inevitable question of whether sustainability can be achieved in an extractive system that externalizes many of its costs. Sustainability, desired by all, is elusive. Yet food is abundant for all but the poor, and commodity surpluses are common in developed countries (Stout and Thompson, 1991).

Weed scientists correctly claim that the widespread use of herbicides has been a significant factor in increasing yields of most crops, but verification of the claim that the highly productive, chemical- and petroleum-dependent system is sustainable, are muted at best. For example, it takes as much energy to run US tractors as is contained in the food produced (Clark, 1975). A common response is: agriculture's task is not to produce energy, but to produce food and that is done very well.

For most US crops, 85% of the acreage is treated annually with herbicides for weed control (Gianessi and Sankula, 2003). These authors and many others claim that "without herbicides, hand weeding and cultivation most likely would replace current practices," that is, herbicides. They estimate these alternatives would cost more than $14 billion annually or more than double what US growers spend on herbicides plus their application. Gianessi and Sankula (2003) reported that for 35 of 40 agricultural crops, yields without herbicides would be reduced 5–7%. They do not claim that herbicides are essential to yield maintenance, but they do claim that their loss would demand much more cultivation and hand weeding to maintain yield. Both of these are expensive and there is no assurance sufficient labor for hand weeding would be available at the required time, thus, yields would decline. More cultivation would also increase soil erosion. Neither option appears sustainable. Gianessi and Sankula conclude that if herbicide use were eliminated, the yield loss of the 40 crops would be $13.3 billion, which is equivalent to 21% of the national production of the 40 crops. Grower income would decline by $21 billion annually. Other studies, cited in Degregori (2001, p. 89), claim losses as high as 70% if all physical,

[4] http://www.usda.gov/nass/pubs/trends.htm (accessed 1991). 2009 data support the claim.

biological, and chemical (not just herbicides) pesticides were eliminated (Oerke et al., 1994, p. 750). Knutson et al. (1990) estimated a 32% reduction in US corn production if all pesticides were eliminated and a 53% reduction if no pesticides and no fertilizer were used. Without any chemical use, soybean production would decline 37%. Peanut production would decrease 78% and wheat 38%, with no chemicals at all. Knutson et al. said that in constant 1989 dollars, consumers would have to spend $228 more per household per year if all pesticide use was eliminated. If the ban were expanded to include fertilizers, household spending would rise by $428 per year. Knutson et al. projected this as a 12% increase in the weekly food bill for middle-income consumers and a 44% increase for the poor. Pimentel (1992) estimated that losses to pests would be 10% higher if no pesticides were used and losses in some crops could approach 100%. In general, each dollar spent on pesticides returns $4 in saved crops (Pimentel). Farah (1994) acknowledged losses up to 40% of potential agricultural production due to pests in developing-country agriculture, but emphasized the adverse effects of pesticides on human health. Environmental health and the increasing problem of pesticide resistance also must be considered. Large loss estimates if pesticides are not used are common in the agricultural literature, whereas human and environmental health effects are not mentioned as frequently. Lehman (1993, 1997) presents moral arguments for reducing (not eliminating) pesticide use in agriculture.

Most farmers in industrialized countries and an increasing number of farmers in developing countries rely on herbicides to manage weeds. The data cited above support the view of most weed scientists that herbicides are important, if not essential, technology if farmers are to continue to be able to feed the world. Their advantages (see Chapter 3) are well known: low cost, safety, efficacy, selectivity, persistence, energy efficiency, profitability, and yield increases. However, the disadvantages are equally well documented and herbicides are indeed a "two-edged sword" (Kudsk and Streibig, 2003). They conclude that "herbicides strongly contribute to sustain and secure yield and are indispensable in modern arable farming." They applaud the continued development of new "chemical hoes" since the 1940s and note that this allowed farmers and weed scientists to regard weed control independently of the whole crop production system. Because weed control was studied independently of the cropping system, weed science became isolated from other agricultural sciences (Kudsk and Streibig). This reinforced agriculture's and by implication weed science's image as an intellectual and institutional vast, wealthy, and powerful island (Mayer and Mayer, 1974).

In Kudsk and Streibig's view, reliance on herbicides "resulted in shifts in the weed flora and the selection of herbicide resistant biotypes." They do not mention associated ethical dilemmas, but they can be identified. They assume, but do not ask if the practices they advocate are sustainable. Their position can be formulated as the invalid syllogism below.

- *Premise*: Present weed management practices will not harm people or the environment.
- *Premise*: Weed management practices that harm people or the environment are wrong.
- *Conclusion*: No present weed management practices are wrong.

A correct and valid formulation of the syllogism is:

- *Premise*: All wrong weed management practices are those that harm people and/or the environment.
- *Premise*: No present weed management practices harm people and/or the environment.
- *Conclusion*: Therefore, no present weed management practices are wrong.

Kudsk and Streibig conclude that present practices (i.e., extensive herbicide use) should be (must be) part of future, sustainable agricultural systems. Their argument appears to ignore or dismiss public concern about the profession's (herbicide users, manufacturers, and researchers) violation of acceptable standards of professional ethics—perhaps best formulated as Do no harm—that led to imposition of "increasingly strict registration requirements in some countries." The political response to public concern about human and nontarget species health and the public's general environmental concern led to imposed regulations (laws) that govern the future of weed science. These things seem minor to Kudsk and Streibig, who conclude by observing that "society at large is, however, not aware of the benefits of herbicides, and there is urgent need to optimize their use." Optimization was not defined, but its components implicitly include developing ways to respond to governmental regulation of herbicides, addressing public concern about pesticide residues in food and water, and minimizing possible adverse environmental effects. They also want to "ensure that herbicides will remain an effective and valuable tool to farmers." Continued herbicide use, in their view, is part of achieving a sustainable agricultural system. Pesticide residues in food and water are real in Kudsk and Streibig's view, but "they do not actually pose a risk to public health, although their presence does cause concern." Maintenance of present agricultural practice is to be done simply by education. The public must be shown through education that the negative view of herbicide use in agriculture is wrong.

Their argument is a scientific one based on good data, which demonstrates environmental concentrations of herbicides are so low that any possible effect is so remote as to be impossible *or* that there is no evidence that existing concentrations have been demonstrated to cause any harmful effect to any living organism. However, this accurate scientific response is just that—a scientific response to nonscientific, moral questions. Evans (1998, pp. 219–220) states the dilemma clearly: Scientists, including agricultural scientists in all their variety, "see themselves as helping to feed and clothe the rapidly growing human population while cherishing the earth." Yet the public sees agricultural scientists as "destroyers of nature, wastrels of water, eroders of land and genetic resources, polluters of the environment, and hand maidens of agribusiness." Thus, the public regards agricultural scientists and those who practice agriculture as having violated expected professional ethical norms. Kudsk and Streibig's argument is similar to that made by many well-intentioned agricultural scientists who, in Evans' view, "seek technological solutions to problems of social and economic inequity." The view is consistent with the scientist's infatuation with more and better science and technology that will solve the problems science and technology created. The claim is that science and technology are not problems, they are required to provide solutions to all societal problems and are necessary to achieve agricultural sustainability. It is, in a less kind view, continuing to do agriculture and agricultural science in the same way to achieve a sustainable system, but expecting a different outcome from the same methods. This has been called a form of insanity by those who observe such behavior.

The claim that improved science and technology will solve the problems science and technology created ignores, or those who make it are unaware of, the Jevons paradox, sometimes called the Jevons effect (Polimeni et al., 2008).[5] It was developed in 1865 by W.S. Jevons, an English economist. The fundamental proposition is that technological progress that increases the efficiency with which a resource is used tends to increase (rather than decrease) the rate of consumption of that resource. Jevons observed that technological improvements that increased the efficiency of coal-use led to increased consumption of coal in a wide range of industries. He argued that, contrary to common intuition, technological improvements could not be relied upon to reduce coal consumption. By reducing the amount needed for a given use, increased efficiency can, and often does, accelerate economic growth, increase the demand for resources, free capital for expansion, lower the relative cost of using a resource which increases demand for the resource. These factors potentially counteract any savings from increased efficiency. The Jevons paradox occurs in agriculture because increased demand for food and fiber causes an increase in overall resource use and that demand lowers prices, etc. For example, the Jevons paradox supports the claim that energy conservation is futile because increased efficiency and lower prices increase fuel use. The paradox applies to agriculture and makes progress toward sustainability more difficult.

As discussed in Chapter 1, appealing to science to solve social and economic problems fails to recognize the difference between rational scientific truth and what one is to do because of it and the demands of personal or subjective truth. Questions from the realm of personal truth frequently use empirical data to bolster the claim that the questions are scientifically and ethically legitimate. An example, no one knows for sure, but it is estimated that there are one to five million cases of pesticide poisoning every year in the world, resulting in 20,000 deaths. The World Health Organization (WHO) estimated that every year three million people in the world (mostly poor people in developing countries) suffer from severe pesticide poisoning. In 1990, it was estimated that 220,000 die annually, mostly in the world's developing countries (WHO, 1990 cited in Pimentel and Greiner, 1997, p. 52). In 2002, Halweil (2002, p. 72) estimated there were at least 20,000 unintentional human deaths and an additional 200,000 suicides. It is reasonable to conclude that no one knows, but that these numbers do not represent all poisonings or deaths because most occur among the poor. Hence, many are not reported and the cause of death is not always known. The data also do not include the many unreported cases of people who are temporarily sick or incapacitated after applying or mixing pesticides. Those affected die quietly, far from the news of the day. Such occurrences do not characterize an ethically desirable or a sustainable system.

[5] I learned of the Jevons paradox in a general October 2006 letter from Wes Jackson, President, The Land Institute, Salina, KS.

Acute effects on human health are an immediate public concern. Chronic effects are of increasing concern, especially relative to the possibility that low levels of synthetic organic chemicals in the environment, at sensitive stages of fetal development, can act as endocrine disruptors that affect the structure and functioning of the mammalian immune system (Repetto and Baliga, 1996; Colburn et al., 1996). Accusations of this kind include empirical and moral questions that ought to be addressed clearly and carefully by the agricultural community. Defensive rejections are the common response.

Over several decades of pesticide development, each new pesticide has contributed to a general trend of products that are safer to the environment, to the user, and to consumers of treated produce (Major 1992). New products are more specifically active (i.e., they do one thing very well), less expensive, easier to use, and compatible with other products and techniques (i.e., suitable for integrated pest management programs). Pesticide formulations have become safer to the user and the environment. Application techniques have improved and off-target spray drift has been reduced. Manufacturers have also developed programs to address surplus pesticide disposal and responsible container disposal or recycling. Taken together, one might expect that these actions and the development of desirable pesticide characteristics and improved application techniques should have led to more public confidence about the manufacture, marketing and use of pesticides (Major), because each appears to contribute to greater sustainability. Exactly the opposite has happened as Jevons paradox predicts. The result reflects the view that—to do more efficiently that which should not be done in the first place is no cause for rejoicing or praise.

Major (1992) concludes that those who recommend and use pesticides must accept the legitimacy of ethical concern, be more open, and become partners in environmental improvement, which, if they are sincere, will often trump increased production. Only then can progress be made toward a sustainable system that maintains and enhances productivity but is environmentally, socially, economically, politically, and morally acceptable and therefore, sustainable.

A few weed scientists have accepted the multiple challenges of achieving agricultural sustainability. Liebman (2001), writing of the need for ecological approaches to weed management, advocates recognition and management use of the many beneficial roles weeds play in agro-ecosystems. In his view, a broad range of ecological processes can be combined with required and new farming practices "to manage weeds more effectively, while better protecting human health and environment quality." Liebman's view is similar to Douglass' (1984) view of sustainability as stewardship. Liebman claims, in contrast to the claims of advocates of modern chemical, energy, and capital intensive systems, that the alternative systems he has developed increase farm profitability by cost reductions and price premiums, while maintaining healthy rural communities. In short, his claim is that ecologically based systems of weed management will be sustainable over time and will be as productive as the present system, which in his view, is unsustainable. Alternative systems do not exclude herbicides, but will not rely on them as the first choice for weed management. New weed management systems will be more reliant on ecological processes such as "resource competition, allelopathy, herbivory, disease, seed, and seedling responses to soil disturbance, and ecological succession." These processes have been known to ecologists for many years, but modern weed management systems have ignored them in favor of readily available, effective chemical methods to manage weed populations. These weed management systems are not sustainable (Douglass, 1984), although they are highly productive. They assume sustainability can be achieved through greater efficiency in agricultural operations and improved technology. It is faith in technological solutions that ignores the fact that many of agriculture's problems have been caused by technology (Rees, 2010).

Mohler et al. (2001), in the conclusion of a book in which Liebman's work appears, write of weed management in a broader context. They suggest that "by reducing the need for herbicides, ecologically based weed management strategies can help farmers reduce input costs, reduce threats to the environment and human health, and minimize selection for resistant weeds." They specifically disagree with the conventional agricultural position (see Avery, 1995; Borlaug, 2001; Waggoner, 1994) that it is only possible to protect land for wildlife and feed a burgeoning world population if modern high-yield technology (pesticides, fertilizers, genetic engineering, energy, etc.) is used widely on the best land. Their view is exactly the opposite of the dominant agricultural view that we must subdue and dominate nature (see White, 1967). That view asserts that it is our task (indeed our obligation) to transform and shape nature to fit human needs. Weeds and other pests must be managed (controlled) to facilitate the transformation of nature so it will be productive of what humans want and need. Mohler et al. (2001) advocate working with nature rather than against it. In their view, to become sustainable, weed management must become an ecologically based rather than a chemical control-based discipline. The natural world is not to be regarded as a mechanical, dead place to be managed by humans for human ends, but as an organic, living place that we can learn from as we try to make it more productive of what we need.

**Highlight 7.2**

*In the end they will lay their freedom at our feet*
*and say to us, make us your slaves, but feed us.*

*Dostoevsky, F. The Brothers Karamazov.*
*Book 5, Chapter 5.*

The word sustainability was first used in the report of the United Nations World Commission on the Environment and Development (The Brundtland report) published in 1987. Their definition was to "meet the needs of the present without compromising the ability of future generations to meet their own needs." The definition implies a commitment to the future. Present generations should strive to live their lives within available ecological constraints because failing to do so means passing the burden of sacrifice on to other people—our's and other's children and grandchildren. Doing this would show little love or respect for our descendants and will be regarded by many as an immoral act.

The quest for sustainability is driven by genuine social concern, that is, we care deeply about our descendants and by two competing philosophical perspectives:

1. Nature has inherent value and sometimes human interest must be sacrificed to ecological or environmental values.
2. Ecological balance is important but only because it has instrumental value for present and future humans.

There is little reason to question if people care about their descendants. Many people regularly make sacrifices for their heirs. There are reasons to question if we care enough for our descendants to change the way we do things.

More critics of developed country agriculture point to the same problems:

- Polluted water,
- Depletion of potentially renewable soil, energy, and water supplies,
- Harmful effects on wildlife,
- Dependence on chemicals that create environmental residues,
- Depopulation and loss of rural communities, and
- Concentration of capital and control within agriculture.

These concerns are driven by a growing awareness of ecological interactions. An awareness of the fundamental rule that it is not possible to do just one thing. Many people believe that in spite of the benefits to some industries, some large farmers, and land-grant universities, the financial, social, environmental, and health costs of modern agriculture are too high.

While it seems important to many to know who to blame, it is not proper to blame agriculture for all of these problems. Wackernagel and Rees (1996) note that coffee drinkers require 25 $m^2$ of what might have been tropical forest to sustain their consumption. The typical American fossil fuel user needs 2–3 hectares (about 5–7.5 acres) of forest somewhere on the planet to absorb $CO_2$ emissions (and that has not been sufficient given rising atmospheric $CO_2$ levels). Typical Americans need 4–9 hectares (10–22 acres) of land somewhere to support their consumer lifestyle. They also note that the American Great Plains are not just out there, somewhere, they are an essential part of the urban ecosystem because they are used to produce food. All developed nations have exceeded their domestic carrying capacity and are borrowing from the ecological system to support their consumer lifestyle (Rees, 2010). If the world's nearly 7 billion people, many of whom do and all of whom probably do, want, the consumer lifestyle of the North American nations, it would require there to four additional planets (Rees). It is clear that we care about our descendants, but not enough to change our lifestyle and achieve a sustainable system that demonstrates how much we care.

Therefore, one is driven to the conclusion that it is nature's instrumental value that is foremost in the human mind, not its intrinsic value, its inherent rights, or the sustainability of the agricultural and other systems imposed on it. The evidence is that humans are willing to sacrifice ecological and environmental values to satisfy current wants and perceived needs and thereby risk the satisfaction and perhaps the survival of our descendants.

There is resistance among farmers and research scientists to this move. Mohler et al. (2001) suggest the resistance is due to five factors:

1. The ease and low risk of failure when herbicides are used explains why farmers continue to use them to manage weeds. Uncontrollable events such as drought, severe sudden storms, poor markets, and so on encourage farmers to seek certainty when they can.
2. Aggressive marketing of agrichemicals. In support of this point, Kroma and Flora (2003) describe the greening of pesticide advertising. They demonstrate how agricultural media (i.e., farm magazines, radio, television, and so forth) have appropriated current societal values in the imagery that accompanies advertisements of agricultural products, including pesticides. Their work demonstrates how pesticide advertising changed from 1940 to 1990 in response to the US sociocultural setting. The industry strategically repositioned itself "to sustain market share and corporate profit by co-opting dominant cultural themes at specific historical moments," by appearing to adopt expected professional ethical norms. Simultaneously, industry advertising avoided or masked environmental and social challenges to pesticide use. Pest and weed management was done by successfully selling farmers new chemical products (Mohler et al., 2001).
3. The externalization of environmental and human health costs. The real costs of environmental pollution, water contamination, nontarget species harm, and harm to human health are not borne by the user, the manufacturer, those who approve registration and use, those who recommend, or those who apply the pesticide. These costs are externalized and borne by society when food is purchased and taxes are paid.
4. The increasing dominance of large-scale, industrial farms. Such farms obtain economies of scale and savings from labor-saving technologies.
5. Government policies that encourage chemical and energy-intensive agricultural practices. Government policies favor use of technology that increases production as opposed to policies that favor integration of production goals with simultaneous achievement of social justice and environmental quality.

Mohler et al. conclude that achieving an ecologically based sustainable farming system is a task that must be shared by farmers, agricultural research scientists, government policy makers, and a public that demands and supports the change. As mentioned at the beginning of this chapter, it is a public responsibility, not exclusively a task for the farmer. However, as Burkhardt (1989) points out so well, that which is everyone's obligation (a collective societal obligation) is often no individual's obligation and little is accomplished.

## The Moral Case for Sustainability

Society is comfortable with ethics preceded by adjectives such as Christian, medical, legal, and more recently environmental. To paraphrase Rolston (1975), the moral noun ethics does not regularly take the scientific adjective agriculture. Ethics and agriculture do not go together easily for two reasons. The first is that the few philosophers who focus on agriculture are not read widely within the agricultural community and their thoughts have not yet had a major effect on conventional agricultural thought. The second reason is that mentioned in Chapter 1: people engaged in agriculture are sure that food and fiber production are among the most ethical things anyone could do and are not therefore, a proper focus for ethical challenges. However, if agricultural practice is to achieve the necessary desired but elusive virtue of sustainability, the focus must include the good end of food and fiber production *and* an examination of the means to that end. The individual claim of virtue and ethical correctness because food production is a worthy goal, a good thing that should be applauded, must be tempered by a collective view of what is good for all. By all I mean all creatures and the environment that supports them. Producing food and fiber are good things. Feeding people is a good thing. But these are not the only good things that must be considered as we try, as we must, to achieve sustainable agricultural systems.

Thompson (1995, p. 15) claims that "agriculture cannot continue indefinitely without an environmental ethic, or at least it cannot continue happily." How to achieve an environmental ethic and agricultural sustainability is a scientific and a moral question. This means to me that all agricultural practitioners, including pest control scientists, need to develop a supportive ethical foundation (see Chapter 4). An environmental ethic must become part of the ethical stance of agriculture and while it will not create, it will significantly aid the quest for, a sustainable system. Creating that ethic will compel review of the adverse consequences of modern agriculture and that review may lead toward development of a universal agricultural/environmental ethic. Agricultural scientists from all disciplines may become that which they have disparaged—environmentalists. That is, agriculturalists, if they are to achieve the universally desirable goal of sustainability, must reinterpret their basic ideas about agriculture and sustainability.[6] Reinterpretation will require careful, difficult thought about growth. What is possible is sustainable development of people and technology. Sustainable growth of anything is not possible. "Sustainable growth is a clear oxymoron" (Daly, 1996, p. 7, also see Daly, 1993). There are limits to growth as pointed out so well by Meadows et al. (1972). "Growth should refer to quantitative expansion in the scale of the physical dimensions of an economic system. Development should refer to the qualitative change of a physically non-growing economic system in dynamic equilibrium with the environment" Daly and Cobb (1989, p. 71). Daly (1996, p. 7) challenges agriculturalist's view, which is shared by many others in our society, that acknowledging limits to growth is intellectually wrong because growth is regarded as the solution to poverty; growth is required to fulfill the moral obligation to feed the world.

Sustainable growth has become a synonym for sustainable development. Growth or quantitative increase in physical size has limits. Development or qualitative change in ability, potential, a communities' character, and the quality of one's life is possible. "Children grow and develop simultaneously, a cancer grows without developing, Earth develops without growing" (Daly, p. 167). A further challenge offered by Daly and Cobb (p. 76) is to consider what those who practice agriculture must do to achieve sustainable development, which is, meeting the needs of the present without compromising the ability of future generations to meet their own needs. To do this, needs must be distinguished from wants and impossible desires. Second, and more difficult, is creating a way to estimate the ability and requirements of future generations to meet their needs.

[6]The idea comes from A.N. Whitehead; fundamental progress has to do with reinterpretation of basic ideas.

But why bother? In a pure utilitarian calculus, the adverse consequences of agriculture may be viewed as being more than offset by the fact that the world now feeds more people a better diet than ever before. In the 1950s, when I was in high school, the world had 2.5 billion people and some were hungry. In 2011, the world has almost 7 billion people and some (perhaps 800+ million) are regularly hungry. But the present capital, energy, and chemical system of agriculture now feeds more than 6 billion, a task that people in 1950 thought, if they thought at all about the world's food supply, was impossible. Thus, the balance of pleasure over suffering is significant. What the agricultural system has achieved must be ranked as among the greatest of scientific achievements. Agriculture, in the view of many, presents no special ethical problems (Thompson, 1995, p. 6) because its real problems pale in comparison to its real achievements. However, in our world of material abundance for some and a similar material abundance wanted by millions of others, a dilemma is that to pursue the utilitarian greatest good for the greatest number may lead to destruction of the resource base on which the greatest good is absolutely dependent (Busch et al., 1995, p. 214). The dilemma is what Durning (1994) calls the conundrum of consumption:

> *limiting the consumer life-style to those who have already attained it is not politically possible, morally defensible, or ecologically sufficient. And extending that life-style to all would simply hasten the ruin of the biosphere. On the other hand, reducing the consumption levels of the consumer society, and tempering material aspiration elsewhere, though morally acceptable, is a quixotic proposal. it bucks the trend of centuries. Yet it may be the only option.*

Through what Busch et al. (1995, p. 214, citing Sagoff, 1988) call the "stepped-up appropriation and commodification of nature" we may destroy nature and "the very culture that provided us with and fostered the idea that there is a 'greater good' than simple satisfaction of preferences or freedom from material wants." Such goals "presuppose the reality of public or shared values ... that are discussed and criticized" (p. 29). These kinds of values should not "be confused with preferences that are appropriately priced in markets" (Sagoff). Thus, agricultural ethical debate is required to aid in decisions when there is tension between the imperative to produce and the values inherent in the need to conserve or protect the resource on which production depends. Such debate will help resolve the conflict between the agrarian philosophy that wants to protect communities and family farms and those who value the economic efficiency of large-scale industrial agriculture to produce cheap, abundant food. An agricultural ethic will help us address the loss of biodiversity from large-scale but highly productive monocultures. An appropriate ethical foundation will guide resolution of the tension but not dictate the answer. It is not feasible to transfer the established tenets of environmental ethics to agriculture, because environmental ethics has consistently given conservation priority over productive use. It is clear that difficulties would result from attempts to make environmental ethics applicable to agriculture. For example, an ethic for wild areas is not applicable to agricultural areas.[7] Agriculture needs its own environmental ethical foundation to achieve sustainability.

[7] Some of these thoughts are derived from Boyd, F. (1997). *Do we need an environmental ethic for agriculture?* Unpublished.

However, one is still left with the questions—what exactly is it that is to be achieved by the quest for sustainability? Why should we work to achieve it? What are the characteristics of a sustainable agricultural system? There is, as mentioned at the beginning of this chapter, a disorienting array of interpretations of sustainability.

### Highlight 7.3

The related terms, "sustainable" and "sustainability" are popularly used to describe a wide variety of activities that are generally ecologically laudable but may not be sustainable. An unambiguous definition of the concept of sustainability must accept that it has to mean, for an unspecified long time. It implies increasing endlessly, which, of course, means that whatever is growing will become infinite in size or scope. If one accepts, as one must, that the earth is finite and its resources and the environment are not infinitely expandable. Then one must accept and understand what Bartlett (1999) calls the most fundamental truth of sustainability: when applied to material things the term sustainable growth is an oxymoron. These principles are the foundation of Bartlett's laws of sustainability, which lead to the ineluctable conclusion that a sustainable agricultural system is "one that meets the needs of the present without compromising the ability of future generations to meet their own needs." Bartlett concludes with his laws of sustainability, some of which are applicable to agriculture's quest for sustainability.

1. Growth in rates of consumption of resources cannot be sustained.
2. In a society such as the United States, as population and consumption of resources grow, it becomes increasingly difficult to become sustainable.
3. The size of a population that can be sustained (the carrying capacity) and the average standard of living of the population are inversely related.
4. Sustainability requires that the population's size be less than or equal to the carrying capacity of the ecosystem in which it exists.
5. The benefits of population growth and growth in rates of resource consumption accrue to a few; the costs are borne by all of society.
6. Growth in the rate of consumption of a nonrenewable resource (e.g., fossil fuel) causes a dramatic and rapid decrease in the life-expectancy of the resource.
7. The benefits of efforts to preserve the environment are easily overwhelmed by added demands placed on the environment by more people who demand more things.
8. Humans will always be dependent on agriculture.

William Jennings Bryan, the Democratic candidate for President of the United States, in 1896 affirmed the essentiality of agriculture, and thus of its sustainability (he did not use the word) in his Cross of Gold speech. He said:

*Burn down your cities and leave our farms, and your cities will spring up again as if by magic; but destroy our farms and the grass will grow in the streets of every city in the country.*

## What is Sustainability?

The International Alliance for Sustainable Agriculture held a conference in 1990 at the Asilomar conference center in California. The Asilomar Declaration[8] for sustainable agriculture was approved by the more than 800 delegates who attended the conference. It begins with the assertion that "the present system of American agriculture cannot long endure." The destructive consequences of regarding agriculture as an industrial-technological process rather than a biological/ecological one (Merrill, 1986) are implicit in the Asilomar Declaration. The challenge is not simply to increase production, in fact, production is not even mentioned as a goal because if sustainability is achieved, production is assured. To sustain is defined in the dictionary sense of—to keep in existence; keep up; maintain or prolong; to keep up without interruption, diminution, or flagging. The challenges of the Asilomar Declaration are:

1. To promote and sustain healthy rural communities.
2. To expand opportunities for new and existing farmers to prosper using sustainable systems.
3. To inspire the public to value safe and healthful food.
4. To foster an ethic of land stewardship and humaneness in the treatment of farm animals.
5. To expand knowledge of and access to information about sustainable agriculture.
6. To reform the relationship among government, industry, and agriculture.
7. To redefine the role of US agriculture in the global community.

In 1992, The Union of Concerned Scientists issued what Rees (2010) calls a "strident assessment" of the state of the planet.

> *We the undersigned, senior members of the world scientific community, hereby warn all humanity of what lies ahead. A great change in our stewardship of the Earth and the life on it is required if vast human misery is to be avoided and our global home on this planet is not to be irretrievably mutilated.*

Social, environmental, agronomic, and economic challenges are included in the Asilomar Declaration. They define agriculture as more than just a productive activity. Sustainable agriculture is to be regarded as the salvation of our souls (Merrill, 1986). "From agriculture we learn that we are not sufficient unto the day—that we do not and cannot have all the answers." Our vast agricultural technological efficiency has given rise to a new question that has not been answered. The question posed by Pettersson (1992) is relevant, "How successful should we be in controlling and manipulating nature or its ecosystems?" A dispassionate person might, at this point, conclude that because humans believe they are the most intelligent creatures on earth, it logically follows that coordinated, widely supported action to achieve sustainability would now be the norm. It is not. Humans occupy or have colonized most of the earth; polluted water, soil, and air; and been the major contributors to the extinction of many other species. These things deny humanity's intent to achieve what most agree is desirable: agricultural sustainability.

To the Asilomar list one could add preservation of nature, which can be achieved by preserving farmland as a buffer between developed urban areas and wild areas

[8] The document is available from The Newman Center at the University of Minnesota.

(Westra, 1998). In accordance with Westra's (p. 28) sixth, second-order principle, an ethic of integrity requires that humans view all activities as taking place within a buffer zone that shields and protects core, ecologically intact, wild areas. "True buffers entail that most natural ecologically evolutionary processes be present, although" such areas may be manipulated for agriculture and forestry. Such manipulation must not impose degradation or disintegrity on the agricultural or care landscape (p. 138).

A simple definition that incorporates sustainability's complexity is that prepared by the Alliance for Sustainability (2004). A sustainable agricultural system must satisfy four criteria:

- *Ecologically sound*: Able to achieve species diversity and be resource efficient to conserve resources, avoid system toxicity, and decrease input costs.
- *Economically viable*: The system must yield a positive net return when resources expended are compared to those returned. In short, it must be profitable for farmers.
- *Socially just*: Resources and power must be distributed equitably so basic needs of all are met and rights are assured. People must be empowered to control their lives. It is in Sen's (1999) term, development as freedom.
- *Humane*: Good farmers are humane. They are kind, tender, merciful, and sympathetic to all life forms, even though the practice of agriculture changes the environment and affects other creatures. Humans have an interdependent relationship with the environment and with animals. Those who raise animals (dairymen, pork, sheep producers, ranchers) know that if they care for their animals, the animals will care for them.

**Highlight 7.4**

Wes Jackson earned a B.A. in biology from Kansas Wesleyan, an M.A. in botany from the University of Kansas, and a Ph.D. in genetics from North Carolina State University. After obtaining his Ph.D. he returned to his alma mater to teach biology. Subsequently he moved to California State University in Sacramento and established their program in Environmental Studies. After becoming a full Professor at California State University, his career diverged dramatically from most those who obtained a doctorate in some field of agriculture and begin an academic career. He left academia and founded the Land Institute in Salina, Kansas.

The Land Institute has studied the problems of agriculture for more than 20 years from a very different perspective than that found in most Land Grant Colleges of Agriculture. Jackson decided that modern chemical, energy, and capital intensive agriculture although highly productive of food and fiber was inevitably unsustainable and doomed to fail. What was needed was a new model for agriculture It was to be a model based on the land ethic of Aldo Leopold:[1]

*A thing is right when it tends to preserve the integrity, stability,*
*and beauty of the biotic community. It is wrong when it tends otherwise.*

The model Jackson chose to achieve the goal Leopold established was the prairie. Many well-established, respected agricultural people scoffed. The comments were similar to: Prairies don't produce food and fiber, they produce prairies. Prairies might be good for some cattle grazing, but not for producing food crops.

Jackson also claimed his goal was to develop a high-yielding, seed-bearing, perennial polycultures. The response from the traditional agricultural community was the, quite reasonable, botanical claim that perennials put their energy into perennial structures (roots, rhizomes, stolons, tubers), not into seed production. Perennation and high seed yield don't go together. Besides, the doubters said, perennial plants that grow on prairies don't produce seeds that humans will eat. They are not palatable; they taste bad.

Ah, but prairies are highly sustainable. Without human disturbance they have survived and remained productive for centuries. Prairies don't have weed, disease, insect, or fertility problems. And, if they do, they recover on their own. Prairies sponsor their own pest control and fertility and don't need to be irrigated. A prairie does all that a sustainable agricultural system ought to do. They do what modern agricultural systems fail to do. A prairie is place to begin to try to learn what nature has to teach.

The Land Institute does what Land Grant Colleges of Agriculture have not done, or, at least, the efforts of colleges to develop a new model are not readily apparent. The Land Institute consults nature, which is regarded as the source and measure of the membership of humans in the natural world. The Institute is designed to develop an agriculture that preserves rather than destroys soil and assures the life of rural agricultural communities.

Scientists at the Land Institute are now confident that they have demonstrated the scientific feasibility of natural systems agriculture. They anticipate that such systems can and ought to be adopted worldwide. The prairie will obviously not be the proper model for all world agriculture. Models must be found in local ecosystems. Diamond (2004, pp. 280–282) provides a clear example of what can be learned from agricultural systems that have survived for centuries but to the western eye appear primitive. For example, New Guineans irrigated sweet potato gardens with vertical ditches that ran down the slope, not across the slopes or along a contour. A European agricultural advisor convinced them to re-orient the ditches with the result that the next heavy rain washed ditches and crop away. These farmers had learned, over centuries, what worked. They had a highly sustainable, productive system that fed people, conserved soil, rotated crops, prevented erosion, added organic compost, and combined annual crops and forests. It was not modern, but it was successful and a model from which modern folk could learn.

The principle that one can learn how to practice agriculture by studying and learning from the natural systems and successful farmers that exist in all world areas should be accepted as a place for all engaged in agriculture to begin to work toward agricultural systems that sustain food production and people.

[1]Leopold, A. 1966. A Sand County Almanac—With essays on conservation from Round River. Ballantine Books, New York. p. 262.

A sustainable agricultural system must be all of these things but one more element must be added—it must be political acceptable. Any system that includes each of the four elements but is not politically acceptable is doomed to fail. Although it is 21 years old, neither the Asilomar Declaration nor the 19-year-old Union of Concerned Scientists (1992) "strident assessment" (Rees, 2010) about stewardship, human misery and irretrievable mutilation have been discussed within the agricultural community. Their scientific and political acceptability are unknown. The assertion that "the present system of American agriculture cannot long endure" is a significant challenge as is the claim about the destructive consequences of agriculture's industrial-technological processes. The challenge is not to the necessity of production. Both statements are a challenge, a request, to the agricultural community, indeed to all, to consider how sustainability is to be achieved so all desirable goals including production will be assured.

## Why must Sustainability be Achieved?

Kirschenmann,[9] a farmer and philosopher, acknowledges that philosophers have the annoying habit of asking questions "that the prevailing culture doesn't like to ask" or answer. Philosophers listen to claims and commonly respond with the frustrating question—"I understand your point, but have you considered ...?" It is the task of philosophy to ask such questions, to explore meaning, even when the questions disturb the prevailing culture. Societies typically try to ignore hard, challenging questions and when that doesn't work, some people who ask hard questions are killed: Jesus was crucified, Socrates was compelled to drink the hemlock juice (he did, although he could have left town), Mahatma Gandhi, Martin Luther King, and Abraham Lincoln were assassinated.

Agriculture and its practitioners have ignored challenging ethical questions and continuing to ignore them can only worsen and weaken agriculture's position in society and the quest for agricultural sustainability. As mentioned at the beginning of this chapter, all are in favor of sustainability. It is universally regarded as a good thing. But the agricultural community has not provided reasons to sustain anything other than production and profit. Modern capital, chemical, and energy dependent agriculture, however, is "neither always full of promise nor profit" (Thomas and Kevan, 1993). For example, production expenses increased more than 100% from 1970 to 1986 while net farm income remained nearly stable, one-fourth of all farm loans were non-performing or delinquent, farm debt to asset ratios increased dramatically, and farm machinery sales dropped (National Research Council, 1989, pp. 90–93). Keeney (2003) claims that the 1970 US policy that proposed that US agriculture should feed the world has failed:

- The world, because as many people are hungry now as were hungry in 1970;
- The United States, because many Americans are hungry and half are overweight; and

[9] http://www.leopold.iastate.edu/fredspeech.html (accessed October 2000).

- Agriculture because large, expensive farm supports have harmed the environment and family farms.

The policy was based on economic rationality and the quest for ever greater profit. It was, in Keeney's view, an agribusiness, not an agricultural vision. It was not and is not sustainable. Tinkering at the margins won't make modern agriculture sustainable or more profitable. Long-term, sustainable agriculture cannot be simply maximizing commodity production per acre (Thomas and Kevan, 1993) and profit. Clearly these are necessary but not sufficient conditions to achieve a sustainable system. They are at best adequate economic criteria that lack an ethical foundation.If production and profit are inadequate criteria of sustainability, one must ask what other criteria must be considered? Thomas and Kevan propose that sustainability must include land practices that "operate at lower levels of purchased inputs" and that overtly embrace agroecology and incorporation of natural processes that allow production to work with rather than against the natural system.

Gale and Cordray (1994) discuss what should be sustained and give nine answers to the question, preceded by four essential questions:

- What is to be sustained?
- Why sustain it?
- How is sustainability measured?
- What are the politics?

Or one might ask, Who benefits and who loses? Table 7.1 shows the matrix, the relationship among the four essential questions and the things one might choose to sustain (Gale and Cordray, 1994). Most of agriculture's concerns are in the first item (dominant product) in Table 7.1. The emphasis of the agricultural establishment has been on the value of producing an abundance of high-value crops. The justification has been the economic efficiency of producing large quantities of desired goods and the necessity of maintaining the flow of the specific agricultural resources the market demands. This position does not implicitly neglect agriculture's social dimensions and the human benefits but surely does not emphasize them. A good society and improved human health are regarded as the inevitable outcome of sustaining crop and livestock production. Ecosystem integrity and diversity (the third and fourth items in the table) are not undesirable goals within the agricultural community, but they rarely, if ever, assume dominance over sustaining production. A major point of Gale and Cordray's (1994) work is that while all agree that sustainability is a good agricultural goal, the debate about what is to be sustained in addition to the yield of high-value products has barely begun within the agricultural community.

The ethical position of those who practice agriculture is that all humans are to be treated equally but animals, at best, only deserve to be treated without cruelty (Table 7.2). Adopting an ethical position that considers the appropriate treatment and rights of other nonhuman animals and other, not always smaller, creatures is presently excluded from agricultures moral realm. Nash's (1977) ethical progression is demanding. It does not demand that all creatures great and small[10] must receive

[10]The title of a bestseller, Herriot, J. (1972). *All Creatures Great and Small.* Bantam Books. Title derived from the well-known hymn All Things Bright and Beautiful by Cecil Alexander.

**Table 7.1** Sustainability Types and Four Defining Questions (Gale and Cordray, 1994)

| **Sustainability Type** | **Four Defining Questions** | | | |
|---|---|---|---|---|
| | **What is Sustained?** | **Why Sustain It?** | **How is Sustainability Measured?** | **What are the Politics?** |
| Dominant product | Yield of high-value products | Economic efficiency | Quantity produced | Maintain flow of narrow resource-specific resources versus broad, diverse resource production |
| Dependent social systems | Social systems communities, families | Lifestyle values | Social system persistence | Local, targeted resource-dependent social systems versus broadly distributed use or preservation |
| Human benefit | Diverse human benefits | Human rights to resource abundance | Range of ecosystem products and uses | Broadly distributed multiple uses versus ecocentrism or resource specialization |
| Global niche preservation | Globally unique ecological | Global human ecosystem | Ecosystem health | "Spaceship earth" versus which niches to maintain |
| Global product | Globally important high-value products | Human need for products even if few areas produce them | Price and supply fit of local products into international market | International comparative advantage versus global exploitation and resource nationalism |
| Ecosystem identity | General types of ecosystems or resources uses | Commitment to general ecosystem diversity | Persistence of global ecosystem diversity | Worth of general ecological characteristics versus market-driven ecosystem conversion |
| Self-sufficiency | Ecosystem integrity | Commitment to ecosystem autonomy and naturalness | Ecosystem integrity without external input | Ecosystem rights and values versus human values and needs |
| Ecosystem insurance | Ecosystem diversity | Ensure against ecological disaster and diversity loss | Vitality and amount of insured resources, resistance to ecological crises | General need for reserved areas versus questions of future need and technological optimism |
| Ecosystem benefits | Undisturbed ecosystems | Respect rights inherent in natural ecosystems | Ecosystem continuity natural evolution | Restorative intervention or ecocentric autonomy versus human dominance and use |

**Table 7.2** An Ethical Progression (Nash, 1977)

| | |
|---|---|
| Future | The intrinsic value of the environment |
| | Life in general |
| | Plants |
| | All other nonhuman animals |
| | Nonhuman mammals |
| Present | Humans |
| | One's race |
| | One's nation |
| | One's tribe |
| The ethical past | Family |
| | Self |

equal treatment. No one thinks sheep should have a right to go to school or geese should have a right to vote. Nash does suggest that the lives of all creatures should be respected and none should suffer unnecessarily. Nash does not say, but for me, his progression suggests that even pests deserve some moral consideration. I cannot prescribe what they deserve, only that agricultures practitioners ought to consider and do not consider what they deserve. The result of careful thought may be part of achieving a sustainable agriculture system.

People in the agricultural community are less challenged now than they were a few decades ago, by discussions of nonhuman animal rights and the ethical treatment of nonhuman animals. When the rights of animals are dismissed as another useless academic debate before debate has begun; little good can follow. The agricultural community is now more willing to give nonhuman animals the rights they routinely and without question give to human animals: for example, the right to a full life, freedom from suffering, the right to raise descendants, freedom from torture. Students struggle with the challenges offered by Singer (2000), who claimed as Bentham[11] did, that all sentient beings should have equal consideration (not including the right to attend religious services or own a bicycle) because all can suffer. The capacity for suffering and enjoyment are necessary and sufficient conditions to say that, without question, nonhuman animals deserve equal consideration. Unnecessary suffering should be stopped. For further discussion, see Chapter 10.

Those who practice agriculture have not progressed to the level of ethical concern prescribed in the famous statement and environmental challenge from Leopold (1947, p. 262):

> *A thing is right when it tends to preserve the integrity, stability and beauty of the biotic community. It is wrong when it tends otherwise.*

The endless, and not undesirable, quest for production in agriculture is rarely concerned with the integrity, stability, or beauty of the biotic community. Agricultural

[11] Jeremy Bentham (1748–1832), an English jurist and philosopher, is best known for his advocacy of utilitarianism.

practice does not inevitably produce ugliness or lead to instability, although it does destroy stability, especially when it is based on uniform (over large areas) monocultural crop production. In fact, cropped fields can be quite beautiful (amber waves of grain). But the beauty is temporary, created only to be destroyed and, one hopes, created again next year. It is not the beauty of permanence, of a stable biological community. It is not a beauty or stability that is created because of an ethical stance similar to the physicians' ethical code: *Primum non nocere*—Above all, do no harm.

The practice of agriculture has changed from something that many people did, often because they had to in order to survive, to something that few people do. Somewhere along the way it also changed from one that working with nature. Farmers were subject to the vagaries of climate and weather. They had no choice. Agricultural practice is still subject to the weather (hail, drought, and severe storms ruin crops), but humans have much greater control than was possible in the early twentieth century. Development of improved ability to control nature occurred with the firm belief that humans had the right (indeed the obligation) to control nature. Agricultural scientists were able to ignore the ecological basis of agricultural practice and, in turn, ecologists ignored agriculture because it was not part of natural processes. Agriculturists knew they were building a sustainable production system. Ecologists knew that was not true, but they did not often speak to each other. As technological solutions to agricultural production problems began to fail (pest resistance appeared, groundwater receded, soil erosion continued, fertilizers and pesticides polluted water, petroleum energy became more expensive, and pesticides affected humans and nontarget species), production became more complicated and less profitable. Sustainability became a highly desired goal, but neither the question of what to sustain nor the question of why achieving it was necessary had been answered.

Sustainability is about the future. We cannot sustain the past and today, each moment is fleeing as we live it. It seems clear that we want to preserve something for tomorrow. What is not clear is what our obligation to the future may be, if we have one at all. It is a common philosophical position that we have no obligation to future generations because it is hard to prove that nonexistent future generations have rights. The argument is that to have a right is to be a present bearer of the right. Because future generations do not exist yet, they cannot bear rights or claims against present people. However, Burkhardt (1989) argues that we can decide that future generations have rights or that we ought to think they do. Those who have them care deeply about their children and grandchildren and may choose to extend their care to generations yet unborn. What we do now in agriculture may affect future generations. We inherited a world that supports us and we can choose to assume the obligation to treat the world in such a way that we pass it to those who follow in as good or better condition than it was in when we received it. We have or we can assume an obligation to sustain and improve the world. At the very least, we can posit that future generations have a right not to be harmed and the right to be helped, which may be equivalent to a right not to be harmed. Therefore, Burkhardt suggests that future people ought to have rights which present people have a duty to ensure. These rights include:

- Protecting the earth's capability to provide sufficient food for, however, many people they democratically decide it is in their interest to support.
- Preserving and developing scientific knowledge and technologies that can assist them in the provision of sufficient supplies of food, clothing, and shelter, subject to their own environmental and cultural values.
- Creating and preserving democratic institutions that promote active participation of all people in addressing whatever problems they might encounter.
- Creating and preserving traditions of moral trust and respect such that values of community and the excellence of human life can be live and promoted.

As Burkhardt says, "it seems clear the earth should not be dead when future generations arrive." That means its productive capacity and the institutions that enable that capacity must be sustained by those who precede them. Agriculture has therefore, an obligation to feed present people and maintain its productive resource base. It also has an obligation to contribute to democratic institutions and caring communities. Burkhardt argues that "any set of practices, policies and institutions that respects the environment, including nonhuman animals, can be morally adequate." Any farming system that feeds people enough, high-quality food is adequate to achieve sustainability. It is not the system or its components that are critical, it is its achievements that will lead to sustainability that are in the interests of all present and future generations and are the obligation of all. Because it is a shared obligation, it is essential that societies collectively define and work to create the kinds of economic, educational and political institutions so that all, including those directly engaged in agriculture, can act on the obligation to achieve sustainability.

## A Concluding Comment About Sustainable Weed Science

During my academic career, I have met many people who on learning that I was a professor would often ask, "What do you teach?" It was and is a good question, and I am eager to answer it. Early in my academic career, I thought it was the wrong question, because I wanted to talk about my research and the importance of weed control. What I wanted to tell people was not what they were asking. I began to wonder if I had the right stuff. Wolfe (1979), writing about the Apollo astronauts, let me know I did not, and told me that some did. "The world was divided into those who had *it* and those who did not. This quality, this *it*, was never named, nor was it talked about in any way." Some had *it*, the right stuff, most did not.

I knew some of my weed science colleagues had the right stuff. They, in Wolfe's words had "the moxie, the reflexes, the experience, the coolness" to think of and answer questions about weed control that most weed scientists thought were the right questions. For me, they were the elite, those who had the right stuff to acquire resources and organize people to ask and answer the right questions. Now, I am retired and in the twilight of my professional life, but still believe some of my superb colleagues had and still have the right stuff, but I am no longer sure they had the right questions.

The most important questions of my early years in weed science were: What is the problem weed? What is its identity? The second question was, How can it be controlled? The second question implicitly also asked, What herbicide will be most effective for controlling the weed, selectively? These are good questions and they are still asked frequently. They are not the most important questions. It is my view that their persistence has, at once, enabled weed science to make important contributions to agricultural productivity while obscuring the right questions. The questions I asked and others still ask are consistent with the dominant production paradigm of modern agriculture. Acceptance of this paradigm has placed increasing production as a high, if not the highest, value, and most weed scientists regard employment of all appropriate modern technology to achieve and maintain profitable production as prudent and correct. The conviction is that the highest level of production or weed control that can be achieved profitably is the best level. The paradigm includes an unexpressed, fundamental assumption of the unqualified right of humans to transform, control, and dominate nature to achieve production of food, which is essential for life. Giampietro (2004, see preface and introduction), from a quite different perspective, suggests agricultural science has failed to alter its operative paradigm to properly address issues of sustainability.

We manipulate the natural world to produce food, and weeds inevitably hinder food production, but the emphasis on control has obscured the right question, which is: Why is the weed where it is? That is to say, what is it about the production system or the way agriculture is practiced that allows a specific weed or weed population to be so successful? The right question is a systemic, holistic one that accepts transformation of nature as a necessary prerequisite to food production, but rejects domination of nature. Transformation of nature may yield weeds; an undesired result. Weed control is not bad or forbidden when the right question is asked. But control is subsumed under vegetation management. The right question, a question of applied ecology, is compatible with the quest for sustainable agriculture and holistic understanding, because it is derived from ecology, a discipline that studies the principles that regulate distribution and abundance of species in communities. Weed scientists, myself included (Zimdahl, 2004), have asked how a certain density and duration of weeds affects crop yield. We have asked these questions to gain an economic answer to a production question. How many weeds are required to reduce crop yield more than the cost of weed control? When we know that answer, we ask how to control the weeds selectively and profitably. The right question will not forbid asking what to do, but it demands that research begin with a *why* question rather than a *what* question. A *why* question leads toward development of a foundational theory to guide weed science. *Why* does something happen? *What* questions are fundamentally empirical and their answers reveal what to do, but not necessarily why a particular course of action is best or why it should be taken. Control questions, those that ask what to do, frequently yield short-term solutions and do not lead to what Berry (1981) calls "a ramifying series of solutions," which are in harmony with the larger patterns in which they are contained. Until the right questions are asked and the characteristics of a production system (the larger pattern) that create opportunities for weeds to succeed are understood, weed scientists will continue to develop and recommend employment of short-term solutions to weed problems. Weed science and successful, sustainable agriculture systems are, or should be, derived from studies in applied ecology. Experiments designed to ask why questions based on the ecological principles that regulate the abundance and distribution of weedy species in disturbed, cropped environments will be asking the right questions to achieve a sustainable agriculture.

## References


Alliance for Sustainability. (2004). Sustainable agriculture defined. http://www.mtn.org/iasa/susagdef.htm (accessed May 28, 2004).

Avery, D. (1995). *Saving the Planet with Pesticides and Plastic: The Environmental Triumph of High-Yield Farming*. Indianapolis, IN, Hudson Institute.

Bartlett, A.A. (1999). Reflections on sustainability, population growth, and the environment—revisited. *Focus* 9(1):49–68.

Berry, T. (1999). *The Great Work—Our Way into the Future*. New York, Bell Tower.

Berry, W. (1981). *The Gift of Good Land: Further Essays Cultural and Agricultural*. San Francisco, CA, North Point Press. P. 137.

Borlaug, N. (2001). Ending world hunger: the promise of biotechnology and the threat of antiscience zealotry. *Trans. Wisconsin Acad. Sci.* 89:25–34.

Burkhardt, J. (1989). The morality behind sustainability. *J. Agric. Ethics* 2:113–128.

Busch, L., W.B. Lacy, J. Burkhardt, D. Hemken, J. Maraga-Rojel, T. Koponen, and J. de Souza Silva. (1995). *Making Nature Shaping Culture: Plant Biodiversity in Global Context*. Lincoln, NE, University of Nebraska Press.

Buttell, F.H. (2003). Internalizing the societal costs of agricultural production. *Plant Physiol.* 133:1656–1665.

Clark, W. (1975). US agriculture is growing trouble as well as crops. *Smithsonian* January:59–64.

Colburn, T., D. Dumanoski, and J.P. Myers. (1996). *Our Stolen Future: Are we Threatening our Fertility, Intelligence, and Survival?—A Scientific detective Story*. New York, Dutton Division of Penguin Books.

Daly, H.E. (1996). *Beyond Growth—The Economics of Sustainable Development*. Boston, MA, Beacon Press.

Daly, H.E. (1993). Sustainable growth: an impossibility theorem. Pp. 267–273 *in* H.E. Daly, K.N. Townsend (eds.). *Valuing the Earth: Economics, Ecology, Ethics*. Cambridge, MA, MIT Press.

Daly, H.S. and J.B. Cobb. (1989). *For the Common Good: Redirecting the Economy toward Community, the Environment, and a Sustainable Future*. Boston, MA, Beacon Press.

Davison, A. (2001). *Technology and the Contested Meaning of Sustainability*. Albany, NY, State University of New York Press.

Degregori, T.R. (2001). *Agriculture and Modern Technology: A Defense*. Ames, IA, Iowa State University Press.

Diamond, J. (2004). *Collapse: How Societies Choose to Fail or Succeed*. New York, Viking.

Douglass, G. (1984). The meanings of agricultural sustainability. Pp. 3–29 *in* G. Douglass (ed.). *Agricultural Sustainability in a Changing World Order*. Boulder, CO, Westview Press.

Durning, A.T. (1994). The conundrum of consumption. Beyond the Numbers: A Reader on Population, Consumption, and the Environment. Washington, DC., Island Press, Pp. 40–47.

Eberstadt, N. (1995). Population, food, and income. Pp. 8–47 *in* R. Bailey (ed.). *The True State of the Planet*. New York, Free Press.

Edwards, S. (1980). Farming's rewards at risk. *Center Mag.* November/December:20–31.

Evans, L.T. (1998). *Feeding the Ten Billion: Plants and Population Growth*. Cambridge, Cambridge University Press.

Farah, J. (1994). Pesticide policies in developing countries: Do they encourage excessive use? World Bank Discussion Paper No. 238. Washington, DC, World Bank.

Gale, R.P. and S.M. Cordray. (1994). Making sense of sustainability: nine answers to 'What should be sustained?' *Rural Sociol.* 59:311–332.

Giampietro, M. (2004). *Multi-Scale Integrated Analysis of Agroecosystems*. Boca Raton, FL, CRC Press.

Gianessi, LP. and S. Sankula. (2003). *Executive summary, The value of herbicides in US crop production*. Washington, DC, National Center for Food and Agricultural Policy.

Halweil, B. (2002). Farming in the public interest. Pp. 51–74 *in* C. Flavin (ed.). *State of the World – 2002*. New York, W. W. Norton & Co.

Keeney, D. (2003). Feed the world: a failed policy. *Leopold Lett.* 15(4):7.

Knutson, R.D., D.R. Taylor, J.B. Penson, and E.G. Smith. (1990). Economic Impacts of Reduced Chemical Use. College Station, TX, Knutson and Associates.

Kroma, M.A. and C.B. Flora. (2003). Greening pesticides: a historical analysis of the social construction of chemical advertisements. *J. Agric. Human Values* 20:21–35.

Kudsk, P. and J.C. Streibig. (2003). Herbicides—a two-edged sword. *Weed Res.* 43:90–102.

Lehman, H. (1997). Environmental ethics and pesticide use. Pp. 35–50 *in* D. Pimentel (ed.). *Techniques for Reducing Pesticide Use*. New York, John Wiley & Sons.

Lehman, H. (1993). Values, ethics, and the use of synthetic pesticides in agriculture. Pp. 347–379 *in* D. Pimentel, H. Lehman (eds.). *The Pesticide Question: Environment, Economics and Ethics*. London, Routledge, Chapman & Hall.

Leopold, A. (1966 [1949]). *A Sand County Almanac*. New York, Ballantine Books. Original Published by Oxford University Press.

Liebman, M. (2001). Weed management: a need for ecological approaches. Pp. 1–39 *in* M. Liebman, C.L. Mohler, C.P. Staver (eds.). *Ecological Management of Agricultural Weeds*. Cambridge, Cambridge University Press.

Major, C.S. (1992). Addressing public fear over pesticides. *Weed Technol.* 6:471–472.

Mayer, A. and J. Mayer. (1974). Agriculture, the Island Empire. *Daedalus* 103:83–95.

Meadows, D.H., D.L. Meadows, J. Randers, and W.W. Behrens. (1972). *The Limits to Growth—A Report for the Club of Rome's Project on the Predicament of Mankind*. New York, Universe Books.

Mepham, B. (1998). Agricultural ethics(1998). *Encyclopedia of Applied Ethics* Vol. 1: San Diego, CA, Academic Press. Pp. 95–110.

Merrill, M.C. (1986). Some philosophical prerequisites for a sustainable agriculture. *In:* Proceedings of 6th International Science Conference of the International Federation of Organic Agriculture Movements, Pp. 83–91. Santa Cruz, University of CA.

Mohler, C.L., M. Liebman, and C.P. Staver. (2001). Weed management: the broader context. Pp. 494–518 *in* M. Liebman, C.L. Mohler, C.P. Staver (eds.). *Ecological Management of Agricultural Weeds*. Cambridge, Cambridge University Press.

Nash, R. (1977). Do rocks have rights? *Center Mag.* November/December:2–12.

National Research Council, (1989). Problems in US agriculture (1989). *Alternative Agriculture*. Washington, DC, National Academy of Science Press. Pp. 89–134 (esp. Pp. 90–93).

Oerke, E.C., H.W. Dehne, F. Schöhnbeck, and A. Weber. (1994). *Crop Production and Crop Protection: Estimated losses in Major Food and Cash Crops*. Amsterdam, Elsevier.

Pettersson, O. (1992). Pesticides, valuations and politics. *J. Agric. Environ. Ethics* 5:103–106.

Pimentel, D. (1992). Environmental and economic costs of pesticide use. *Bioscience* 42:750–760.

Pimentel, D. and A. Greiner. (1997). Environmental and economic costs of pesticide use. Pp. 51–78 *in* D. Pimentel (ed.). *Techniques for Reducing Pesticide Use. Economic and Environmental Benefits*. New York, John Wiley and Sons.

Polimeni, J.K., M. Mayumi, Giampietro, and B. Alcott. (2008). *The Jevons Paradox and the Myth of Resource Efficiency Improvements*. Earthscan, London, UK. P. 184.

Pretty, J.N. (1995). Sustainable agriculture in the 21st century: challenges, contradictions and opportunities. Brighton Crop Protection Conferenc—Weeds. Pp. 111–120. British Crop Protection Council. Surrey, U.K.

Rees, W. (2010). What's blocking sustainability? Human nature, cognition, and denial. *Sustainability: Sci. Practice, Policy* 6(2):13–25.

Repetto, R. and S.S. Baliga. (1996). *Pesticides and the Immune System: The Public Health Risks*. Washington, DC, World Resources Institute.

Rolston, H. (1975). Is there an ecological ethic? *Ethics* 85:93–109.

Sagoff, M. (1988). *The Economy of the Earth: Philosophy, Law and the Environment*. Cambridge, Cambridge University Press.

Schaller, N. (1993). Farm policies and the sustainability of agriculture: Rethinking the connections. Policy Studies Report No. 1, from the H.A. Wallace Institute for Alternative Agriculture.

Sen, A. (1999). *Development as Freedom*. New York, A.A. Knopf.

Singer, P. (2000). All animals are equal. Pp. 28–46 in Writings on an Ethical Life. New York, HarperCollins Publication, Inc. Originally published in *Animal Liberation*. 1975. New York, Random House.

Standaert, M. (2003). World's farmers struggle with globalization issues. wysiWg://79/http://www.enn.com/news/2003-12-31/s_116 (accessed January 5, 2004).

Stout, B.A. and P.A. Thompson. (1991). Beyond the large farm. Pp. 265–279 *in* P.B. Thompson, B.A. Stout (eds.). *Ethics and Research Goals in Agriculture* Vol. 1: Boulder, CO, Westview Press.

Thomas, V.G. and P.G. Kevan. (1993). Basic principles of agroecology and sustainable agriculture. *J. Agric. Environ. Ethics* 6:1–19.

Thompson, P.B. (1995). *The Spirit of the Soil: Agriculture and Environmental Ethics*. New York, Routledge.

Union of Concerned Scientists (UCS). (1992). 1992 World Scientists Warning to Humanity. Union of Concerned Scientists, Cambridge, MA.

Wackernagel, M. and W.E. Rees. (1996). Our Ecological Footprint—Reducing Human Impact on the Earth. Stony Creek, CT, New Society Publishers.

Waggoner, P.E. (1994). *How Much Land Can Ten Billion People Spare for Nature?* Ames, IA, Council for Agricultural Science and Technology.Task Force Report No. 121.

Westra, L. (1998). *Living in Integrity: A Global Ethic to Restore a Fragmented Earth*. Totowa, NJ, Rowman and Littlefield.

White, L. (1967). The historical roots of our ecological crisis. *Science* 155:1203–1207.

Wolfe, T. (1979). *The Right Stuff*. New York, Farrar, Strauss, and Giroux. P. 24.

Zimdahl, R.L. (2004). *Weed-Crop Competition—A Review*, 2nd Ed. Ames, IA, Blackwell Publishing.

# 8 Biotechnology

*I am thy creature, and I will be even mild and docile to my natural lord and king if thou wilt also perform thy part, that which thou owest me. Oh, Frankenstein, be not equitable to every other and trample on me alone, to whom thy justice, and even thy clemency and affection, is most due. Remember that I am thy creature; I ought to be thy Adam but I am rather the fallen angel, whom thou drivest from joy for no misdeed. Everywhere I see bliss, from which I alone am irrevocably excluded. I was benevolent and good; misery made me a fiend. Make me happy and I shall again be virtuous.*
*'Begone! I will not hear you. There can be no community between you and me; we are enemies. Begone, or let us try our strength in a fight, in which one must fall.'*

***Shelley (1818)***

Chapter 3 suggests that biotechnology has changed the debate about the criteria used to determine the acceptability of any agricultural technology by introducing a new question: Do we need it? The question could also be moral rather than just scientific or economic, by asking, Ought we to do it? In the above quote from Mary Shelley's novel *Frankenstein,* published in 1818, all will recognize the monster Dr. Frankenstein created as the parable of perverted science, of science gone bad (Warner, 1994, pp. 29–30). Shelley's more important message is Dr. Frankenstein's response (the last three lines above) to his monster's plea. Shelley recasts the monster in the image of Dr. Frankenstein, its creator. The monster is Frankenstein's brainchild, his double and serves to define his creator. Her lesson is: scientists might make a monster, but be incapable or unwilling to take responsibility for what they have created, a concern echoed by Kneen (1999).

The dilemma is one that plagues biotechnology: ought we to do it? Will biotechnology create another monster and will its creators then be unable or unwilling to accept responsibility for the bad behavior of their creation, or is biotechnology the next great scientific step that will benefit all? Surely the new technology shows vast promise of crops that are more pest and drought resistant, more nutritious, higher yielding, and other desirable traits to be mentioned. In Chapter 5, Kirschenmann and Youngberg (1997) suggest that in spite of the early, heated controversy, which has diminished in recent years, biotechnology will be the primary force that shapes tomorrow's agriculture. The world's farmers [(in order of hectares planted) in the US, Brazil, Argentina, India, Canada, and China (none in the EU)], planted 148 million hectares (366 million acres[1]) of biotech

[1] Because it is the accepted world standard, hectares will be used. Hectares × 2.471 = acres. Acres/2.471 = hectares.

**Agriculture's Ethical Horizon. DOI: 10.1016/B978-0-12-416043-9.00008-8**

(genetically engineered—GE) crops in 2010.[2] Biotech crops have been adopted more rapidly than any other agricultural technology. In 1996, the first year of availability, 1.7 million hectares were planted, 64 million in 2003 (CropBiotech Net, 2004), 134 in 2009, and 148 in 2010. The area planted continues to increase, at least 7% per year. The US grows half of all GE crops. Six countries grew GE crops in 1996, 18 in 2003, 23 in 2007, and 29 in 2010. In 2010, there are 10 available GE crops.[3] The primary crops and the percent which is GE in the US are: Soybean 93, HT[4] Cotton 78, Bt cotton 73, HT corn 70, and Bt corn 63. Herbicide tolerance is incorporated in 63% and Bt-insect resistance is in 18% of worldwide GE crops. According to surveys conducted by USDA in 2001–2003, most farmers (79%) adopting GE corn, cotton, and soybeans indicated they did so to increase yields through improved pest control. The second most cited aim was to save management time and make other practices easier (15–26%, except for Bt corn, which was much lower); the third reason was to decrease pesticide costs (9–17% of adopters).

## The Debate

The debate about biotechnology is similar to others that have occurred about agricultural technologies (e.g., pesticides). It seems that neither scientists nor technology developers (who may also be scientists) have learned much from the earlier debates about new agricultural technologies. Previous debates have included competing industrial concerns, university research scientists, a confused, but perhaps not very concerned, public, and questionable or absent ethics. DeGregori (2001, p. 125) states the scientific perception of the problem: "public discourse is being driven by emotional language." The biotechnology debate is beset with emotion and ambiguities (Zwart, 2009). It may refer to modification of living organisms or to a specific set of recent (1990s[5]) technologies capable of bioengineering and genetic modification. The definition chosen reveals one's position (bias?). Zwart claims that biotechnology is part of the human condition and, especially in agriculture, modifying plants, animals, and the environment has become an essential part of agricultural practice. Since agriculture began, humans have practiced genetic modification by selecting seeds to plant from plants perceived to have more desirable characteristics (primarily higher yield).

[2] These data and several following points are from 1. Genetically Modified Crops Only a Fraction of Primary Global Crop Production. http://www.worldwatch.org/node/5950 (accessed March 20, 2011), 2. Executive Summary: Global Status of Commercialized Biotech Crops/GM Crops. http://www.isaaa.org/resources/publications/briefs/42/executivesummary/de....htm (accessed March 20, 2011), and 3. www.ers.usda.gov/Briefing/Biotechnology/chapters1,2,and%203.htm (accessed March 20, 2011).

[3] Alfalfa, canola, cotton, maize, papaya, potato, soybean, squash, sugarbeet, sweet pepper. GE alfalfa, primarily Roundup Ready™ alfalfa, created by Monsanto has been the subject of intense debate. It was approved by the USDA in 2005. Opponents quickly filed a lawsuit, and in 2007 a federal judge ordered more review. The USDA issued its final report in December 2010. Frenzied activity and more discussion followed. USDA Secretary Vilsack argues that without any action, courts will dictate the future of GE alfalfa and other crops.

[4] HT, herbicide tolerant; Bt, Bacillus thurigiensis, incorporated for resistance primarily to European corn borer and corn rootworm.

[5] Herbicide tolerant crops were first available to farmers in 1996. Their use expanded rapidly. Scientists at Monsanto successfully introduced a foreign gene into a plant cell for the first time in 1982 (Hsin, 2002).

For more than a century plant breeders have used art and science to change the genetics of plants. They employ techniques ranging from simply selecting plants with desirable characteristics and crossing them by selective pollination with others of the same species, and more recently, complex molecular techniques. Because pests evolve and the environment changes, breeding new crop varieties that are higher yielding, more resistant to pests, drought resistant or regionally adapted is important to ensure food security. Genetic engineering (the biotech revolution) as it has evolved since the 1970s has permitted selective modification of plants and other organisms with greater levels of precision (Zwart). It includes the previously impossible incorporation of genes from one, unrelated species into the genome of another species.

**Highlight 8.1**

The Anheuser-Busch company announced in 2005 that it would not buy rice grown in Missouri if genetically modified drug-containing crops (crops modified to produce a pharmaceutical product) are allowed to be grown in the state (Hahanel 2005). *(I have been unable to determine if this is still true.)* Anheuser-Busch, headquartered in St. Louis, is the largest buyer of rice in the United States. Ventria Biosciences requested permission from the state of Missouri to grow 200 acres of rice that had been genetically modified to produce human proteins from which drugs could be made. The intent of the bipoharming project was to lower the cost of drug manufacture by using plants to produce a drug. The most common fear is cross contamination of non-genetically modified crops, grown nearby. Biopharming projects have been growing for nearly a decade, although many are concerned that there has not been sufficient scientific study of the risk and safety of such crops.

Borlaug (2001) argues biotechnology is scientific progress and "must not be hobbled by excessively restrictive regulations." If we are to meet the needs of the 8.3 billion people that may be on our planet by 2025, Borlaug claims that conventional technology and biotechnology will be not just needed, they will be essential. He also claims that "extremists in the environmental movement, largely from the rich nations or the privileged strata of society in poor nations, seem to be doing everything they can to stop scientific progress in its tracks." *The Economist* (2004a) agrees with Borlaug's view and opines that "scaremongering by western green activists has discouraged investment in such valuable research." Borlaug acknowledges agriculture's debt to the environmental movement because their efforts have led to "legislation to improve air and water quality, protect wildlife, control the disposal of toxic wastes, protect soil, and reduce loss of biodiversity." However, now the anti-biotechnology extremists have gone too far and their policies will have "grievous consequences for the environment and humanity." It is mildly ironic that Borlaug claims this in spite of the fact that those in the environmental movement are continuing to pursue improvement of the same things for which he thanks them.

The scientific view is "science can determine a fact, that these facts represent objective reality, and that values or beliefs play no role in determination of facts" (Barker and Peters, 1993, p. 5). Science is objective and value-free, as it should be. It is not the scientist's task to create or change social, economic, or political policy according to Barker and Peters. Objective science is driven by curiosity about the natural world, the mission of the employing institution, and the demands of the funding that enable the research. Scientists attempt to understand and explain the natural world and technology applies scientific findings to the world. In general, science has been regarded by the public as good and technology was judged to be good or bad depending on how it was used (Boulter, 1997). However, science, in general, has moved from being viewed as an unalloyed public benefit (e.g., better living through chemistry) to being regarded with suspicion, if not outright distrust. Scientists were used to be seen as being guided by a wholesome curiosity and a search for the elusive truth. Now the public is wary. It is well known that science, similar to many other human activities, is strongly influenced by social, economic, and political pressures (Boulter). It is equally well known that some bad consequences, scientists said were unlikely, have happened (see Chapter 3). For example, there are pesticides in some drinking water supplies and food, nitrates pollute water, space shuttles fail, and airplanes fall from the sky. Do scientists deign to think about where their cleverness as opposed to their wisdom is taking us (Brower, 2010)? How often does the moral question, ought we to do it, enter the scientific/economic realm?

Public and scientific debates about biotechnology often appear to be dominated by polar opposite views. The debaters don't really speak to each other. The disputes are frequently based on scientific facts—often selected facts—but the disputants nearly always disagree over the story (Charles, 2001). All stories are, in some sense, true, especially when one knows the preconceptions of the storyteller. The dispute is over the goodness of the characters (their virtue or lack thereof), the plot (why is this happening?), the editing (what facts count?), and how it will all end. Charles provides examples of the relevance of the story's (its myth) metaphor by contrasting deeply embedded stories that affirm the possibility of progress, discovery and creativity, problem solving, and expanding the boundaries of human possibility with other stories (the countervailing myths), which affirm that technology is unpredictable, threatening, and a product of human folly that believes it can (and should) control the world.

What one hears or reads in these conflicting stories is often not a reasoned debate of the issues. It is a presentation and defense of one of the polar views: biotechnology is good and required to feed the expanding human population vs. biotechnology is bad, for a variety of reasons, especially the unpredictability of future effects and will not help feed people. There are notable exceptions. The purpose of this chapter is to present and explore the debate (some of the stories) about biotechnology. It is intended to be an exploration not a resolution.

## Technological Problems

Giampietro (1994) notes that it "is dangerous to convey to the general public the idea that problems faced today by humankind can be fixed without changing cultural and political paradigms, remaining in the business-as-usual mode, just by working out a better technology." Genetic engineering, he acknowledges is like other technologies, in that its useful or harmful effects will be determined by how it is used. We humans tend to analyze and address the problems that bother us now, fix those problems when possible, and ignore long-range problems because, well, they are long range and there are so many other problems demanding immediate attention. Ecological problems are frequently long-range in the sense that they don't appear quickly, cannot usually be addressed by quick, short-term solutions, and the best solutions require a long-range, historical/ecological perspective. Biotechnology is regarded by many critics as one of the quick fixes that ignores long-range potential problems while focusing on short-term benefits. Without a long-range perspective many solutions create a new set of problems that require more quick fixes but do not yield true solutions (Giampietro). The solutions that proponents of biotechnology suggest seem to be the kind that Berry (1981) deplored because so many can make the problem worse. Berry advocated solving for pattern and identified three kinds of solutions to any problem. The first solution (that he deplored) causes a ramifying series of new problems, the second immediately worsens the problem, and the third, most desirable solution, creates a ramifying series of solutions. The dependence of modern agriculture on technical solutions that depend on declining (but perhaps not as rapidly as the doomsayers suggest) supplies of petroleum energy to increase food production is not in the latter category. In economic terms such dependence may be right but it may not be sustainable ecologically (Giampietro) or defendable morally.

Thompson (1987) frames the problem well by suggesting that facts and values about biotechnology are neither "readily separable in the regulatory tangle" nor in the public mind. Citizens are aware of some of the issues but scientists do not speak in language that is readily understandable and the media often seems to exacerbate prevailing polar views of what the story is about. Thompson suggests that the casual reader of available news can easily draw the conclusion that the scientist's story of agricultural biotechnology may be just as likely to defend personal interests as it is to present unbiased judgments. Thus, concerned citizens may shift their focus from the scientific issue to the values of the scientist or the scientific community, which is frustrating because these values are rarely explicit. Therefore, citizens tend to focus on the characteristics and presumed values of the scientist, what Charles (2001) called the goodness of the characters, and the values of the scientific community, which also are rarely explicit.

Lovins and Lovins (2000) point out that with the good goal of feeding the growing human population, agricultural biotechnology is replacing nature's wisdom with human cleverness. Nature and natural systems are not guides but obstacles. Nature is to be restructured to meet what all agree is a good goal: feeding people. Lovins and Lovins article is in the center of a longer article by Joy (2000),[6] who no one should identify as an opponent of new technology. Joy argues that "new technologies like genetic engineering ... are giving us the power to remake the world" and we risk

[6]William N. Joy is a computer scientist and a co-founder of Sun Microsystems. Among other things, he is well known for his article in Wired magazine (Issue 8.04, April 2000)—Why the future doesn't need us.

"failing to understand the consequences of our inventions while we are in the rapture of discovery and innovation." He identifies this as a common fault among scientists and technologists.

## Regulation

Andrews (2002) and Lovins and Lovins (2000) identify a critical commonality among the products of agricultural biotechnology: "they are different enough to patent but similar enough to make identical food." The accepted definition of a transgenic organism (a more precise term) is one that contains an inserted gene sequence that could not have been acquired by natural or artificial (i.e., human facilitated) hybridization. Those who create and sell transgenic (i.e., genetically modified) crops have convinced the Food and Drug Administration (FDA), the Environmental Protection Agency (EPA), the US Department of Agriculture (USDA) (overlapping regulatory agencies) that the inserted genes are products of nature (they are naturally created) and therefore the resultant plant or animal is no different from their nonengineered cousins. The regulatory agencies regard them as "substantially equivalent." However, at the same time, they have persuaded the US Patent Office that the genetically engineered crops are not products of nature (the characteristics could not have been acquired naturally) and are different enough to be patented (Andrews, 2002). Mann (1999) discussed his view of the "lack of a rigorous regulatory framework to sort out the risks inherent in agricultural biotech." He suggests the lack exists because there is no agreement on what should be regulated. Thus, we find the European Union's more rigorous insistence on labeling so people have the freedom to choose what they want to eat or not eat, is what Borlaug (2001) calls environmental extremism and what many scientists call an emotional response without any scientific foundation. Others call the EU position an irrational (i.e., nonscientific) barrier to free-trade.

Food from biotech crops is similar enough to non-biotech crops to eat, because the US regulatory community has accepted the concept of substantial equivalence to assess the risk of genetically modified (GE) food. Substantial equivalence involves quantitative chemical analytical methods of evaluation to determine if key nutrients or anti-nutrients in plants to be used for food or feed have been changed by genetic modification (Anonymous, 2004b). The analysis considers essential vitamins, minerals, fatty acids, carbohydrates, amino acids, naturally occurring toxicants such as erucic acid and glucosinolates in canola or solanine in potato. Analyses may also consider allergenic proteins in common foods such as soybean and wheat (Anonymous, 2004b). To determine substantial equivalence, health and regulatory officials consider natural variation, the changes that may be introduced by processing, and the food's intended use. The tests are what Pouteau (2000) calls a reducing concept because they ignore the effects of much of the production and processing of food that occurs between harvest, store, and table.

The concept of substantial equivalence was first described in 1993 and implemented in 1998 by the OECD (Organization for Economic Cooperation and Development) (OECD, 1993; Pouteau, 2000). It has been accepted for regulation

**Table 8.1** Opinions vs. Experimentation in the Referred Literature on GE Food Safety

| Base Phrase for Medline Database Search | Total Number of Identified Citations | Number of Citations Specifically Related to the Question | |
|---|---|---|---|
| | | Citations Reporting Experimentation | Citation of Opinion without Experimentation |
| Toxicity of transgenic foods | 44 | 1 | 7 |
| Adverse effects of transgenic foods | 67 | 2 | 16 |
| Genetically modified foods | 101 | 6 | 37 |

Adapted from Domingo (2000) by Clark and Lehman (2001)

of GE foods by the United Nations Food and Agriculture Organization (UN/FAO), the World Health Organization (WHO), and by the US government. According to Pouteau, the process used to determine substantial equivalence was never intended to be a substitute for true safety evaluation. It is a chemical analytical process that compares the composition of GE food or feed with a non-GE food or feed that humans or animals already consume with no evidence of harm. It has been used widely for decision making about the safety of GE foods. Clark and Lehman (2001) provide citations in support of and in opposition to GE crops and then cite Domingo (2000) to show that widespread use of GE crops has not been preceded by "rigorous, scientifically defensible analysis of benefits and harms." The quantitative, chemically supported concept of substantial equivalence has been used to determine safety for users and the environment. Clark and Lehman used Domingo's data (Table 8.1) from the Medline database[7] to illustrate that opinion rather than scientific evidence has dominated safety determination. Thus, according to the work cited by Clark and Lehman, nine mostly rodent-based studies were (in 2001) all the peer-reviewed, experimental data available on the safety of GE food. They conclude that widespread commercial adoption of GE technology has not been preceded by a "rigorous and scientifically defensible assessment of benefits or harms for consumers or primary producers, for the Third World, for biodiversity, or for the environment." They call substantial equivalence "argument by analogy." If GE crop X is chemically similar to non-GE crop Y and X has not caused any obvious harm to humans or animals that consume it, then Y must also be safe. There is no independent, peer-reviewed, scientific evidence of safety for anything. The use of substantial equivalence as the primary determinant of acceptability and presumed safety is a commercially approved political/regulatory decision that is not based on any requirement for scientific evidence of safety. Lovins and Lovins (2000) comment that our "technical ability has evolved faster than social institutions" and is not far off the mark. In their view, "skill has outrun wisdom."

Millstone et al. (1999) called substantial equivalence a pseudo-scientific concept because it is a commercial and political judgment masquerading as if it were

[7] http://www.ncbi.nlm.nih.gov/pubmed/

scientific. It is, moreover, inherently anti-scientific because it was created primarily to give regulators an excuse not to require extensive biochemical or toxicological tests. It therefore serves to discourage and inhibit potentially informative scientific research. It is the reason there have been no safety studies of GEO foods, no post-market monitoring, no labels, no laws, no agency coordination, and no independent review.

The regulatory system contributed to, if not created, the polar views mentioned above. Advocates point to achieved and proposed benefits and downplay assumed risks. They argue that substantial equivalence is accepted by regulators charged with assuring safety and that there is no scientific evidence of harm to anyone (Clark and Lehman, 2001; DeGregori, 2001). Those who object take the view that absence of evidence is not evidence of absence, because possible effects may take years to appear and the right studies are not being done or even planned. The claims sound very similar to those made about organo-chlorine insecticides and pesticides in general in the 1960s and 1970s and similar to the defensive claims made by manufacturers of cigarettes and polychlorinated biphenyls (PCBs) (Clark and Lehman, 2001). Caution is appropriate, given the example of chlorinated fluorocarbons (CFCs) and their well-documented effects on atmospheric ozone. They were used widely because the best scientific evidence showed they were simple, inert chemical compounds that could not cause any environmental harm. But they did. In spite of past evidence of scientific fallibility the proponents of biotechnology are sure the regulatory system provides appropriate regulation and adequate protection of the public. Experience tells us that biological entities have a way of doing unpredictable things which are not necessarily what we want them to do. Biochemists know that pyrrole evolves into porphyrin rings that create chlorophyll in plants and hemoglobin in animals. The evolution is no longer a mystery. The question was predictable before the scientific work was done. It gives one pause because the doctrine of substantial equivalence assumes that the complex chemistry will not be altered at all. It is logical to conclude that careful oversight is required. But we haven't proven to be very good at that.

Pouteau (2000) recommends inclusion of substantial equivalence as the first level of evaluation. It would be succeeded by her concept of qualitative or nonsubstantial equivalence. The concept includes country of origin, methods of production, mode of harvest, methods of conservation, and methods of processing. Some of these can be analytical but most, in her opinion, would escape detection by the substantial equivalence screen. Pouteau then advocates that products be evaluated for their ethical equivalence. This equivalence includes ethical criteria for environmental sustainability, socioeconomic acceptance (i.e., sustaining human resources), effects on wealth allocation, and socio-cultural effects (i.e., effects on freedom, respect for individual identity, respect for religious and philosophical beliefs). These, as is often true of many ethical criteria, are demanding, non-quantitative, vague standards that the scientific community and general society may resist accepting. Hubbell and Welsh (1998) claim that existing GE varieties, while potentially improving agricultural sustainability relative to existing chemical-based systems, fail to enable truly sustainable agriculture. In their view, the primary reason is that genetic traits and regulatory systems that have a higher potential for promoting sustainable agriculture have been precluded from development. A few examples:

- Development of GE crops that do not require agricultural chemicals,
- Application of GE technology to minor (orphan) crops,
- Development of traits that are self-limiting and enhance use of sustainable practices,
- Substitution of technology that is less environmentally damaging,
- Incorporation of or enhancing nitrogen fixing capability, and
- Converting annuals to perennials.

One must note that Pouteau is French and perhaps inevitably reflects the EU view that there should be a public forum for evaluation of the social consequences of any technology that has the potential to affect a lot of people (Williams, 1998). For example, it is likely that farmers with less access to credit and who grow specialty crops for smaller markets will be less able to access or benefit from GE (National Academies, 2010). The extent of these and other social effects are not well understood. European opposition to GE food appears as strong as ever, despite increasing scientific dissent. Although it seems simplistic, biotechnology applied to food can affect all of us because we all must eat. Short of the US Congress, there is no forum for evaluating the social consequences of any technology (Thompson, 2000b). In Europe, this is not true. There the social consequences of technological innovations are debated and the debate can affect a technology's fate (Thompson). In Europe, the public accepts that technological choices are simultaneously political choices and that they play a major role in determining the kind of society that results (Middendorf et al., 1998). It seems that in Europe, talk is not cheap. It can make a difference because debate about social consequences means something (Thompson). In the US, it may be just talk, albeit in good academic journals, newspapers, and magazines, but nevertheless, it is often just talk without social consequences. I am inclined to favor those who favor careful oversight. The evidence of scientific fallibility is persuasive (see Chapter 3) and supports the view that regulating systems and regulators have not been very good at providing oversight that assures the public that all is well and will continue to be so.

## Arguments in Favor of Agricultural Biotechnology

Proponents of biotechnology and genetic engineering of plants or animals see the techniques as a continuation of rather than a radical departure from what scientists have been doing for decades—manipulating the genome of plants and animals by selective breeding. Humans have been selecting plants for higher yield, greater height, drought tolerance, ease of harvest, and other desirable characteristics for a long time. Similarly, for a long time, particular animals have been chosen for breeding because they gave more milk, produced more beef, were more docile, and so forth. Now, proponents say they are just continuing to do the same things with techniques that allow changes to be achieved faster and more efficiently with better, more predictable results. The technique, in its simplest terms, is insertion of a gene

**Highlight 8.2**

The fourth in a series of farm-scale trials of genetically modified food crops (oilseed rape, sugar beet, and maize), conducted over four years, showed, once again, the GM crops may be harmful to wildlife, e.g., birds, insects, and wild flowers. The problem was not specifically with the crop or its genetic modification. It was caused by the herbicides used to control weeds. Commercial firms (e.g., BASF and Monsanto) have withdrawn and cited the lengthy, continuing, never-ending, and unfriendly regulatory struggle as the reason. Killing common broad-leaved weeds, which is exactly what the herbicides were designed to do, was cited as a major concern by environmental groups.

Elsewhere the news is quite different. Seven provinces in China now grow genetically modified poplar trees. The trees have been modified by incorporation of genes from the common bacterium *Bacillus thuringiensis* (Bt) and thus sponsor their own insect control without spraying other insecticides. Chinese scientists are also working to modify larch and walnut trees (see *The Christian Science Monitor* March 10, 2005, pp. 14, 15). This is the first large-scale growth of genetically modified trees in the world and may herald China's competitive entry into the lumber and paper industries. Perhaps because of inevitable environmental criticism and government required regulatory approval, both of which are not as critical in China, no other country has planted GE trees on this scale.

One view is that faster-growing trees that don't require intensive pest management with insecticides will produce more biomass for fuel, capture more carbon, release more oxygen, and be environmentally favorable. The Chinese are ready to capture these advantages. They are not as concerned, and perhaps not concerned at all, about pollen drift, genetic drift, or the possibility of the modified trees becoming weedy, invasive species. A value judgment has been made that the advantages outweigh the potential disadvantages and it is full speed ahead.

Chinese scientists have also shown that when two strains of genetically modified rice are grown together in the field on small farms, they produce higher yield (approximately 9%), use less pesticide, and are better for the health of the farmers (who do not use as much pesticide) when compared to farmers who grow non-GE strains of rice. One GE rice strain contains genes from *Bacillus thuringiensis* that encode a protein that paralyzes the digestive system of insects. The same gene is widely used in cotton and maize. The second rice strain uses a gene from the cowpea (*Vigna unguiculata*) that produces a protein that inhibits the activity of trypsin, a major digestive enzyme, but it acts only in insects. Full government approval has not been given, but the benefits seem to outweigh potential disadvantages.

sequence that could not have been acquired naturally. The first major achievement of importance to agriculture occurred in 1988 when Pioneer Hi-Bred patented the first viable, replicable transgenic corn plant. The point is that scientists have simply switched tools to accomplish the same goals achieved by traditional breeding techniques. What is happening is similar to what happened in agriculture when hybrid corn seed, the tractor, synthetic fertilizers, and synthetic organic chemical pesticides were developed and introduced (Bailey, 2002b). As time passed and no major disasters occurred, the public attitude (assuming the public notices agricultural developments) toward these changes shifted toward acceptance. Similar changes have occurred, albeit slowly and not unanimously, in the public's attitude toward the use and results of genetic engineering. Although the claim is common, there is no evidence that GE crops have had any adverse effects on human health but fear that they do persists. Norman Borlaug, creator of the Green Revolution in agriculture, was enthusiastic about the potential of genetic engineering of plants. He viewed the claim of risks as rubbish, whereas benefits were potentially endless. From the beginning (1944) of his career in Mexico, his moral stance was that hunger had to be addressed, feeding people could not wait. Genetic engineering, which developed after he retired, was the next step to accomplish the goal. It is what agriculture must and ought to do.

It is clear that the technology is here to stay. Few will argue that modern agricultural techniques have resulted in no harm (e.g., pesticides, fertilizers, irrigation), but those who favor agricultural biotechnology argue strongly that the benefits to date have been greater than the risks and the future potential is great. For example, present achievements include:

- Plants that contain their own insecticide eliminate the need to apply organophosphate-based insecticides.
- Plants resistant to a herbicide make weed management simpler and more complete.
- China has approved a GE maize variety, developed by Chinese researchers, that should make better pig feed. GE cotton reduced pesticide use for Chinese farmers.
- Oji of Japan, a paper manufacturer, has transferred a carrot gene into eucalyptus trees so they will grow well in acidic soil (Economist, 2004a).

Future achievements may include:

- Pigs and potatoes that produce human proteins for medical use (none have been approved).
- A vaccine for foot and mouth disease of cattle may come from alfalfa.
- Peanuts, soybeans, wheat, and other crops in which genes that produce allergenic effects have been silenced or removed.
- Cotton cultivars are being developed with improved processing characteristics (colored cotton).
- Canola with high beta (ß) carotene (an antioxidant) content.
- Plants that produce specialty chemicals—pharmaceuticals (rabies vaccine from corn).

I told students in my weed science class that if they wanted to become famous there was one certain way—discover a way to control parasitic weeds in crops. Parasitism occurs in 17 plant families. There are three important genera: witchweed (*Striga* nine species), broomrape (*Orobanche* five major and five minor species), and

Dodder (*Cuscuta* 14 species) (Parker and Riches, 1993). Witchweed is the scourge of African agriculture (Mann, 1999). Wherever parasitic weeds appear, they are devastating to crop production. There are herbicides that control parasitic weeds but most are not selective, that is, they may kill or negatively affect the crop. Scientists at the International Center for Maize and Wheat (CIMMYT) in Mexico have studied development of herbicide-resistant sorghum. In the early 1990s, Pioneer HI-Bred developed a transgenic sorghum that resisted parasitism by *Striga*. While this research is good and holds great promise there is a significant problem. Parasitic weeds infest crops in developing countries, which are often grown on small farms. The farmers are poor and cannot afford to buy expensive seeds. When seed is developed in the private rather than the public sector, developers, understandably, want to make a profit. The moral obligation to help those in need, to narrow the disparity between the citizens of developed and developing countries is, if not neglected, a low, or even absent, value. Another aspect of the problem is that the seed has to be purchased each year leading to farmer's dependence on large corporations. Proponents of GE crops see them playing a major role in solving the food production problem of developing countries. If opponents succeed in blocking further development they may very well hurt the poor of developing countries. It is, as so many issues in agriculture are, a moral dilemma. Most developed country farmers will survive and prosper regardless of the outcome of the debate. Poor farmers in developing countries may, quite literally, not survive.

Chapter 3 claimed that many of the past charges about the evils of agricultural technology have been wrong. Evidence has frequently been exaggerated to support a false claim of danger to humans or the environment. The case that modern agricultural technology has harmed humans because they must eat is much less easy to support than the claim that modern technology has harmed the environment on which agriculture depends. The environmental harm caused by modern agriculture is real and must be diminished. Regardless of whether one thinks modern agriculture is an unsustainable disaster or a system that just needs smoothing of its rough edges, Wildavsky's (1997) evidence should give us pause as we examine the claims for and against agricultural biotechnology. It is indisputable that since the early 1950s, the earth's human population has more than doubled. It is also true that there are too many people [more than 1 billion, one-seventh of the earth's people (Pappas, 2011)], who are hungry each day and probably most of the time. To permit this to continue, if it can be prevented, is wrong. However, it is also true that the world's agricultural systems now feed more people a better diet than ever before. The world's farmers feed more than twice as many people as were alive in the early 1950s. That outstanding achievement is due, in large measure, to the technology used in modern farming. Farming methods and associated technologies, so often maligned by critics, have saved millions from starvation. Technological achievements in agricultural and public health ought to be applauded because without them, many people would not be alive and more of the six billion who are now fed, would be constantly at the edge of starvation.

It surely is true that the world has more people than can be supported at a Western middle-class standard of living. The claim is beyond the scope of this book and will not be debated here. But it must be considered by those in agriculture who are committed to the moral responsibility to feed those who are here and those who will be here in the decades ahead. The foundational moral stance is—all humans have a right to eat just because they are human and alive. However, no government has a policy that ensures all will be fed, that there is a right to eat. The firm moral obligation is always secondary to economic and political realities. The agricultural community accepts that humans have a right to eat and the duty to produce food. It must be noted that acceptance of the moral responsibility to produce food for people does not include an obligation to make sure the hungry are fed. Distribution of abundance, elimination of storage losses, and food wastage are another's obligation. If grain production is used as the standard, there is enough food produced in the world annually to fulfill the basic nutritional needs of all people. The moral obligation to be sure the hungry are fed,[8] if it exists, is ignored.

A major advantage of agricultural biotechnology is that by any definition it is what one calls "cutting edge science." It is new, fascinating, and rapidly advancing. Because it is so new, one cannot know what may be discovered or what applications may appear. It is certain that the strategies, the scientific methods, for genetic manipulation that are developed will be applied to several different crops and animals (Herrera-Estrella, 2000). Genetically engineered virus, insect, or disease resistance or delayed ripening techniques could benefit several crops in the regions of the world where they are grown. Herrera-Estrella also points out an infrequently mentioned advantage of biotech crops—in theory, adopting such a crop does not require a farmer to change the farming system. Changes in the farming system may be desired, but they are hard to achieve. If yield increases or other desired advances can be achieved without major system changes, adoption will be rapid, as past adoption of technology has been (Kalaitzandonakes, 1999).

This new way of doing what has been done for years—manipulate the genome of crop plants and farm animals—will help, and in the view of many, may be essential, if the world's growing population is to be fed (Borlaug, 2001). Other than increased production, potential benefits include reductions in pesticide use from incorporation of genes for *Bacillus thuringiensis* (Bt) into several crops to control insect pests and incorporation of resistance to some herbicides into several crops to reduce the need for other, less environmentally favorable, herbicides. For example, trials of modified Bt maize in China have reported yield increases of up to 23% (Anonymous, 2003) and reduced insecticide use. China's efforts have focused primarily on development of insect and disease resistance in crops to reduce the need for pesticides and thereby increase farm income (Pearce, 2002). It must also be noted that reduction in pesticide use led to a rapid increase of non-target secondary insects.

Human need has driven development of some advances in agricultural biotechnology. A good example is incorporation of genes for β carotene production into the rice genome to make Golden Rice (Potrykus, 1999). β carotene is a precursor for the production of Vitamin A, which is essential to prevent blindness. Nash (2000)

[8] My reference and stimulus is Matthew 25:31–46. "For I was hungry, and you gave Me something to eat; I was thirsty, and you gave Me drink; I was a stranger, and you invited Me in; naked, and you clothed Me; I was sick, and you visited Me; I was in prison, and you came to Me."

reported that one-third (2+ billion) of the world's people depend on rice for 50% of their daily caloric intake. TIME's July 31, 2000 cover claimed that "This rice could save a million kids a year." Nash (2000) said that as many as 10% of the world's people are afflicted with some form of Vitamin A deficiency that can lead to blindness. There may be as many as 1 million children who die annually from Vitamin A deficiency and up to 350,000 become permanently blind. Golden Rice was created to address this world humanitarian catastrophe. To date, because one must consume large quantities of golden rice to satisfy the daily Vitamin A requirement (Recommended daily allowance—RDA = 900 mg), golden rice has not achieved the success early publicity promised. It is the first but surely not the last scientific advance to introduce or enhance the nutrient content of crops. This is unquestionably a good thing to do. As many as 30% of the world's people may suffer from iron deficiency (McGloughlin, 1999) and scientists have worked to increase the iron content of rice cultivars (Potrykus, 1999; Goto et al., 1999). Humans need some of 40 nutrients. Of these, the supply of four is critical and chronically short—iron, zinc, iodine, and Vitamin A. It has not been done, but increasing the content of these in crops would be a major contribution to those, that Fanon (1963), speaking of a different historical change, called The Wretched of the Earth.

In addition to cotton cultivars with improved processing characteristics, cotton provides a more interesting example. Cotton seed is a rich source of protein, which is usually unavailable because an extensive refining process is required to remove gossypol, a chemical toxic to mammals and chickens. Gossypol helps protect the plant from attack by insects and microbes. If gossypol is removed, as plant breeders have, there would be a cheap abundant form of protein, but the plants would be destroyed by insects. Enter genetic engineering. A group at Texas A&M University found a way to turn off the genes that stimulate production of gossypol, while the rest of the plant keeps its natural defenses. Forty-four million metric tons of cotton seed with 23% protein are produced annually. Detoxified cotton seed could meet the protein needs of 500 million people (Anonymous, 2009). Who will object to that achievement?

The promises of genetic modification are, if anything, even more spectacular. Toenniessen et al. (2003) advert that "over the past half century genetic improvement of crops (via traditional plant breeding) combined with improved agronomic practices have benefitted billions of poor people in developing countries." Plant biotechnologies are already providing millions of farmers in developing countries with increased yields, reduced insecticide cost, and fewer health risks. Benefits have been gained through the improved control of the previously uncontrollable, devastating witchweed (*Striga asiatica* L.) a parasite of maize. Better human nutrition has been achieved by increasing the mineral content of crops. We may soon see cereal and other crop cultivars that are tolerant of drought or extremes of heat or cold. Soon, we are told, there will be crop cultivars that are tolerant of (able to grow in) soils high in aluminum, alkalinity, and other previously toxic conditions. Crops might grow well in soils and under conditions where they previously could not grow or be grown with assurance of a good yield. Borlaug (2001) knew the promise of biotechnology was crops with improved fertilizer use efficiency, disease control or resistance, and nutritional quality. These could become reality, not just dreams. A major scientific achievement will be transfer of genes that make corn and small grain crops (e.g., wheat, barley, millet, sorghum) into legumes capable of making their own nitrogen. These things will benefit not only crops and farmers in the developed world, but also those in developing countries as illustrated by the work of Wambugu (2000) in Kenya, who developed, by genetic modification, sweet potatoes, resistant to feathery mottle virus, which can reduce yields 20–80% (Wambugu, 1999). Sweet potato is an important crop on small farms and a staple of the Kenyan diet. Sahai (1997) argues that one must balance the possible risks of agricultural biotechnology against the risk of doing nothing to improve food production in the world's developing countries. In their view, the risk of doing nothing is a much less defendable moral stance.

Incorporation of genes in plants that will immunize those who consume the plant product against cholera or diarrhea, common diseases in the world's developing countries (Arakawa et al., 1998; Tacket et al., 1998) is well underway but not yet commercialized. McGloughlin (1999) claims that development of edible vaccines delivered in locally grown crops that are dietary staples, could "do more to eliminate disease than the Red Cross, missionaries, and United Nations task forces combined, at a fraction of the cost." If she is correct, that claim alone is a clear moral argument to pursue biotechnology. Transgenic livestock may play a role in producing new medications for treatment of human diseases. Livestock may be modified to actually produce "recombinant proteins (including therapeutic proteins and antibodies) for treatment of human diseases" (Keefer et al., 2007). In this work, as in many similar projects, "safety and effectiveness are of paramount importance, it is also critical to establish economic feasibility." The Council for Agricultural Science and Technology (CAST, 2002) paper discusses the unique and vital potential of the work and mentions, but does not dwell on, ethical issues such as animal welfare, food safety, and environmental effects. The authors never ask whether or not the work ought to be done. The conclusion claims that "education regarding the advantages and challenges associated with this new technology is the key to public understanding" and, by implication, acceptance. It is the agricultural tactic: If the public will listen to us they will understand the value and safety of what we do and what we propose. The converse is not part of the tactic.

In the fourth century, Aristotle commented on the phenomenon known as bioluminescence. Soon it may be possible to put the enzyme luciferase into other organisms and make them glow. Some might glow brighter rather than fainter as the environment becomes more toxic. Potential ecological benefits include adding luciferase to plants and tying it to plant stress genes to allow plants to express their health. Crops might glow in areas where insects were attacking or irrigation was needed (Economist, 2011). It sounds like science fiction, but those who object on moral grounds must develop reasons why it is wrong.

Even more science fiction can be found in the goals of engineers, who are working to develop microbes that generate cheap petroleum from plant waste or engineering microbes to generate components of natural gas. Those who object, on moral grounds, must develop reasons why this is wrong.

Genetic modification introduces new proteins never before encountered by the mammalian digestive system. That may be true, but there is no evidence to support the fear such a claim creates. For example, 75% of all cheeses contain chymosin, which is, and has been for some time, produced with genetically engineered bacteria. The mammalian digestive system is quite versatile and all the proteins that have been introduced have been non-toxic and sensitive to heat, acidic and enzymatic digestion (Thompson, 2000b).

Proponents also argue that genetic engineering will reduce costs to farmers, be benign or better for the environment as opposed to modern chemical and energy intensive agricultural technology, and because of extensive testing will not be a threat to the health of humans or other species. A major development has been the spectacular rate of adoption of herbicide-resistant crops, especially in the US. They offered/promoted several advantages. Weed control is less complicated because one or two applications of a single herbicide will, in many cases, achieve nearly complete control of a broad spectrum of weeds without or with minimal crop injury (Gianessi and Carpenter, 2000). Unfortunately, the rapid spread of herbicide-resistant weeds has resulted in the opposite effect: weed control is acknowledged as the most serious management problem in GE cropping systems. Adoption of herbicide-resistant crops could improve soil and war quality because less or no tillage is required for weed control, which reduces soil erosion. The greater simplicity of weed management is attractive to farmers because management decisions are easier. Multiple, expensive applications or combinations of herbicides will not be required. Soybean growers saved more than $200 million in herbicide costs in 1998 (Gianessi and Carpenter) by using fewer, less toxic, or less persistent herbicides (Ervin et al., 2000). Farmers thereby achieved greater flexibility in coordinating weed control with other required crop production tasks. Finally, because the most common herbicide for which resistance has been engineered in several crops is glyphosate (Roundup™), there is no concern about carryover to succeeding or rotational crops. Glyphosate is not environmentally benign, but it is one of the most environmentally favorable herbicides ever developed and is much more so than the herbicides that had been used for weed control in what are now called Roundup Ready™ crops. It is rapidly and completely adsorbed to soil, has only contact activity, and no soil residual activity. It degrades rapidly with a field half-life of 47 days (Ahrens, 1994).

In the early phases of adoption of GE crops, Canada halved agricultural pesticide use. In the US, pesticide use decreased by 1–2% from 1996 to 1998. Due to herbicide use increasing 20% and 27% in 2007 and 2008 and thus an increase of 383 million pounds of herbicides applied in the US over the first 13 years of commercial use of GE crops. The increase is primarily due to the emergence and rapid spread of herbicide-resistant weeds (Benbrook, 2009, p. 3).

Skepticism about the potential of GE crops remains but many scientists and other share the opinion that they will be better for the environment and help meet the moral obligation to feed all in the twentieth century. In short, biotechnology has developed rapidly and is nowhere near reaching its potential. Scientists have not been working in the field long and it takes time to develop and bring good things to market. "The potential upside of genetic modification is simply too large to ignore—and therefore environmentalists will not ignore it. Biotechnology will transform agriculture, and in doing so will transform American environmentalism" (Rauch, 2003).

Rauch does not deny that it includes risks but claims, without identifying them, that traditional crossbreeding does also. Government regulation is required so the risks (gene escape, invasive or destructive species) can be anticipated and controlled. However, the benefits far outweigh the known and unknown risks. Rauch, echoing Avery (1995) and Waggoner (1994), suggests GE crops will be major factors in the quest to grow all the food the world needs while reducing the human ecological footprint, returning cropland to wilderness, repairing damaged soils, and restoring ecosystems. The poor in the world's developing countries will be poorer if the current hysteria of the environmentalists is allowed to prevail (DeGregori, 2001, p. 119).

Agriculture in the world's developing countries must be improved, but it does not have to be modernized, in the sense that it must become more like the presently unsustainable systems of the developed world (Chrispeels, 2000). If developing-country agriculture can be improved, with biotechnology's help, skip the high input, chemical, and energy intensive system of the developed world and proceed to a truly sustainable system, that will be a good thing all can applaud. Biotechnology will do much to satisfy what Fedoroff and Brown (2004, p. xiii) cite as the challenge of the coming decades: "To limit the destructive effects of agriculture even as we continue to coax more food from the earth." In their view, that task is made less daunting by the new knowledge of genetic modification and new methods—if used wisely. They see molecular approaches to plant and animal modification (improvement) as the best way to increase food production in an environmentally responsible way.

A good test of the acceptance of genetic modification is farmer use. The Economist (2004b, p. 64) reports that the notion that the world's farmers have to be coerced, deceived, or bribed to grow GE crops is demonstrably false. In 1997, the world grew 1.4 million hectares (ha) of GE cotton and 7.2 million ha in 2003. China authorized commercial planting of GE cotton in 1997. Chinese farmers grew 1.5 million ha in 2001 (30% of the total cotton area). China approved planting of GE rice and maize in 2010. It is important to note that both were developed by Chinese researchers, without funding or help from Western companies (Economist, 2010). India and Brazil are developing new GE crops. Only 100,000 ha of GE cotton were grown in India from 2003 to 2004. The area planted was expected to grow quickly (Economist, 2004b, p. 64). In 2009, India, a major world cotton producer, authorized use of GE cotton, developed in the public sector. When farmers see a clear benefit in dollars, yuan, or rupees, they rapidly adopt new technology including GE crops.

Many of these technological achievements have demonstrated significant improvement over previous technology, but they are also subjects of serious concerns and debate (Graff et al., 2004).

## Arguments Opposed to Agricultural Biotechnology

The case against is more nuanced than the case for agricultural biotechnology. This could be because many of the arguments in favor are promises of what is to come and those against are fears of what may already be. It could be as DeGregori (2001, p. 112) claims, that it is hard to use the scientific method to create public understanding of complex issues when one faces opposition from "skilled propagandists working for strongly motivated ideological groups." His claim implies that biotechnology issues are scientific and can be addressed and reconciled with scientific data. The view is typical of the arrogance of scientists who view all issues as scientific and claim that the science that supports their view is sound while the science others use is bad (flawed) (Kirschenmann, 2003). Perhaps the case against is as simple as the apparent fact that we humans often find it easier to be against something than to develop strong, persuasive reasons to be for it.

It could be that opposition stems from a loss of public faith in science because in spite of all that science has brought us, it also has often failed (see Chapter 3). The public's suspicion of pesticides began with *Silent Spring* (Carson, 1962), a carefully documented, thoughtful, passionate polemic against pesticides. Carson's book is widely regarded as one mark of the beginning of the US environmental movement. The public's suspicion has been reinforced by the agricultural community's defense of pesticides of all kinds and the claims that they did not harm wildlife or contaminate the environment when used according to label directions, until they had to admit pesticides did both. Agribusiness firms claimed and still claim that when pesticides are used in accordance with label directions they will not contaminate food, but they do. When nitrates were found in groundwater at harmful levels, the agricultural community claimed the source could not be fertilizer, but it was and is. Such claims were only reinforced by the more general suspicion of science that resulted from concern about thalidomide, antibiotics in meat, the herbicide atrazine in surface and well water, groundwater mining for irrigation, the effect of chlorofluorocarbons on the earth's ozone layer, and the link between mad-cow disease and Creutzfeldt-Jakob disease in humans (Kirschenmann, 2001).

Critics are aware that most (not all; see Wambugu, 2000) of the biotechnology products now available to the agricultural market are designed to enable producers to come closer to the physiological yield ceiling. Raising the ceiling has been elusive (Ruttan, 2001). Available products can be characterized as product protecting or market extending techniques rather than plant or animal physiological innovations. Many claim that if one truly wants to feed the world, it can be done best by using traditional plant breeding, improved soil and crop management techniques (Ruttan), integrated pest management systems designed for specific ecological and cropping systems, and some of the presently available GEs (e.g., Bray, 1994; Altieri, 2000). Voosen (2011) claims that molecular marker assisted breeding is becoming the conventional breeding process. It relies on inventorying clusters of genes that have been identified as having some influence on desirable traits. The new plants are not classified as GE because the first step in their development is pollination. Molecular marker breeding does not involve inserting foreign genes, but it would be impossible without the techniques of genetic engineering used to evaluate gene function. Plant breeding with molecular marking uses genetic knowledge, but does not create a GE plant.

In short, a lot of yield enhancement can be accomplished with what is known combined with new knowledge. New solutions with foreign genes are neither required nor the best way to spend limited resources. Biotechnological solutions may be good, appropriate solutions to production problems but they are, by definition, scientific/technical solutions to the important, morally just, task of feeding people. Others claim that delaying introduction of GE crops condemns many in the world to starvation. The best solutions to hunger problems, especially in developing countries should not dismiss technology but they must consider underlying social and economic problems that may be the true causes of hunger (Altieri, 2000). For example, Lappé (2002) claims that hunger is "not caused by a scarcity of food but by a scarcity of democracy." Horst (2010) and Gorman (2006) advert that what Africa needs to achieve food security is:

- Techniques for and teaching about caring for seriously degraded soil.
- Improved fertilizer supply and application equipment. Farmers in sub-Saharan Africa use an average of 8, whereas US farmers use an average of 100 kg/ha. It is discouraging to learn that Africa's fertilizer use declined 2.3% in 2008–2009 (Styslinger, 2011).
- Improve the status of women who do most of the farming of food crops.

These needs are well known. None are easy to accomplish. Success will take time just as Green Revolution technology did and it will have the added benefit of contributing to maintenance and reinforcement of essential social, financial, and biological foundations. Genetic engineering of seed is not to be abandoned, but, alone, it is not the solution to Africa's agricultural problems. It should be included as one element of Africa's agricultural development.

It could be true that those opposed to agricultural biotechnology are indeed simply anti-science zealots (Borlaug, 2001) and skilled propagandists who use scientific evidence only when it supports their cause (DeGregori, 2001, pp. 112–113). Nevertheless it is important to explore the elements of the case against biotechnology in agriculture.

Opposing, often quite passionate, arguments can be fairly characterized as progressing through three phases of what Thompson (1987) calls uncertainty arguments. The first phase raises a succession of doubts about the safety of the practice itself. Secondly one finds arguments that raise doubts about the reliability of methods for assessing risk. Finally, opponents question the motives, integrity, and reliability of scientists (the goodness of the characters) who have conducted risk analyses. Without evaluating the claims made, these kinds of arguments are illustrated by Bailey and Lappé (2002) and Altieri (2001). They include claims that agricultural biotechnology:

1. Cannot feed the world and may not even help to do so.
2. Is likely to harm human health.
3. Will harm the environment.
4. Is another in a series of technological solutions that will not lead to a sustainable agricultural system and will have negative ecological effects.
5. Will increase the role and dominance of chemical agribusinesses and their seed producing subsidiaries.

## *Feeding the World*

The primary point is that the transgenic crops have not been created to feed the world. They have been created because they were patentable and would improve profits for the industrial company that created them (Lovins and Lovins, 2000). Creation has been largely in the private realm by agribusiness companies and their association with university researchers. Development, with some exceptions (e.g., China and India) has been done primarily by the developed world's capitalist, industrial, multinational corporations that sell seed and pesticides, directly linked to new biotech products. Patenting and the fact that it is not in the immediate interest of any multinational corporation to develop products that won't lead to improved sales and profits means widespread public benefit is not an immediate or obvious goal. Patenting has led the Food and Agriculture Organization of the United Nations to warn of a growing "molecular divide" between industrial and developing countries. For example, sorghum is not a major crop in the world's developed countries as it is in Africa, where potential profits are low. Orphan crops (sorghum, millet, pigeon pea, chickpea, and others) have received little or no attention. Orphan crops do not offer significant opportunities for profit. There is an unreconciled conflict between the clear moral obligation to feed the poor and the proprietary nature and profit orientation of developers of GE crops. Such scientific endeavors are in the category of what ought to be done. They are morally good, but immediate and even long-term profit is far from assured. Multinational corporations are not primarily aid giving organizations and the direction their research has taken is neither surprising nor malevolent.

The relevant question is, will transgenic research benefit the world's poor? It is highly likely that the answer is—Yes, it will. Drought tolerance is recognized as one of the most important targets for crop improvement programs. Biotechnology has been identified as the best way to achieve the goal of maize drought tolerance. The African Agricultural Technology Foundation (AATF) is leading a public-private partnership called Water Efficient Maize for Africa (WEMA). The partnership includes national agricultural research programs in Kenya, Mozambique, South Africa, Tanzania, and Uganda, plus CIMMYT and Monsanto. Monsanto will provide proprietary germplasm, advanced breeding tools and expertise, and drought tolerance transgenes. Maize varieties developed by the project will be distributed to African seed companies through AATF, without royalties, and they will be made available to small holder farmers. In this case, the research is incomplete, but biotechnology is seen as the solution, not the problem.

The green revolution was created by public centers (the CGIAR's[9] international agricultural institutes) that focused on providing technology, without cost, to national agricultural programs for further development and local adaptation (Chrispeels, 2000). Altieri (2000) and Altieri and Rosset (1999) argue that the claims of the proponents of genetic engineering that it will increase world food production, enhance food security for all, and reduce dependence on chemicals and thereby reduce agriculture's environmental problems are false. The basis of their claim is that the world's food problems are not due to a lack of food but to poor distribution of

[9] CGIAR, The Consultative Group for International Agricultural Research. The CG system is the governing body for 15 international centers, including IRRI—The International Rice Research Institute, CIMMYT—Spanish—Centro Internacional Por Maize y Trigo—English—The International Center for Wheat and Maize, and ICRISAT—The International Center for Research in the Semi-Arid Tropics.

presently adequate production. They see serious environmental risks, little chance of helping small farmers in developing countries, and an increased dependence on pesticides, none of which, in their view, will lead to sustainable agricultural systems. Farmer's costs will rise because large, multinational seed and chemical pesticide manufacturers will control the cost and supply of both. Initially, production systems that include herbicide tolerant or insect resistant crops may well be simple and lead to lower costs. Altieri and Rosset suggest that ecological problems and pest resistance will follow and make such systems less attractive, more expensive, and less sustainable. They conclude that available agroecological approaches can solve the problems biotechnology says it will solve "in a more socially equitable and in a more environmentally harmonious manner." Altieri and Rosset do not deny that biotechnology can solve some problems but insist other more ecologically appropriate solutions are available or can be developed easily.

This view is in sharp contrast to that of Wambugu (2000) who argues that GE technology can mean higher yields for Africa and could literally weed out poverty by reducing the need for women to weed crops by hand. GE crops will increase food productivity by reducing pest problems and that will make Africa more self-sufficient and bring down the price of food. She advocates the work should be done by local scientists who understand the need and the culture. Ability is not in question, capability with limited resources is.

Golden Rice, discussed above, was created to address the fact that many people, including many children, go blind or die each year in the world's developing countries because they do not have enough Vitamin A in their diet. In contrast to the TIME article by Nash (2000), Pollan (2001) calls Golden Rice the great yellow hype. Pollan argues that Golden Rice offers much more promise to biotech companies than it does to the world's rice-eating poor. This is another case where it is too early to know for sure. Golden Rice has been created and scientists are working to incorporate the trait for production of β-carotene into other rice cultivars that have desirable yield, texture, and olfactory characteristics that are so important to rice eaters. There is also exploration of a way to make Golden Rice white, which would make it more appealing to those for whom rice must be white. These new varieties have not been released to farmers, thus it is too early to tell how great or small their contribution may be. In any case, Golden Rice cannot be a complete solution to Vitamin A deficiency in diets of poor people. The importance of rice cannot be over stated. Two to three billion people depend on rice for about 50% of their daily caloric intake. The average Asian adult rice eater may consume 300 g of rice each day. That much Golden Rice will provide 8% of the recommended daily intake of Vitamin A. Obviously it cannot solve the problem, it can only help. It will directly help people rather than just make a plant resistant to a herbicide or able to provide its own insect control. Pollan suggests Golden Rice is a solution to the industry's public relations problem. Others (Altieri, 2000; Altieri and Rosset, 1999) argue that Golden Rice is not a solution at all. The solution, in their view, is to teach people the benefits of eating what is already abundant—wild and cultivated leafy greens rich in vitamin A. These things are not part of the Asian peasant's diet but yellow rice is not either (white rice is). One will cost money, the other while not free is available at much lower cost.

### *Harm to Human Health*

The US regulatory system combines the following:

- The Animal and Plant Health Inspection Service (APHIS) determines agricultural and environmental safety of proposals for field evaluation of genetically modified organisms,
- The EPA is charged with assessing environmental risks through its mandate to regulate pesticides, and
- The FDA, which is responsible for the safety and labeling of all food and animal feed (Thompson, 2000).

Each agency uses a science-based approach, which means that a transgenic crop will be approved for the market (it can be sold, planted legally, and consumed) if there is no firm scientific evidence that it causes harm. In contrast, the European Union (EU) has adopted the precautionary approach that maintains that a transgenic plant can only be approved for the market if there is firm scientific evidence that it does *not* cause harm. The difference is that in the US, a transgenic product is innocent until proven guilty and in the EU the opposite is true; a much more demanding standard. The EU process shifts the burden of proof from the government regulatory agency (in the US, it must prove harm to deny an application) to the applicant (usually a private firm, which must prove no harm and establish benefit) (Buttel, 2003). The EU process also rejects the concept of substantial equivalence that holds that "if a new or modified food or food component is determined to be substantially equivalent to an existing food, then further safety or nutritional concerns are not expected to be significant" (OECD, 1993). The EU also has a social effect standard that goes beyond the regulatory criteria of product efficacy, human safety, and environmental and non-target species safety that are accepted by the EU and the US (Buttel, 2003).

Criticisms of the US regulatory system (Ervin et al., 2000) focus on four major deficiencies:

1. The need for more participation by ecologists.
2. An environmental agency (not an agricultural agency; USDA) should be the lead agency for ecological assessment.
3. The entire process should be more transparent to the public.
4. Values and socioeconomic criteria should have equality with scientific criteria. A value driven system of evaluation would tend to focus on questions of public welfare including human and environmental questions, relative benefits and risks, and appropriate caution (Bailey, 2002a; Buttel, 2003).

Perhaps the essence of the claims surrounding possible harm to human health is that consumers who worry about unknown, potentially dangerous technology, do not have a choice and ought to. It is clear that Americans and Europeans are both concerned about their health and ways to protect it now and in the future. Ecological effects are important in both cultures and there is a rising concern about the ever more dominant links between industry, government, and academia. Over all of these shared concerns, one must recognize the very different cultural and historical forces that exist and help to explain the different reactions to GEs in the two cultures. Japan is a third example of concern. The Japanese prohibited import of GE rice and wheat and threatened to ban all US exports. The EU has made similar threats. They do not import GE rice and are concerned that stored non-GE rice had been contaminated by experimental strains of GE rice.

There is no question that people's health can be affected by what they eat (e.g., malnourishment and resultant obesity due to eating too much of the wrong food in developed countries) and by the fact that diets inadequate in basic nutritional needs lead to poor health. The potential for allergic reactions in people exposed to new proteins in genetically modified food is of concern to many. They ask: If I do not know the origin of what I am eating how can I avoid eating what may be harmful to me? To date there have no incidents of allergic reactions to genetically modified food. This is not because of blind luck on the part of those who genetically modify crops. The lack of incidents is because scientists have progressed in identifying and categorizing protein allergens (Fedoroff and Brown, 2004, p. 189). Allergenic proteins are usually abundant in a food, survive cooking, are quite stable, and resist metabolism by stomach acids. There are several examples of modification projects that have been stopped because allergens were found, noted, and caused the project to be stopped or altered. The potential negative health effects of an allergen unknown to the consumer is not debatable.

Health, it is claimed, can also be affected negatively if there is an inability to choose what one wants to eat. There are many sets of guidelines designed to assist people in choosing a healthy diet. However, if it is not possible to tell what one is eating because the product is not labeled, some element of free choice has been eliminated. That bothers many people. The relevant question is not is the food safe. It is, because I cannot tell, do I want to eat this food? Thompson (2002b) points out that for many people, eating is not just an instrumental act (I must eat to live), it is a spiritual act that ties one to the community, the producer, and the material world. For some, eating is a sacramental act, a religious rite. One does not just eat, one must eat correctly and give thanks (Thompson, 2010; Chapter 6).

One of the basic tenets of a democratic society is that citizens must be consulted. They should have a choice about how they are to be governed and how the society is to be managed. In a democracy, citizens are not subject to the whim of a dictator, they are consulted because their opinions matter. What one eats ought to be a matter of choice and the decision is not purely peer-reviewed science (Thompson, 2002b) which may show that a food is substantially equivalent to another food and therefore OK to eat. Citizens of a democracy are quite sure they have certain inalienable rights and choosing what they want to eat is among them. If I don't want to eat fast-food because I think (I could be wrong) that such food is harmful to my health, I don't have to. I can, in Thompson's words, opt out of that choice. It is my right to do so. With transgenic crops and animals there is no ability to opt out because such products are not labeled. The EU's regulators regard labeling of all transgenic foods as the right measure to protect public health and choice, whereas the US opposes labeling as being too costly, unneeded, and impractical (see Hilleman, 2002, Sherlock and Morrey, 2002, and Halloran and Hansen, 2002). Proponents argue that labeling is not required because the products are not different; they are substantially equivalent. They may be right, but I may still want to be free to choose what I will eat and I cannot if it is not labeled. As Thompson (2002b) notes, "the right of exit is a necessary condition of a just society." The right does not extend to a personal ability to demand that my society provide me with anything I want. But if I am required to do what I may not choose to do because the freedom of choice has been taken away, that is a moral wrong. The right of exit, the ability to opt out, should not demand that all produce scientific evidence of no danger. The right means only that as a citizen of a democratic society we all have been given the privilege to choose based on religious, political, and cultural considerations and these need not be based, indeed, by definition, they cannot be based on scientific evidence. The freedom to choose is fundamental in a democratic society and a refusal to label based on scientific evidence or political decisions that may be unknown to consumers, not understood if available, or unavailable, denies that freedom. A key problem that characterizes debates about agricultural technologies and is part of the biotechnology debate is not simply whether to accept or reject it. The problem is knowing who is making the decisions that may affect me.[10]

## Harm to the Environment

The fear is often expressed by the word Frankenfood. This term expresses the public's horror about the monster created by Dr. Frankenstein, who was incapable or unwilling to take responsibility for what he had created. Based on what has happened so far, it seems unlikely that genetically modified foods will be unsafe to eat. A great fear is environmental. What might happen if a genetically modified fish or plant escaped and started breeding in the wild? The answer may be nothing untoward will happen. No really bad effects on humans have happened so far. No one knows and no one can predict with 100% reliability what the future holds. However, it is not unreasonable to conclude that careful monitoring of field practices and post-market studies may be key to determining what happens. It is reasonable to conclude that under the present regulatory regime, if there is harm to humans, other species, or the environment the industrial companies that developed the technology will not be held responsible and unfortunate consequences will be treated as externalities.

Most transgenic crops (71%) have been planted in the US, 17% in Argentina, and 10% in Canada (Ervin et al., 2000). Ervin et al. claimed that 40 transgenic crops had been planted in eight developed and four developing countries. More than 20% of the area was soybeans. Potential risks include uncontrolled gene flow via pollen transfer to wild relatives, transfer of incorporated resistance (herbicide, insect, or disease) to wild relatives or weeds from the same genus, gene transfer to microorganisms, a reduction of *in situ* crop genetic diversity, dispersal and invasion of GE plants into ecosystems and adverse effects on unrelated, perhaps beneficial, non-target organisms (Ervin et al., 2000; Kremer, 2004). Few large-scale field studies have been done to evaluate and verify these risks. Firbank and Forcella (2000) did a careful study in the UK and in Minnesota to ask if genetically modified crops affect biodiversity.

[10]Thompson, P.B. and K. David. 2005. Unpublished book prospectus—What can Nano Learn From Bio?

They accepted that normal farming practices have led to a decline in farmland biodiversity, especially in the UK. The advent of genetically modified crops has led to a decrease in weed diversity, but some weed species escape control by germinating after a contact herbicide has been applied and others are tolerant of the herbicide. They studied weed seed production as it might affect bird populations and concluded that "it was simply too soon to tell." Their conclusion justifies some of the fear of those who object to GE technology—there has not been enough time to determine the risks. Answers to important questions, such as effects on wildlife, cannot be obtained quickly. Gains from use of genetically modified crops are private, whereas losses reduced food production and environmental contamination are public.

A recent report (not a published, peer-reviewed scientific study) supports the argument that it is too soon to tell what may happen. The essence of the report (Spotts, 2004) is the radical concept that a gene or a set of genes from a single species can affect an entire ecosystem. That is, individual genes can have major effects on plant or insect communities and community composition. A few examples support the hypothesis. Schweitzer et al. (2004) reported that differing condensed tannin concentration is genetically based and is the best predictor of ecosystem-level processes. Condensed tannin inputs from foliage of different types of cottonwood trees explained 55% to 65% of the variation in soil net nitrogen mineralization. They concluded that plant genes had "strong, immediate effects on ecosystem functioning." Bailey (cited in Spotts) showed that beavers, who are well known for creating landscapes, showed a preference for cottonwoods, determined by tannin concentrations—the lower the tannin level, the more beavers liked the wood. Wimp et al. (2004) showed that the genetic diversity in cottonwoods, a dominant riparian species, affected the arthropod community of 207 species. Cottonwood genetic diversity accounted for nearly 60% of the variation in arthropod diversity and effectively structured arthropod diversity. Whitham et al. (2003) presented evidence that heritable genetic variation within dominant or keystone species has community and ecosystem consequences. Their work demonstrates what they call "extended phenotypes." They propose that there can be an effect of genes on system organization at levels higher than the population that demonstrate community heritability. These studies suggest that the greater the genetic diversity in individual cottonwoods, the greater the diversity in the insect community associated with a given tree. A reasonable conclusion is that individual genetic factors may have a more pronounced effect on communities than was heretofore believed. It thus follows, although it has not been proven, that while an inserted single gene for herbicide or insect resistance is one among millions in a complex ecosystem, it still may have profound effects. As Firbank and Forcella (2000) noted, it is too soon to know what the effects may be or what they may mean for ecosystem stability.

Ervin et al. (2000) conclude that if biotechnology is to achieve its full potential for environmental benefit and avoid human or environmental catastrophe, national and international governmental regulatory oversight is required. They recommend a "cautious approach" that includes two elements. The first is the necessity of increased investment in public research and development that, in their view, will ensure that the "neglected environmental aspects of transgenic crops receive adequate attention."

It is their view that public investment will lead to a "comprehensive monitoring system" that will achieve proper evaluation of the crops and their environmental effects. Private companies do not have incentives to invest in public goods (e.g., adequate street lighting in urban areas, mass transit systems) and will not do so unless coerced. Investment in public goods does not increase profit or return to stockholders. Public research and evaluation is required if public rather than private good is to be maximized.

Secondly, Ervin et al. (2000) advocate a change in the US regulatory process and its governing agencies. It claims that there are many demonstrable shortcomings of the present three agency system. There are gaps in regulatory coverage and a failure of the agencies "to conduct comparably rigorous reviews with original scientific data." Environmental scientists, their data, and their perspectives are often just left out.

A frequent objection is that the release of a transgenic organism into the environment is especially dangerous because the organism cannot be recalled. Plant escape via seed movement, pollen transfer, or intentional or unintentional human intervention is inevitable. Once an organism escapes, given nature's complexity, one cannot predict what will happen. Will ecological diversity be affected? Will an organism be more ecologically fit and become a serious pest or will it die because it is not? One cannot always know, so caution is wise. As Commoner (2002) points out, "What the public fears is not the experimental science but the fundamentally irrational decision to let it out of the laboratory into the world before we truly understand it." When we think we understand, it is called science. When one learns to use it, it is called technology and it is the frequently weak link between understanding and use that the public often fears. A report in the *New York Times* (Anonymous, 2004a) illustrates the concern. Genes from Roundup-resistant creeping bentgrass developed by Monsanto and Scotts migrate much farther than anyone had thought was possible. The initial estimate by Scott's scientists was pollen travel up to 1000 ft. Work by EPA scientists showed pollen travel to sentinel plants of the same species as far as 13 mi and to wild relatives nearly 9 mi away. The concern is that the herbicide resistance genes may spread to wild relatives and complicate the task (i.e., may make it impossible) of controlling weeds with Roundup (glyphosate) herbicide in many landscapes.

When transgenic plants are demonstrated to be substantially equivalent (see section on regulation above) it is not required that the applicant show that the plant "actually produces a protein with same amino acid sequence as the presumably inserted protein" (Commoner, 2002), although this seems to be a reasonable regulatory request. If the plant behaves as predicted or desired (it expresses a Bt protein or resists glyphosate) then it is substantially equivalent, even if that is not precisely described. Commoner's objection points out the implicit assumption that it is a plant's (or any organism's) genome, its DNA as expressed in its genes, that determines what it is and how it will behave. Molecular biologists have what Commoner calls a false central dogma: "molecular structure is the exclusive agent of inheritance of all living things." The dogma ignores nurture and the environment and their role not in gene creation but in gene expression.

Neither side has good, long-term evidence but both have a view on environmental release of transgenic organisms. Proponents argue that the risks are negligible and largely the creation of environmental groups that create public fear (e.g., Hindmarsh, 1991, p. 200). Critics suggest that there may be pandemics caused by widespread release of created pathogens or weeds and the possible creation of major ecological imbalances. Available evidence compiled in a study by the Council for Agricultural Science and Technology (2002) suggests that commercialized GE corn, soybean, and cotton cultivars yield environmental benefits (e.g., reduced pesticide use and water requirements, and increased yield) and are not environmental problems. The evidence to support these claims is not strong. Hamer and Anslow's study (2008) showed reduced yields for cotton and soybeans. When improved insect, disease, or weed management is obtained it is "consistent with improved environmental stewardship in developed and developing nations" (CAST, 2002).

A concern of those opposed to biotechnology is the potential for loss of genetic diversity as genetically modified crops occupy more and more of the world's arable land. Ecologists assert that large monocultural areas are ideal environments for large-scale infestations of insects, diseases, and weeds. Large monocultures also create the risk of the loss of genetic diversity as farmers abandon traditional varieties to adopt the new. The claim is the same as that which followed widespread adoption of Green Revolution varieties in the 1960s. The risk is loss of the genetic diversity, which, once lost, cannot be recreated. The international crop research centers of the CG system have a responsibility to preserve germplasm found in traditional varieties and each has done so. Large monocultures are at risk of large-scale pest infestations, but traditional germplasm is being sought and preserved whenever possible. Even though such varieties may not be planted regularly, the unique germplasm has not been lost. Skeptics worry that control will pass from developing countries, where much of the germplasm is found, to multinational corporations that will find just the genes they need to make yet another modified crop, patent them, and then sell it to farmers in the country from which the gene came. This is entirely possible but the issues raised are not simply questions about biotechnology. They are scientific (Do those who create new varieties, have so-called "breeders" rights?), political (Who controls germplasm?), and moral (Who ought to control germplasm?) questions.

## *Transgenic Technology and Sustainable Agricultural Systems*

Technology is never without its disadvantages. Australians reported the first case of glyphosate resistance with rigid ryegrass (*Lolium rigidum* Gaudin) in 1996. Initially these geographically separate reports of resistance (Table 8.2) were interesting but were not regarded as threatening. From 1996 to 1999, six cases had been reported, five with rigid ryegrass (four in Australia, one in California) and one with goosegrass in Malaysia. Horseweed [*Conyza canadensis* (L.) Cronq.] was found to be resistant to glyphosate in 2000 in Delaware. It was assumed that glyphosate resistance was a problem, but only a minor one. This changed rapidly. Through April 2011, 21 weed species in 25 US and 5 Australian states and 15 countries have been reported to be resistant to glyphosate.[11]

[11] http://www.weedscience.org/In.asp

**Table 8.2** Reports of Glyphosate Resistance

| Year | Number of Countries |
|---|---|
| 1996 | 1 |
| 1997 | 2 |
| 1998 | 1 |
| 1999 | 3 |
| 2000 | 1 |
| 2001 | 4 |
| 2002 | 7 |
| 2003 | 10 |
| 2004 | 6 |
| 2005 | 12 |
| 2006 | 20 |
| 2007 | 17 |
| 2008 | 14 |
| 2009 | 8 |
| 2010 | 11 |

Genetic modification has not used nature as a model. Nature has been viewed, as it has been in much agriculture development, as a place over which we have dominion. It is our charge, indeed our opportunity and obligation, to subdue it and thereby make it productive for human benefit (White, 1967). The task is to improve nature, not learn from it. There will be disadvantages but technical skill will solve the problems as they arise just as it has in the past. Science will march on, but it may not achieve a sustainable agricultural system. As noted earlier, Lovins and Lovins (2000) argue that our technical ability has evolved much faster than the social institutions required to deal with technical advances. There is growing evidence that transgenic technology in developed countries has not increased farmer's yields or profits, improved food quality, or improved the environment (Cox, 2002). Cox also claims, based on unpublished data from Goodman and Carson of North Carolina State University, that development of transgenic plant varieties is much more expensive (perhaps up to 10 times) than traditional plant breeding. It is, in his view, "not equipped to solve complex problems."

Not only is it unlikely that transgenic technology will feed the world, it is highly likely (Cox) that its ascendancy and higher costs will lead to the loss of small farms. The economic effects of GE on small farms or growers of non-GE crops are essentially unknown. It is also highly likely that the dominance of GE will destroy traditional plant breeding programs that have made significant contributions to feeding the world. Genetic modification has been accomplished or is under development in at least 24 crops each of which is resistant to one of nine herbicides. The dominant one is Roundup Ready™ technology from Monsanto. Concerns include the transfer of resistance to related weedy species that could become more vigorous competitors in the absence of other weedy competitors. But it is unknown if these related weeds will be more or less fit in the environment in which they must compete. Many call these plants superweeds, but they are not super in the sense of uncontrollable. They are resistant to the herbicide to which the crop is resistant but they are not resistant to other common control techniques such as cultivation or other readily available herbicides. This does not mean that engineered herbicide resistance will fail. It does mean that farmers will once again have to rely on the approach known as integrated weed management rather than on one herbicide to do all weed management. A related objection is that while the herbicides for which resistance has been engineered are comparatively environmentally benign, other companies, to maintain or increase market share, may use genetic engineering to create resistance to herbicides that are much less environmentally favorable. It is also claimed (Hindmarsh, 1991) that the incorporation of genes to create pest (insect or disease) resistance will inevitably exert strong selection pressure that favors that small segment of any pest population that is naturally resistant to the incorporated toxin. As experience with chemical pesticides has shown, this is undoubtedly true. The claim is that such selection will, because of more rapid adoption and widespread use of the resistant cultivars, cause pest resistance to occur more rapidly. More than 1000 glyphosate resistant soybean cultivars are available in the US in 1999 (Lawton, 1999). Some argue that rapid adoption of GEs has led to increased pesticide use and reliance on them, more rapid development of herbicide resistance, and more pesticide residues in food (Cummings, 2005). Hamer and Anslow (2008) claim that GE crops cost farmers and governments more money than they make. The consequence is that the technology may fail rapidly, but it will be assumed that more technical fixes will solve the problems technology caused. The technological treadmill will have to go faster just to keep up. A common response of the technological optimists is that the new genetic techniques will permit alteration of the genes themselves to make the toxin more effective (Hindmarsh, 1991). One is sure this will be possible, if it has not already been done.

It is likely that acceptance and rapid adoption of HR crops and the attention they receive by weed scientists will delay development of non-chemical weed management techniques (Duke, 1998). The University of Nebraska Agbiosafety website[12] reported that 17 crops in 2004 and 22 in 2011 had been genetically modified with 73% (2204) and 82% (2011) receiving a gene from an external source (primarily a bacterium). Nearly all of the genetic modification was done to achieve herbicide or insect resistance. These chemical- and petroleum-based crops are still reliant on certain herbicides and they will inevitably force development of resistance to those herbicides. Herbicide-resistant crops (HRCs) have led to outcrossing of the resistance trait to weeds and crops,[13] a change in the spectrum of weeds to be controlled, the development of tolerant crops, and the need for new techniques to manage resistant weeds and crop volunteers. If a sustainable agricultural system is to be achieved, many feel it cannot be done by continuing to follow the path of the present chemical, energy, and capital dependent system that has caused so many of today's agricultural problems. HRCs, being more of the same kind of technology, won't lead to sustainablility, they will only lead to more problems.

[12] http://agbiosafety.unl.edu/education.shtml (accessed June 2004 and August 2011).

[13] Conventional plantings of non-GEO crops, including corn, soybeans, canola have been found to be contaminated with low levels of DNA originating from GE varieties of the crops.

Radosevich et al. (1992) linked biology, technology, and ethics to frame questions about HRCs early in the technology's development. They proposed that if such questions were addressed early, the technology would have a more solid scientific and moral foundation. Martinez-Ghersa et al. (2003) revisited the same questions ten years later to determine their current relevance. They found that after a decade of development, HRCs (which the earlier paper called Herbicide Tolerant Crops, HTCs) had not increased crop yield, had made weed management simpler (more efficient), but had not significantly reduced weed management costs. There has been uneven adoption of the technology around the world due to rejection of the technology by some countries and the resistance of agribusiness to invest in countries where intellectual property rights are not protected. Martinez-Ghersa et al. conclude that there is still insufficient ecological information available to affirm that there is or is not ecological and agricultural harm from use of HRCs. Science is still not certain about answers to questions concerning development of herbicide-resistant weed populations, the development of so-called superweeds in other crops, transfer of inserted genetic constructs into wild relatives, and the long-term effects of reduction in species diversity on agroecosystems. In the moral realm, the answer to the question of who really benefits from and who or what may be harmed by the technology is still unknown (Martinez-Ghersa et al.). The need for continued ethical discussion exists although it and new biological questions (production of allergens or toxins, genetic erosion, effects on non-target organisms, plants modified to produce pharmaceutical products) are often secondary in the continued testing and release of new HRCs.

Herbicide-resistant crops allow farmers to use reduced or no-tillage cultural practices more effectively and eliminate use of herbicides that are more environmentally harmful than glyphosate and glufosinate, the most common herbicides to which resistance has been created. Both are applied post-emergence to the crop, are non-selective and have minimal environmental effects. But continued use of any herbicide will cause weed population shifts and speed the selection of resistant weed genotypes. Development of resistance in natural weed populations was hypothesized by Gressel and Segal (1978), but there were no reports of resistance by 1992 (Radosevich et al., 1992). As reported above, resistance has been confirmed for 21 species/biotypes in 17 countries (e.g., Powles et al., 1998; Dill et al., 2000; Lee, 2000; VanGessel, 2001; Pringnitz, 2001; Owen, 1997). Resistance is a continuing concern. In 2011 (see Chapter 3) 357 weed biotypes, 197 species (115 dicots and 82 monocots) in 430,000 fields were resistant to herbicides from 20 mode-of-action groups (see footnote 11).

Although incorporation of herbicide resistance in crops has permitted easier control of many previously recalcitrant weed species, there is concern among weed scientists about evolution (selection for) of resistance to presently useful herbicides and population shifts to naturally resistant weed species. As mentioned, gene flow from resistant crops to wild relatives or closely related species may not be important (Gressel, 2000), but should not be ignored. Several major crops do not have close relatives in the areas where they are grown (e.g., corn, soybean, cotton) and gene transfer should not be of concern. However, in 2001 native maize in Mexico (the origin of modern corn) had suffered some level of genetic pollution from US GE corn.[14] Some crops do have closely related species with which they share habitats: canola is related to several species of *Brassicaceae* weeds, potato is in the same family as *Solanaceous* weeds, wheat shares part of its genome with jointed goatgrass (*Aegilops cylindrica* Host), rice interbreeds with red rice (*Oryza sativa* L.) (Langevin et al., 1990). Glufosinate resistance has been transferred from rice to red rice (Sankula et al., 1998).

van den Belt (2003) claims that GE crops have become a symbol for "all that Europeans don't like about modern agriculture." He agrees with Hilleman (2001) that the European debate is neither simply emotional nor about food safety, it is a proxy for a larger debate about how agriculture should be done. Europeans view biotechnology as something that may destroy wilderness, and the natural environment (as exemplified by farmland) as they know it (Hilleman, 2001). Gould (1985) characterized the debate well. He contrasted "immediate and practical" with "distant and deep" issues. Immediate and practical issues included concern about "potent and unanticipated effects," many of which are mentioned above. Distant and deep concerns are clearly the EU focus. They ask, "What are the consequences, ethical, aesthetic, and practical, of altering life's fundamental geometry and permitting one species to design new creatures at will, combining bits and pieces of lineages distinct for billions of years?" Gould and the EU are not campaigning for abolition but rather for proper development and use that give adequate and proper consideration to the health, environmental, and sustainability issues discussed above.

## *The Role of Large Companies*

Biotechnology is relatively new in weed science and although HRCs have been rapidly adopted much is still to be learned about their effects on weed science and weed management techniques (Duke, 1998). The science has been market driven toward development of transgenic crops that allow use of patented broad spectrum herbicides that contribute to corporate profit (Gressel, 2000) and crop production. To date, not much has been done to use the potential of biotechnology to develop weed management systems that are not dependent on chemicals. These could include enhancing crop competitiveness for nutrients, light, or water or by exploiting natural allelopathy (Gressel). Gressel also suggests that biotechnology could be used to modify weed populations to make them less competitive and to make hypervirulent biocontrol agents that are safe but not able to spread (they are self-limiting). These innovative ideas show biotechnology's potential but such achievements may only occur when research is publicly funded rather than profit driven. Profit is not evil, but the quest for profit inevitably leads research in directions that may not be environmentally, socially or politically desirable. The primary GE weed management technology (incorporation of herbicide resistance) has changed the herbicide used but has not reduced inclusion of herbicides as the essential element of weed management programs. Similarly, no GE crops have reduced the need for petroleum-based

[14] Environmental Milestones. 2011. A chart that traces key events in the quest for sustainability from 1962 to 2010. World Watch Institute. Washington, D.C.

fertilizer to achieve (perhaps assure is a better word) yield goals. GE technology is best adapted to large scale, industrial, monocultural agriculture, which, depending on the facts one accepts and the view they represent, may or may not be the best way to feed the world, protect human health and the environment and achieve long-term sustainability. There are notable exceptions to this view (see *Arguments in favor of biotech,* above, and Borlaug, 2001; Sahai, 1997; Tonniessen et al., 2003; and Wambugu, 1999, 2000). Acceptance by big-farms has led editorial writers of the *Economist*, a magazine that traditionally supports capitalist, industrial entrepreneurs, to worry that GE crops will inevitably destroy small farms and lead to control of a significant portion of the natural world (Anonymous, 2010). They go on to say that the fact (no source is given) that "90% of the farmers growing GE crops are comparatively poor and in developing countries is sinister not salutary. Monsanto's dominance in America's soyabean market, seems to suggest a goal of world domination."

At this point it is worthy of note that in some quarters, it is fashionable to excoriate the pesticide chemical industry as one that takes advantage of farmers and the environment. My experience has been that people in the pesticide chemical industry are capitalists and idealists. They are driven by the quest for profit but are optimistic that their work may benefit the world. It has made and will continue to make significant contributions to agriculture's moral obligation to feed the world. Modern agriculture would not have achieved what it has without the aid of the research and discoveries of the pesticide chemical industry.

The agricultural view is that genetically modified crops are essential to the moral obligation to feed to world and they will be lost without technological advancements that solve the problem of weed resistance, provide clear benefits, and are part of a sustainable system. Improved technology will solve the problems technology created. An ethical foundation that guides consideration of potential health, environmental, sustainability, and social effects is absent.

## The Moral Arguments

Moral claims are clear in the agricultural biotechnology debate, but while they are always present they may not be recognized or they are ignored because it is so easy to neglect them beneath the flurry of competing empirical and economic questions. It appears that little has been learned from the similar, if not nearly identical, controversy over pesticides that began with publication of Carson's book in 1962.

To illustrate the problems with questions of what one ought to do, I turn to the wisdom often found in the Calvin and Hobbes cartoon series by Watterson (1993):

The cartoon opens with Calvin filling a balloon with water.
Calvin: "In order to determine if there is any universal moral law beyond human convention, I have devised the following test."
Calvin: "I will throw this water balloon at Susie Derkins unless I receive a sign within the next 30 seconds that this is wrong."
Calvin: pauses to note that—"It is in the universe's power to stop me. I'll accept any remarkable physical happenstance as a sign that I shouldn't do this."

Calvin: Looks up and says, "Nothing's happening, five seconds to go!"
Calvin: Streaks away in search of Susie, saying—"Time's up. That proves it. There is no moral law! Wheee!"
Calvin: With a look of vicious glee, hits Susie on the head with the water balloon.
Susie: Immediately races after the retreating Calvin who screams for help.
Calvin: Lies smashed on the ground with stars swirling about his head, and laments, "Why does the universe always give you the sign AFTER you do it?"

The cartoon clearly states the case for those opposed to the rapid adoption of agricultural biotechnology simply because we can do it. Opponents know the signs of trouble have not arrived, but they are sure they will.

For many people, the morality of biotechnology must deal with the risk of playing God. That is there are some things humans are not supposed to do, especially as Calvin points out, when we don't know or cannot anticipate the consequences. Prometheus was not supposed to give us fire and Pandora was not supposed to open her box. Other people are concerned about possible effects on human health, the health of other species, or the environment. Many questions are essentially empirical:

- Are GEOs safe to eat?
- Will GEOs harm the environment?
- Will the technology help create an agricultural system that is more or less sustainable?
- Will GE technology make it more possible to feed the world's growing population?

It is important to understand the question being asked and the value positions held by the questioner and those who offer answers. While these and similar questions appear to be primarily empirical, each has moral dimensions, which should be considered when answers are offered. Moral issues have been dealt with well by several authors cited in this chapter (e.g., Bailey and Lappé, 2002; Burkhardt, 2000, 2001; Comstock, 2000b; Gendel et al., 1990; Sherlock and Morrey, 2002, and Thompson, 2010). Without any pretense of presenting complete moral arguments, I conclude this chapter with a brief presentation of five important, unresolved moral questions related to GE technology:

1. Labeling
2. Effects on family farms
3. Academic–industry relationships
4. Transgenic pharming
5. The precautionary principle

## Labeling and Biotechnology in the US and the EU

The EU has mandated labeling of any agricultural commodity containing at least 0.5% GE content. US developers and regulators have opposed labeling and expressed regret over the EU's more restrictive stance which will affect the annual $6.3 billion EU market for US agricultural commodities. Both positions are shifting (Pew initiative, 2003). The EU appears ready to approve corn modified with the Bt gene for consumption (consumers can eat it when imported) but not for production (farmers cannot grow it) (Meller and Pollack, 2004). A significant portion of the argument centers not on danger to human health but on danger to wildlife. In the US, approximately 28% of the land is cultivated, and one can find abundant farming and separate wildlife habitat. In Europe farming and wildlife are intimately linked, for example, in the UK, 80% of the land is cultivated (Williams, 1998). Because enough studies have not been done to determine the effect on wildlife, caution is demanded before widespread adoption of transgenic crops is allowed. A risk is that by being too cautious, European companies and scientists will fall behind in the world competition for a potentially huge market. On the other hand, Monsanto, a major developer of transgenic crops, announced that it would not market the world's first herbicide-resistant wheat because of opposition, not from the EU, but from American and Canadian farmers who feared losing billions of dollars in export earnings because European and Japanese consumers would not buy transgenic rice or wheat (Pollack, 2004).

A common view among agricultural scientists is that the nasty environmentalists argue that risks presumed to be real are not understood and are alleged to threaten human health and diminish biodiversity. Even though the ensuing arguments frequently lack scientific credibility the debate has discouraged consumers from eating what some regard as perfectly safe products and has led the food industry and transgenic manufacturers to abandon promising technology. This common view misses many of the concerns that motivate public skepticism or outright rejection of biotechnology. The public concern does not always result from or is it alleviated by measurement of scientific risk. A risk-benefit analysis that is science based often cannot address the concerns that result from nonscientific questions. The public issues are frequently moral and science based answers don't immediately address moral concerns. The moral concerns include such things as (Comstock, 1989):

- What are our duties toward the natural environment?
- What are our political and economic responsibilities?
- What are our ideals and attitudes about communities?
- Is it right to respect individual rights?
- Ought we to help the disadvantaged (i.e. the poor and unemployed in developed nations and residents of developing nations)?
- Can capitalism and technology enframe nature and eliminate our dependence on the natural world?

The questions are similar and can be summarized as—What is it that we ought to do and why is it right? It is a moral question. Given the increasing public suspicion of the morality of large corporations, of agribusiness, and the awareness of the economic and social power of large businesses (see Drutman and Cray, 2004), democratization and transparency of biotechnology seem essential to gaining public approval and support (Strauss, 2003).

For any technical/scientific development issue some questions should be asked during the development process, prior to commercial release. For genetic modification in agriculture there are at least five relevant questions (Comstock, 2000a).

1. What agricultural problem does the genetic modification address?
   i. Does it actually address the problem that proponents say it will solve? Biotechnology is a product of industrial capitalism that emphasizes private profit, short-term control of nature, and neglect of short- and long-term social and environmental consequences

(Middendorf et al., 1998), if the latter are even acknowledged. Production is the problem addressed even as other problems appear. Middendorf et al. suggest this is because we live in a society that has had nearly absolute faith, but now has a growing suspicion, that all human problems may not be solved by technological advances.

2. What harm may follow?
   i. What is the nature of the harm? Who will be harmed and how is the harm distributed among those who may be affected? Proponents of biotechnology emphasize the benefits of increasing the amount and quality of food available without dealing with issues of access to the food or its distribution to those who are hungry (Middendorf et al., 1998)
3. What is known?
   i. Scientific facts must be separated from scientific opinions and what is not known should be revealed.
4. What are the options?
   i. For most agricultural problems there is an array of alternatives to biotechnology. All options should be explored and their relative advantages and disadvantages should be explored.
5. What are and what ought to be the guiding ethical principles? It is likely that the guiding ethical principles will be among those presented in Chapter 4. There is always a guiding ethical principle but it is frequently implicit. It must be made explicit.

The agricultural community has tended to focus on how agricultural biotechnology will increase production of agricultural commodities because increasing production is seen as the best, if not the only, way to feed a growing population. Secondarily the emphasis has been on efforts to improve the quality (nutritional composition, storability) of commodities, and on reducing the cost of food. Other groups, regarded as opponents of biotechnology by the agricultural community, have tended to focus on potential harm, what is known and not known, and the value of discussion of other options. Questions such as what are the potential effects on human and animal health and about effects on environmental quality are primary. These groups also ask about the distribution of effects with questions about social justice and fairness to all stakeholders and the different implications of the advent of a technology on first and third world countries (Burkhardt, 2001). It is easy for either group to give primary emphasis to its important questions and dismiss or ignore other questions. Thompson (2000a) discusses the philosophical debates surrounding rational choice and informed consent with respect to the food we must eat. Advocates of biotechnology emphasize minimization of risk which they think is already minimal and can never be completely eliminated. The more cautious suggest that in a democracy, all have a right to choose what they will consume and an assumption of probable harm is sufficient to require labeling even if the likelihood of harm is, from a scientific perspective, very low. Proponents are confident that labeling will encourage poorly informed individuals to misinterpret labels and reject food that is not harmful. In Thompson's (2000a) view, GE food, already feared by many, will be stigmatized by labeling, without any scientific basis.

For example, the developers of herbicide-resistant crops have correctly pointed out the crop production benefits of greater simplicity and surety of weed control by using herbicides that if not environmentally benign are more environmentally favorable than the herbicides they replace. Moral questions about human duties to the natural environment, political and economic responsibilities to others, and our obligation to future generations (Dekker and Comstock, 1992) have not been part of agricultural evaluation of herbicide-resistant crops.

## Effects on Family Farms

Burkhardt (2000) identifies three moral critiques of agricultural biotechnology that deal with possible effects on family farms. They are potential harm to:

1. An important political–economic entity,
2. A cherished symbol if not the embodiment of basic American moral values, and
3. A solution to long-term natural resource problems.

It is probably correct to say that the vast majority of the American public like family farms support their continued existence, and, unless they have read Berry's *The Unsettling of America -Culture and Agriculture* (1977), have no idea what is happening to them. Most Americans do not know that the idyllic family farms they envision, if they think of farms at all, are in deep economic trouble and are disappearing rapidly (see Halweil, 2004).

At the time of the American Revolution, the vast majority of Americans were farmers. In 1900, 60% of the US population lived in rural areas. America had 5.7 million farms. According to USDA data, in 2001 America had 2.1 million farms and about 0.7% of the population was engaged in full-time farming (fewer people than are in US prisons). Average farm size gradually increased to 436 acres in 2001. Nearly 1.5 million US farms are less than 80 acres and are supported by day jobs held by men and women. That is, such farms are not viable economic entities, although each produces agricultural commodities, few make enough profit to support a family. Of these farms, around 1.3 million are part-time, residential, or retirement farms and more than 60% of retail farm sales are captured by 163,000 large farms. More than 60% of the large farms are bound in some value linkage with a large corporation and are not what one would define as a family farm. It is quite within the realm of possibility that as few as 25,000 commercial farms will remain in the US by 2030. Corporate owners include petrochemical companies, restaurant chains, and urban investors (Burkhardt, 2000). Burkhardt posits that biotechnology is another step the industrialization of agriculture and if its effects were as apparent to the majority of the American public as they are to farmers and residents of rural communities supported by farming, US agricultural policies that have contributed to the demise of family farms might change.

The industrialization of agriculture is a force that is destroying family farms. Therefore, while important moral questions should be addressed they seem to be insignificant when arrayed against the forces that impel the continued development of chemical, energy, and capital intensive agriculture. From a political–economic perspective the question of whether or not policies should be developed to encourage the continued existence of family farms represents, in Burkhardt's (2000) view, "a test of the fairness or justness of democratic, market-based societies." Similarly, from a cultural or moral-value perspective, Burkhardt wonders if the values (e.g., helping others, community life, responsibility, stewardship) associated with traditional family farms continue to be viable and if the survival of family farms may serve as a test of the moral or spiritual health of modern society. Finally, with regard to moral obligations to future generations, Burkhardt asks if the environmentally sustainable practices many family farms have employed in the past can be employed in modern industrial farming and if this is a test of the legitimacy of current production practices. The family farm, Burkhardt says, is "a metaphor for the good life, ethically conceived," rather than simply something people do to make money. Biotechnology is business oriented, materialistic, and not supportive of the values inherent in family farms or of basic human needs. The demise of family farms in the US is a fact and many see it as an inevitable, but not undesirable outcome of the quest for higher production and efficiency. The family farm is viewed as an anachronistic entity that does not fit in the modern world. Family farms are economically inefficient and should be allowed to fade away because family farmers cannot or will not adapt. In the US, the view is—that is just the way it is, get used to it!

That view is correct when the family farm is seen as simply an inefficient production enterprise. It should disappear just as the corner, full-service gas station and the Mom and Pop grocery store disappeared. They could not compete in a changing economic world. However, there is another view. Many European nations see family farms as we used to see them—as the backbone of valuable rural communities. They are not just production units, they are social and cultural units that have value. US agricultural support policy rewards production, whereas policies in many European nations (e.g., Austria, France, Germany, and Switzerland) support people. It is a common European agricultural policy to regard family farms secondarily as production units. First they are essential components for maintenance of rural communities and rural life (see Goldschmidt, 1998). They are the key link in maintenance of rural communities (villages) which are regarded as important, if not essential, to maintenance of national cultural richness (Friedman, 1999, p. 239). Family farms are also supported because the farmers care for the land; farmers are stewards and are rewarded for their stewardship of the land. Their stewardship of the land is supported because, among other things, the farm landscape and the maintenance of the rural communities and their farms, assures that Austria will look like Austria is supposed to look for Austrians and tourists.

Thompson (2010, p. 86) cautions that we should not let nostalgia divert us from the important questions. Family farms may have all the virtues outlined above, but the arguments for preserving them are obliged to consider more. Achieving a sustainable food system is, as all seem to agree, a good goal. Those who think preservation of family farms is best way to achieve sustainability and feed people, must develop arguments and reasons for supporting this view. In Thompson's view, it is not sufficient to "simply assume that family farms promote environmental quality" or can feed the world. There is no compelling reason to hold fast to the view that family-values arguments can legitimately be "the stalking horse[15] for people who fear social change."

[15]A stalking horse is a horse or the figure of a horse used to conceal the hunter stalking an animal. Something used to hide one's true purpose.

## *Academic–Industry Relationships*

An early, influential publication raised ethical concerns about herbicide-resistant crops and their threat to development of sustainable agricultural systems (Goldberg et al., 1990). The report emphasized the likelihood of environmental harm and what they regarded as inappropriate allocation of funds toward biotechnology and away from programs designed to develop sustainable agricultural systems. The report also was highly critical of Land Grant Universities that were vigorously pursuing biotechnology which, in the author's view, would primarily benefit large corporations and harm family farms (see Thompson, 2000c). Goldberg et al. claimed that development of herbicide-resistant crops would not deliver what was needed: "an economically viable and sustainable agriculture that uses safe and ecologically sound pest management strategies."

Blumenthal and Campbell (2000) raised the same issue that Goldberg et al. raised. After several years, it is still not apparent to many engaged in agricultural biotechnology that the issue of effects of biotechnology research on the academic community is important. There is a long tradition of research cooperation and mutual benefit between the academic and industrial communities. In the simplest terms, industry has been a source of funds and the academy has been a source of talent. Universities accept funds and research ideas from the industrial world and return research results and ideas, which often lead to profit based on the transfer of intellectual effort from the academy. Academic scientists frequently serve as consultants to industrial firms, an activity permitted and encouraged by universities. Beginning in the 1990s, government funding for universities began to decline. Some of this may have originated in the idea developed during the Reagan Presidency (1981–1989) that those who benefit from higher education ought to pay the costs. This was a departure from the older idea that higher education was beneficial to those who received it and to society. All benefitted as more became educated. This latter view was part of the motivation for large government expenditures on the GI bill after World War II that enabled returning military people to go to college. Now the costs of doing research are high, the equipment required is expensive, and federal and state support for university research is declining. Scientists are compelled to spend a great deal of time and effort finding research support (often called proposing to work), for without it they will not be able to publish and they will therefore, perish. Research funds are often easy to obtain from supporting industries although they may be accompanied by intellectual direction, which defines specific tasks and goals. That is the source of the funds determines what will be done and the academy thereby loses some of its independence concerning what research will be done.

Blumenthal and Campbell (2000) identify several desirable ties between industry and public research institutions that have benefitted society. Graduate student education has been facilitated by the money, sponsors have much more rapid access to research results, and technology transfer to industry is enhanced. The risk that is especially apparent with biotechnology is that industry sponsors may undermine the university's reputation for objectivity (Blumenthal and Campbell). The quest for external funding has increased as government support for higher education has declined. Rhodes (2001, p. 136) reported that California built 21 new prisons and one new university between 1980 and 2000. Its university system was once regarded as the best in the world, but the state now spends more to incarcerate its criminals than it does to educate its youth. The share of California's budget that goes to higher education fell from 12.5% in 1990 to 8% in 1997 while support for prisons increased by 4.5%, an amount equal to the loss in support for higher education. Similar examples exist in several US states where declining tax revenues and fixed expenses have compelled cuts in the share of state funds available for higher education. The US has as many prisoners as it has graduate students.

These observations do not mean that good scientific research is not being done in universities. That is not true. The trend may mean that the research may be applied to a short-term goal that enhances benefit to the supporting industry. There may be a reduction in scientific openness and more direction of research toward what is popular or can be funded rather than toward what ought to be done. This does not present any immediate threat to the conduct of science in universities or to their central educational mission, but it does raise important questions about the future direction of research. Will the only questions asked be those for which a sponsor can be found? Will money or intellectual curiosity drive university research? Given the potential dominance of corporate sponsorship who will the university really serve?

## *Transgenic Pharming*

Biopharming, which uses genetically modified plants to produce pharmacuetical products is under development in many places. It promises plentiful, less expensive supplies of therapeutic proteins for disease treatment (Byrne, 2002, DeGregori, 2001, for opinions see—Kohoutek, 2004, Lamb, 2004, and Lampman, 2000). The near future holds the promise or fear of plants engineered to produce vanilla and soybeans that produce palm or coconut oil, which developed nations may no longer have to import. Bananas containing hepatitis B vaccine and Golden Rice with enhanced β carotene (the precursor of Vitamin A) production have been engineered. Plants and animals may be developed to produce specific pharmaceutical products more cheaply than they can be produced in a lab. Plants that can be grown easily in the US will be engineered to produce vanilla, palm oil, coconut oil, or one of many other plant products now produced in the tropics and imported. The scientific achievement, while enormously impressive, does not include a compelling reason to ask—Is this what ought to be done? The achievement will be good for US agriculture but will likely be bad for farmers and the economies of developing countries that are dependent on export of tropical plant products. It is very likely that if it can be done, it will be, without asking if it ought to be. If it is considered, the moral question pales in comparison to the scientific achievement and its economic benefits to ... well, that *is* the moral question.

Transgenic pharming may allow pharmaceutical companies to use plants and animals much like a laboratory. Human hemoglobin in pigs and amino acids in sheep have already been produced. An enzyme lacking in humans with the genetic disorder Pompe's disease has been produced in rabbits (Middendorf et al., 1998). Is the animal's well-being affected or is it only human need that matters? Ought we to use animals as chemical factories? If we can do it, ought we? Why? Why not? As pointed out by the *Economist* (October 2004b) the biggest hurdle for all non-food GE products is not the technology, it is public opinion.

## The Precautionary Principle

Raffensperger and Tickner (1999) defined the principle: "when an activity raises threats of harm to human health or the environment, precautionary measures should be taken even if some cause and effect relationships are not fully established scientifically."

The essence of the precautionary principle is—if one is not sure what may happen, caution is the proper course of action. In its simplest terms it is—look before you leap. It implicitly includes the admonition—Better safe than sorry. Caution prevents Calvin's inevitable problem. Kriebel et al. (2001) explain how the precautionary principle "highlights this tight problematic linkage between science and policy." It places the responsibility for ill effects on the developers of the technology—the creators of the hazard—not on its users.

> *Scientific studies can tell us something about the costs, risks, and benefits of a proposed action, but there will always be value judgments that require political decisions. The scientific data used for making policy will nearly always be limited by uncertainty. Even the best theory and data will leave much that is not known about estimates of risks, benefits, or costs. In conducting their research, scientists must make assumptions, choices, and inferences based on professional judgment and standard practices, that if not known by the public or policy makers, may make scientific results appear to be more certain and less value laden than is warranted. Although there are some situations in which risks clearly exceed benefits no matter whose values are being considered, there is usually a large gray area in which science alone cannot (and should not) be used to decide policy. In these gray areas, status quo activities that potentially threaten human and environmental health are often allowed to continue because the norms of traditional science demand high confidence in order to reject null hypotheses,*[16] *and so detect harmful effects. This scientific conservatism is often interpreted as favoring the promoters of a potentially harmful technology or activity when the science does not produce overwhelming evidence of harm. The precautionary principle, then, is meant to ensure that the public good is represented in all decisions made under scientific uncertainty. When there is substantial scientific uncertainty about the risks and benefits of a proposed activity, policy decisions should be made in a way that errs on the side of caution with respect to the environment and the health of the public.*

Kriebel et al. clearly include the presence of value/moral judgments in scientific study and the policy recommendations that may follow. Scientists have a social responsibility to consider all dimensions of their work including value judgments.

[16] There is no validity to the specific claim that two variations (treatments) of the same thing can be distinguished by a specific procedure.

Dundon (2003) is correct in his assertion that

> *it is astonishing to see serious players in agriculture maintaining that one does not have to look before leaping unless one has solid demonstration that a cost effective looking is called for. If someone wishes to impose a risk on me for his benefit, it is his task to demonstrate that the risk is minimally likely to happen. And then the choice is still mine. What part of that is hard to understand?*

The precautionary principle has only been applied in environmental policy since the 1970s and is explicitly incorporated in the environmental policies of several countries but not in US policy (Raffensperger and Barrett, 2002). The United Nations biosafety protocol, adopted in 2000, includes a precautionary approach to international trade. It requires exporters to obtain consent from destination countries before shipping GE crops. There are three essential parts:

1. If there is reason to believe that harm may result from an action or a technology and if there is scientific uncertainty about the harm, measures to anticipate and prevent harm are necessary and justified.
2. Harm to humans, other creatures, or the environment all count, but not equally. There is no uniform global policy on application of the principle. It is harm that counts most and it counts more than profit or technological innovation.
3. Openness and transparency are required as is adherence to the democratic principle that consent of the governed is required when great changes are made.

Advocates of agricultural biotechnology claim that sufficient caution is being practiced in all phases of biotechnology development. In fact, many regard the process as over-regulated and at risk because of excessive regulation. Others, of course, see the entire process as under-regulated with insufficient precaution. Only economic potential and future profit matter. People's fear of potential, but unlikely, effects on small farms, communities, and the environment will be neglected if profit can be made. These contrasting polar views are real, if a bit exaggerated. If dialogue can be conducted to bring the polar opposites closer together it may lead to resolution of many of the moral dilemmas surrounding agricultural biotechnology. A conclusion is that government should regulate to assure public and environmental health and industrial firms should accept that even if things are safe for some time, accidents will happen. Neither seems to describe reality.

Eating is a biological necessity, a daily ritual, and a cultural experience. It is something all creatures must do. Come, let us break bread together. How could people who must eat not be concerned about what and how they eat and what is being done to their food? Agricultural biotechnology has been and will continue to be a scientific success story. It remains to be seen if it will also be cultural success.

Prometheus gave us fire and we have been burned. Pandora opened her box and the only thing left in it was hope. Eve encouraged Adam to eat the fruit from the forbidden tree. Children often do exactly what they are told not to do. We are risk takers and that is often how we learn about risks and opportunities. We explore, we risk, we fail, we harm, but always we learn and try not to make the same mistakes again.

# References


Ahrens, W.H. (1994). Glyphosate. *WSSA Herbicide Handbook.* Champaign, IL, Weed Science Society of America. Pp. 149–152.

Altieri, M.A. (2000). Biotechnology: a powerful distraction from solving world hunger. *Diversity* 15:24–26.

Altieri, M. (2001). *Genetic Engineering in Agriculture: The Myths, Environmental Risks, and Alternatives.* Oakland, CA, Food First Books.

Altieri, M.A. and P. Rosset. (1999). Ten reasons why biotechnology will not ensure food security, protect the environment and reduce poverty in the developing world. *AgBioForum* 2:155–162.

Andrews, L.B. (2002). Patents, plants, and people: the need for a new ethical paradigm. Pp. 67–79 *in* B. Bailey, M. Lappé (eds.). *Engineering the Farm: Ethical and Social Aspects of Agricultural Biotechnology.* Covelo, CA, Island Press.

Anonymous. (2003). Amaizing. *Economist.* November 8, 369, p. 78.

Anonymous. (2010). Attack of the really quite likeable tomatoes. *Economist.* February:16.

Anonymous. (2009). Edible cotton. *TIME.* September 14:54.

Anonymous. (2004a). The travels of a bioengineered gene. *New York Times Editorial,* Section A. September:28.

Anonymous. (2004b). *Substantial equivalence in food safety assessment.* Council for Biotechnology Information (accessed May 6, 2004).

Arakawa, T., D.K.X. Chong, and W.H.R. Langridge. (1998). Efficacy of a food plant-based oral cholera toxin B subunit vaccine. *Nat. Biotechnol.* 16:292–297.

Avery, D. (1995). *Saving the Planet with Pesticides and Plastic: The Environmental Triumph of High-yield Farming.* Indianapolis, IN, Hudson Institute.

Bailey, B. (2002a). A societal role for assessing the safety of bioengineered foods. Pp. 113–124 *in* B. Bailey, M. Lappé (eds.). *Engineering the Farm: Ethical and Social Aspects of Agricultural Biotechnology.* Covelo, CA, Island Press.

Bailey, B. (2002b). Preface. Pp. xiii–xix *in* B. Bailey, M. Lappé (eds.). *Engineering the Farm: Ethical and Social Aspects of Agricultural Biotechnology.* Covelo, CA, Island Press.

Bailey, B. and M. Lappé (eds.). *Engineering the Farm: Ethical and Social Aspects of Agricultural Biotechnology.* Covelo, CA, Island Press.

Barker, A. and B.G. Peters. (1993). Science, policy and governments. Pages *in* A. Barker, B.G. Peters (eds.). *The Politics of Expert Advice: Creating, Using and Manipulating Scientific Knowledge for Public Policy.* Pittsburgh, PA, University of Pittsburgh Press.

Benbrook, C. (2009). Impacts of genetically engineered crops on pesticide use: The first thirteen years. The Organic Center. *Union of Concerned Scientists, Critical Issue Report.* www.organic-center.org (accessed August 2011).

Berry, W. (1981). Solving for pattern(1981). *The Gift of Good Land.* San Francisco, CA, North Point Press. Pp. 134–145.

Berry, W. (1977). *The Unsettling of America: Culture and Agriculture.* New York, Avon Books.

Blumenthal, D. and M. Campbell. (2000). Academic industry relationships in biotechnology, overview. Pp. 1–9 *in* T. Murray, M. Mehlman (eds.). *Encyclopedia of Ethical, Legal, and Policy Issues in Biotechnology.* New York, John Wiley & Sons.

Borlaug, N. (2001). Ending world hunger: the promise of biotechnology and the threat of antiscience zealotry. Pp. 25–33 *in* F.H. Buttel, R.M. Goodman (eds.). *Of Frankenfoods and Golden Rice: Risks, Rewards, and Realities of Genetically Modified Foods* 89: Transactions of Wisconsin Academy of Science, Madison, WI.

Boulter, D. (1997). Scientific and public perception of plant genetic manipulation—a critical review. *Crit. Rev. Plant Sci.* 16:231–251.

Bray, F. (1994). Agriculture for developing nations. *Sci. Am.* July:30–37.

Brower, K. (2010). The danger of cosmic genius. *Atlantic* December:48–52,54, 56, 57, 60, 62.

Burkhardt, J. (2000). Agricultural biotechnology, ethics, family farms, and industrialization. Pp. 9–17 *in* T. Murray, M. Mehlman (eds.). *Encyclopedia of Ethical, Legal, and Policy Issues in Biotechnology*. New York, John Wiley & Sons.

Burkhardt, J. (2001). The genetically modified organism and genetically modified foods debates: Why ethics matters. Pp. 63–82 *in* F.H. Buttel, R.M. Goodman (eds.). *Of Frankenfoods and Golden Rice: Risks, Rewards, and Realities of Genetically Modified Foods* Vol. 89: Transactions of Wisconsin Academy of Science.

Buttel, F.H. (2003). Internalizing the societal costs of agricultural production. *Plant Physiol.* 133:1656–1665.

Byrne, P. (2002). *Bio-Pharming*. Fort Collins, CO, Colorado State University Cooperative Extension. Crop Series—Production No. 0.307.

Carson, R. (1962). *Silent Spring*. Boston, MA, Houghton Mifflin Co.

Charles, D. (2001). Telling the story. *Trans. Wisconsin Acad. Sci.* 89:15–23.

Chrispeels, M.J. (2000). Biotechnology and the poor. *Plant Physiol.* 124:3–6.

Clark, E.A. and H. Lehman. (2001). Assessment of GE crops in commercial agriculture. *J. Agric. Environ. Ethics* 14:3–28.

Commoner, B. (2002). Unraveling the DNA myth: the spurious foundation of genetic engineering. *Harpers* 304(1821):39–47.

Comstock, G. (2000a). Brief for the royal commission on genetic modification of New Zealand 8 October 2000. AgBioView-owner@listbot.com. October 23, 2000.

Comstock, G. (1989). Genetically engineered herbicide resistance, part one. *J. Agric. Ethics* 2:263–306.

Comstock, G. (2000b). *Vexing Nature? On the Ethical Case Against Agricultural Biotechnology*. Boston, MA, Kluwer Academic Publishers.

Council for Agricultural Science and Technology (CAST). (2002). Comparative environmental impacts of biotechnology-derived and traditional soybean, corn, and cotton crops. Executive Summary, CAST, Ames, IA.

Cox, T.S. (2002). The mirage of genetic engineering. *Am. J. Altern. Agric.* 17:41–43.

CropBiotech Net, (2004). *Global status of commercialized transgenic crops: 2003*. Ithaca, NY, ISAAA. (accessed January 2004).

Cummings, C.H. (2005). Trespass: Genetic engineering as the final conquest. *World Watch Mag.* 18(1):24–35.

DeGregori, T.R. (2001). *Agriculture and Modern Technology*. Ames, IA, Iowa State University Press.

Dekker, J. and G. Comstock. (1992). Ethical and environmental considerations in the release of herbicide resistant crops. *J. Agric. Human Values* 9:31–43.

Dill, G., S. Baerson, L. Casagrande, Y. Feng, R. Brinker, T. Reynolds, N. Taylor, D. Rodriguez, and Y. Teng. (2000). Characterization of glyphosate resistant *Eleusine indica* biotypes from Malaysia. Proceedings of 3rd International Weed Science Cong. Foz do Iguazu, Brasil. Int. Weed Sci. Soc.

Domingo, J.L. (2000). Health risks of GE foods: many opinions but few data. *Science* 288:1748–1749.

Drutman, L. and C. Cray. (2004). *The People's Business: Controlling Corporations and Restoring Democracy. The Report of the Citizen Works Corporate Reform Commission.* San Francisco, CA, Berrett-Koehler Publishers.

Duke, S.O. (1998). Herbicide resistant crops—their influence on weed science. *J. Weed Sci. Technol. (Zasso-Kenkyu, Japan)* 43:94–100.

Dundon, S.J. (2003). Agricultural ethics and multifunctionality are unavoidable. *Plant Physiol.* 133:427–437.

Economist (2004a). Empty bowls, heads and pockets. July 31:12.

Economist (2004b). The men in white coats are winning, slowly. October 9:63, 64, 66.

Economist (2010). Taking root—The spread of GE crops. February 27:70–71.

Economist (2011). How illuminating. March 12:17–19.

Ervin, D.E., S.S. Batie, R. Welsh, C. l. Carpentier, J.I. Fern, N.J. Richman, and M.A. Schulz. (2000). *Transgenic Crops: An Environmental Assessment.* A Report from the H. A. Wallace Center for Agricultural and Environmental Policy at Winrock, Int. Morrilton, AR.

Fanon, F. (1963). The Wretched of the Earth. New York, Grove Press, Inc.

Fedoroff, N. and N.M. Brown. (2004). *Mendel in the Kitchen—A Scientist's View of Genetically Modified Foods*. Washington, DC, Joseph Henry Press.

Firbank, L.G. and F. Forcella. (2000). Genetically modified crops and farmland biodiversity. *Science* 289:1481–1482.

Friedman, T.L. (1999). *The Lexus and the Olive Tree*. New York, Farrrar Straus Giroux.

Gendel, S.M. A.D. Kline, D.M. Warren, F. Yates (eds.). *Agricultural Bioethics: Implications of Agricultural Biotechnology*. Ames, IA, Iowa State University Press.

Giampietro, M. (1994). Sustainability and technological development in agriculture: a critical appraisal of genetic engineering. *BioScience* 44:677–689.

Gianessi, L.P. and J.E. Carpenter. (2000). *Agricultural Biotechnology: Benefits of Transgenic Soybeans. A Report.* Washington, DC, National Center for Food and Agricultural Policy.

Goldschmidt, W. (1998). Conclusion the urbanization of rural communities. Pp. 183–198 *in* J.H. Thu, E.P. Durrenberger (eds.). *Pigs, Profits, and Rural Communities*. Albany, NY, State University of New York.

Goldberg, R., J. Rissler, H. Shand, and C. Hassebrook. (1990). *Biotechnology's bitter harvest: herbicide tolerant crops and the threat to sustainable agriculture. Biotechnology Working Group.* See WorldCat listing: http://www.worldcat.org/title/biotechnologys-bitter-harvest-herbicide-tolerant-crops-and-the-threat-to-sustainable-agriculture/oclc/21355061

Goto, F., T. Yoshihara, N. Shigemoto, S. Toki, and F. Takaiwa. (1999). Iron fortification of rice seed by the soybean ferritin gene. *Nat. Biotechnol.* 17:282–286.

Gorman, C. (2006). Seeds of hope. *TIME*. September 25:59–60.

Gould, S.J. (1985). On the origin of specious critics. *Discover*, January:34–42.

Graff, G., M. Qaim, C. Yarkin, and D. Zilberman. (2004). Agricultural biotechnology in developing countries. Pp. 417–437 *in* C.G. Scanes, J.A. Miranowski (eds.). *Perspectives in World Food and Agriculture*. Ames, IA, Iowa State Press.

Gressel, J. (2000). Molecular biology of weed control. *Transgenic Res.* 9:355–382.

Gressel, J. and C.A. Segal. (1978). The paucity of plants evolving genetic resistance to herbicides: possible reason and implications. *J. Theor. Biol.* 75:349–372.

Hahanel, S. (2005). Brewer threatens boycott. *Fort Collins Coloradoan from the Associated Press*. April 13.

Halloran, J. and M. Hansen. (2002). Why we need labeling of genetically engineered food. Pp. 221–229 *in* R. Sherlock, J.D. Morrey (eds.). *Ethical Issues in Biotechnology*. New York, Rowman and Littlefield Pub., Inc.

Halweil, B. (2004). *Eat Here: Reclaiming Homegrown Pleasures in a Global Supermarket.* New York, W. W. Norton & Co.

Hamer, E. and M. Anslow. (2008). 10 reasons why genetically modified crops cannot feed the world. *Ecologist* 38(2):43–46.

Herrera-Estrella, L.R. (2000). Genetically modified crops and developing countries. *Plant Physiol.* 124:923–925.

Hilleman, B. (2002). Differing views of the benefits and risks of agricultural biotechnology. Pp. 111–126 *in* R. Sherlock, J.D. Morrey (eds.). *Ethical Issues in Biotechnology*. New York, Rowman and Littlefield Pub., Inc.

Hilleman, B. (2001). Polarization over biotech food. *Chem. Eng. News* 79:59.

Hindmarsh, R. (1991). The flawed "sustainable" promise of genetic engineering. *Ecologist* 21:196–205.

Horst, S. (2010). Africa needs a brown (not a green) revolution. *Christian Science Monitor*. July 5:34.

Hsin, H. (2002). Bittersweet harvest—the debate over genetically modified crops. *Harvard Int. Rev.* Spring:38–41.

Hubbell, B.J. and R. Welsh. (1998). Transgenic crops: engineering a more sustainable agriculture. *Agric. Human Values* 15:43–56.

Joy, B. (2000). Why the future doesn't need us. *Wired* April:238–262.

Kalaitzandonakes, N. (1999). A farm level perspective on agrobiotechnology: how much value added and for whom? *AgBioForum* 2:61–64.

Keefer, C.L., J. Pommer, and J.M. Robl. (2007). *The Role of Transgenic Livestock in the Treatment of Human Disease*. Ames, IA, CAST.Issue Paper 35. 12 pp. Council for Agricultural Science and Technology

Kneen, B. (1999). *Farmageddon: Food and the Culture of Biotechnology*. Gabriola Island, New Society Publishers.

Kirschenmann, F. (2001). Questioning biotechnology's claims and alternatives. Pp. 35–61 *in* F.H. Buttel, R.M. Goodman (eds.). *Of Frankenfoods and Golden Rice: Risks, Rewards, and Realities of Genetically Modified Foods.* Vol. 89: Transactions of Wisconsin Academy of Science.

Kirschenmann, F. (2003). What constitutes sound science? www.leopold.iastate.edu. 15 pp.

Kirschenmann F. and G. Youngberg. (1997). *Letter from the President and the Executive Director. Annual Report, H. A. Wallace Institute for Alternative Agriculture*. Greenbelt, MD. P. 2.

Kohoutek, B. (2004). The curative crop: Colorado prepares for its first dose of plant-made pharmaceuticals. *The Rocky Mountain Bullhorn*. April 8–14:14–16. www.rockymountainbullhorn.com.

Kriebel, D., J. Tickner, P. Epstein, J. Lemons, R. Levins, E.L. Loechler, M. Quinn, R. Rudel, T. Schettler, and M. Stoto. (2001). The precautionary principle in environmetnal science. *Environ. Health Perspect.* 109:871–876.

Kremer, R. (2004). Introduction. *J. Env. Quality* 34:805.

Lamb G.M. (2004). Are there drugs in my cornflakes? *The Christian Science Monitor*. March 11:14–15.

Lampman, J. (2000). Engineering the future. *The Christian Science Monitor*. Pp. 11–13.

Langevin, S.A., K. Clay, and H.B. Grace. (1990). The incidence and effects of hybridization between cultivated rice and its related weed, red rice (*Oryza sativa* L.). *Evolution* 44:1000–1008.

Lappé, F.M. (2002). Afterword: the biotech distraction. Pp. 157–159 *in* B. Bailey, M. Lappé (eds.). *Engineering the Farm: Ethical and Social Aspects of Agricultural Biotechnology*. Covelo, CA, Island Press.

Lawton, L. (1999). Roundup of a market. *Farm Industry News*. February:4–8.

Lee, L.J. (2000). A first report of glyphosate-resistant goosegrass (*Eleusine indica* (L.) Gaertn.) in Malaysia. *Pest Manag. Sci.* 56:336–339.

Lovins, A.B. and L.H. Lovins. (2000). A tale of two botanies. *Wired.* April:247.

Mann, C. (1999). Biotech goes wild. *Technology Review.* July/August:37–43.

Martinez-Ghersa, M.A., C.A. Worster, and S.R. Radosevich. (2003). Concerns a weed scientist might have about herbicide-tolerant crops: a revisitation. *Weed Technol.* 17:202–210.

McGloughlin, M. (1999). Ten reasons why biotechnology will be important to the developing world. *AgBioForum* 2(3/4):163–174.

Meller, P. and A. Pollack. (2004). Europeans appear ready to approve a biotech corn. *New York Times.* May 15:2–3.

Middendorf, G., M. Skladany, E. Ransom, and L. Busch. (1998). New agricultural biotechnologies: the struggle for democratic choice. *Monthly Review.* July/August:85–96.

Millstone, E., E. Brunner, and S. Mayer. (1999). Beyond substantial equivalence. *Nature* 401:525–526.

Nash, M.J. (2000). Grains of hope. *TIME.* July 31:39–46.

National Academies, (2010). *The Impact of Genetically Modified Crops on Farm Sustainability in the United States. Report in Brief.* Washington, DC, The National Academies.

OECD—Organisation for Economic Cooperation and Development (1993). *Safety Evaluation of Foods Derived by Modern Biotechnology—Concepts and Principles.* Report Paris, France.

Owen, M.D.K. (1997). North American developments in herbicide-tolerant crops. Proceedings of British Crop Protection Conference. Brighton, UK British Crop Protection Council. 3:955–963.

Pappas, S. (2011). *Global Chronic Hunger Rises Above 1 billion. Vital Signs 2011—The Trends that are Shaping our Future.* Washington, DC, Worldwatch Institute. Pp. 92–95.

Parker, C. and C.R. Riches. (1993). *Parasitic Weed of the World: Biology and Control.* Wallingford, CAB International.

Pearce, F. (2002). Reaping the rewards. *New Scientist* 173:12.

Pew Initiative (2003). American's knowledge of genetically modified foods remain slow and opinions on safety still split. info@pewagbiotech.org (accessed September 18, 2003).

Pollack, A. (2004). Monsanto shelves plans for modified wheat. *New York Times.* May 11, Pp. C1 and 8.

Pollan, M. (2001). The great yellow hype. *New York Times Magazine.* March 4:13–14.

Potrykus, I. (1999). Vitamin-A and iron enriched rices may hold key to combating blindness and malnutrition: a biotechnology advance. *Nat. Biotechnol.* 17:37.

Pouteau, S. (2000). Beyond substantial equivalence: ethical equivalence. *J Agric. Environ. Ethics* 13:273–291.

Powles, S.B., D.F. Lorraine-Colwill, J.J. Dellow, and C. Preston. (1998). Evolved resistance to glyphosate in rigid ryegrass (*Lolium rigidum*) in Australia. *Weed Sci.* 46:604–607.

Pringnitz, B.A. (2001). *Issues in weed management for 2002.* Extension Pub. PM 1898. Iowa State University Extension Serv. 15 p.

Radosevich, S.R., C.M. Ghersa, and G. Comstock. (1992). Concerns a weed scientist might have about herbicide tolerant crops. *Weed Technol.* 6:635–639.

Raffensperger, C., J. Tickner (eds.). *Protecting Public Health and the Environment: Implementing the Precautionary Principle.* Washington, DC, Island Press.

Raffensperger, C. and K. Barrett. (2000). In defense of the precautionary principle. Pp. 161–163 *in* B. Bailey, M. Lappé (eds.). *Engineering the Farm: Ethical and Social Aspects of Agricultural Biotechnology.* Covelo, CA, Island Press.

Rauch, J. (2003). Will Frankenfood Save the Planet? *The Atlantic Monthly* 292 (3) October:103–108.

Rhodes, F.H.T. (2001). *The Creation of the Future: The Role of the American University.* Ithaca, NY, Cornell University Press.

Ruttan, V. (2001). Biotechnology and agriculture: A skeptical perspective. Pp. 83–92 *in* F.H. Buttel, R.M. Goodman (eds.). *Of Frankenfoods and Golden Rice: Risks, Rewards, and Realities of Genetically Modified Foods* Vol. 89: Transactions of Wisconsin Academy of Science, Madison, WI.

Sahai, S. (1997). Developing countries must balance the ethics of biotechnology against the "ethics" of poverty. *Genet. Eng. News* 17(10):4.

Sankula, S., M.P. Braverman, and J.H. Oard. (1998). Genetic analysis of glufosinate resistance in in crosses between transformed rice (*Oryza sativa*) and red rice (*Oryza sativa*). *Weed Technol.* 12:209–214.

Schweitzer, J.A., J.A. Bailey, B.J. Rehill, G.D. Martinsen, S.C. Hart, R.L. Lindroth, P. Keim, and T.G. Whitham. (2004). Genetically based trait in a dominant tree affects ecosystem processes. *Ecol. Lett.* 7:127–134.

Shelley, M. (1818). *Frankenstein*, 1st Ed. London, Lackington, Hughes, Harding, Mavor, and Jones.

Sherlock R., J.D. Morrey. (2002). Part III. Food Biotechnology. *Ethical Issues in Biotechnology.* New York, Rowman and Littlefield Publishers, Inc. Pp. 183–189.

Spotts, P.N. (2004). Life under one tree's rule? *Christian Science Monitor*, October 14:13 and 16.

Strauss, S.H. (2003). Genomics, genetic engineering, and domestication of crops. *Science* 300:61–62.

Styslinger, M. (2011). *Fertilizer Consumption Decline Sharply. Vital Signs 2011—The Trends that are Shaping our Future.* Washington, DC, Worldwatch Institute. Pp. 70–72.

Tacket, C.O., H.S. Mason, G. Losonsky, J.D. Clements, M.M. Levine, and C.J. Arntzen. (1998). Immunogenicity of a recombinant bacterial antigen delivered in transgenic potato. *Nat. Med.* 4:607–609.

Thompson, L. (2000). Are bioengineered foods safe? *FDA Consumer* 34:1–5. (accessed July 2004).

Thompson, P.B. (2010). *Agrarian Vision—Sustainability and Environmental Ethics.* Lexington, KY, University of Kentucky Press.

Thompson, P.B. (1987). Agricultural biotechnology and the rhetoric of risk: some conceptual issues. *Environ. Prof.* 9:316–326.

Thompson, P.B. (2000a). Agricultural biotechnology, ethics, food safety, risk, and individual consent. Pp. 17–26 *in* T. Murray, M. Mehlman (eds.). *Encyclopedia of Ethical, Legal, and Policy Issues in Biotechnology.* New York, John Wiley & Sons.

Thompson, P.B. (2000b). Discourse ethics for agricultural biotechnology: Its limits and its inevitability. *Sci. Eng. Ethics* 6:275–278.

Thompson, P.B. (2000c). *Food and agricultural biotechnology: incorporating ethical considerations. A Report Prepared for the Canadian Biotechnology Advisory Committee.* Available at http://www.agriculture.purdue.edu/agbiotech/Thompsonpaper/Canadathompson.html (accessed September 20, 2002).

Thompson, P.B. (2002b). Why food biotechnology needs and opt out. Pp. 27–43 *in* B. Bailey, M. Lappé (eds.). *Engineering the Farm: Ethical and Social Aspects of Agricultural Biotechnology.* Covelo, CA, Island Press.

Toenniessen, G.H., J.C. O'Toole, and J. DeVries. (2003). Advances in plant biotechnology and its adoption in developing countries. *Curr. Opin. Plant Biol.* 6:191–198.

van den Belt, H. (2003). Debating the precautionary principle: "Guilty until proven innocent" or "Innocent until proven guilty"? *Plant Physiol.* 132:1122–1126.

VanGessel, M.J. (2001). Glyphosate-resistant horseweed from Delaware. *Weed Sci.* 49:703–705.

Voosen, P. (2011). Biotech without foreign genes. *The Land Report* No. 99, Spring, Salina, KS. The Land Institute, Pp. 17–21.

Waggoner, P.E. (1994). How much land can ten billion people spare for nature? *Council for Agriculture Science and Technology Task Force Report* No. 121, Ames, IA. CAST, 64 pp.

Wambugu, F. (2000). Feeding Africa. *New Scientist* 165:40–43.

Wambugu, F. (1999). Why Africa needs agricultural biotech. *Nature* 400:15–16.

Warner, M. (1994). *Six Myths of Our Time.* New York, NY, Vintage Books. 135 pp.

Watterson, B. (1993). *Calvin and Hobbes cartoon. The Days are Just Packed.* Kansas City, KS, Andrews and McMeel. P. 118.

White, L. (1967). The historical roots of our ecological crisis. *Science* 155:1203–1207.

Whitham, T.G., W.P. Young, G.D. Martinsen, C.A. Gehring, J.A. Schweitzer, S.M. Shuster, G.M. Wimp, D.G. Fischer, J.K. Bailey, R.L. Lindroth, S. Woolbright, and C.R. Kuske. (2003). Community and ecosystem genetics: a consequence of the extended phenotype. *Ecology* 84:559–573.

Wildavsky, A. (1997). *But is it True? A Citizens Guide to Environmental Health and Safety Issues.* Cambridge, MA, Harvard University Press.

Williams, N. (1998). Agricultural biotech faces backlash in Europe. *Science* 281:768–771.

Wimp, G.M., W.P. Young, S.A. Woolbright, G.D. Martinsen, P. Keim, and T.G. Whitham. (2004). Conserving plant genetic diversity for dependent animal communities. *Ecol. Lett.* 7:776–780.

Zwart, H. (2009). Biotechnology and naturalness in the genomics era: plotting a timetable for the biotechnology debate. *J. Agric. Environ. Ethics* 22:505–529.

# 9 Alternative/Organic Agricultural Systems

*In the end they will lay their freedom at our feet*
*and say to us, 'Make us your slaves, but feed us.'*

*F. Dostoevsky,* The Brothers Karamazov, *Book 5, Chapter 5*

In 2009, *The Economist* magazine asked, Whatever Happened to the Food Crisis? In 2008–2009, farmers harvested 2.3 billion tons of cereals, the largest crop ever, supplies surged, but due to the global recession, demand stagnated. The index of food prices fell 40% between July and December. The 2009 cereal crop was only slightly less than the previous year. In 2010 yield increased 1.5%. Due to severe drought in the US southern states (e.g., Alabama, Arkansas, Colorado, Louisiana, New Mexico, Nebraska, Oklahoma, Texas), the 2011 winter wheat harvest will be at its lowest level in 5 years. Abundant grain reduces import demand and lowers international prices, which are expected to increase in 2011–2012. This exacerbates the food problem for those in at least 29 developing countries where local food insecurity, low production, continued high prices, and lack of access to food combine to cause continued hunger or famine. The rational conclusion is that the modern chemical, capital, and energy-intensive

**Highlight 9.1**

The source of agricultural scientific research funding indicates that the objectives of a doubly green revolution are highly likely to be consistent with and support the methods of modern agriculture. In 1986, of the $3.3 billion the USA invested in agricultural research, 54% came from the public and 46% from the private sector. In 2010, seed and agrochemical companies, which are often the same company, provided 72% of total agricultural research funds (State of the World, 2011). The same change has occurred in public funding of higher education. This is not at all surprising when one considers that in the 1980s state government provided about two-thirds of the cost of tuition for a full-time student and the student (and/or their parents) paid about one-third of the cost to attend a public land-grant university. Now, students pay two-thirds and the state one-third. The situation for state support of research at public universities is similar. Once upon a time, public universities were state funded, then they became state supported, now it is not an exaggeration to say they are only state located. It is not surprising that researchers and research objectives have become more closely aligned with the interests and objectives of the source of funds.

**Agriculture's Ethical Horizon. DOI: 10.1016/B978-0-12-416043-9.00009-X**

agricultural system is very successful and able to meet the world's present and future demand for food, but has failed to do so in several African and Asian countries. There is some future uncertainty due to rising oil prices and the influence of capital, but within the agricultural community, optimism prevails. *The Economist* notes that food prices are still high and there is a genuine mismatch of supply and demand. The era of cheap food may have ended. In the 1990s, it appeared that most agricultural problems were either solved or would soon be solved in developed countries. Yields were rising, all kinds of pests[1] were controlled, and fertilizers were abundant and effective.

Proponents of the highly productive modern agricultural system argue that high prices make feeding the world more difficult. Therefore, more needs to be done to increase supplies, and the way to do that is to spread modern farming to those countries that have a food deficiency. The dominant view among international development agencies, plant breeders, companies that supply agricultural inputs, and university research scientists is that the green revolution was a marvelous success and a second one is needed. The need is for what Conway (1997) calls a doubly green revolution. It must, as did the green revolution of the 1960s, be based on strong scientific support and agricultural policies that facilitate transfer of research results to food deficit nations. Many who are concerned about feeding 9 billion or more people are skeptical, if not hostile, of the claim the modern system can accomplish the goal. The system, in their view, is not sustainable. It is surely possible for modern agriculture to produce more; however, more production is necessary but not sufficient to achieve agriculture's morally accepted goal of feeding the world—ending hunger and malnutrition (De Schutter, 2011), in view of the world's growing population. Existing global hunger cannot be alleviated or eliminated unless marginalization of people and societies, inequality, and social injustice are reduced and eventually eliminated (De Schutter) and access is improved. These, although worthy goals, are not regarded as things that the agricultural community, whether alternative/organic or modern, can, or is designed to, accomplish.

The present (modern is a synonym) highly productive, but fragile and probably (perhaps surely) not ecologically sustainable or socially acceptable system is well defined. It is dominantly monocultural, capital, chemical and energy dependent, and destructive of the environment, rural communities, and small farms. Its dominant ethical stance is as mentioned above—a moral obligation to feed the world. That is to be achieved through ever-increasing production. Alternative systems are not well defined. They are variously called agroecological/ecological, biodynamic, conservative, organic, regenerative, or some combination of these terms. The several names indicate that the definitions and the goals of these systems are as complex as agriculture is. They generally share the view that modern agriculture produces food that is tasteless, nutritionally inadequate, and environmentally disastrous. However, there is little or no scientific evidence that food produced under these systems tastes better or is nutritionally better than food produced by modern agriculture. Organic food commonly has lower nitrate and protein content (Woese et al., 1997), and there is some evidence that organic vegetables are higher in antioxidants.[2]

[1] Here and elsewhere, pest includes disease organisms, insects, and weeds.

[2] Antioxidants may be produced in response to insects feeding on the plants.

The green revolution, while viewed as successful in many quarters, has failed. It has done more environmental damage and brought fewer benefits, especially to poor farmers, than projected (Economist, 2011). Pollan's Omnivore's Dilemma (2006) begins with the question: "What should we have for dinner?" In the view of advocates of a different, an alternative, agriculture system, the appropriate question is: "Will there be anything for dinner?" Those who regard modern agriculture as the only way to feed the world respond by saying that the agricultural system(s) proposed by the people who discount the achievements of modern agriculture will not and cannot feed the world. If these systems become more productive, they may feed North America or Europe, but not the world! Therefore, those who advocate abandoning the modern, very successful system will have to decide who will starve (or will compel others to decide). Criticism of modern intensive farming, it is claimed, is a privilege of those who are rich and well fed (Economist, 2011). When one considers the source of criticism, the claim is difficult to refute.

At this point, it is worth noting that Buchanan et al.'s (2010) report on agricultural productivity, published by the Council For Agricultural Science and Technology (CAST, 2010), does not offer any comments on alternative or organic agriculture. The CAST view is that the best way to ensure food for all is to support an increase in funding for agricultural science, which will ensure continued productivity of the modern agricultural system. Similarly, Federoff et al. (2010) advocate "bringing existing agronomic and food science technology and know-how to people who do not yet have it." In other words, the way to ensure all are fed is to continue to develop the technology that supports modern agriculture including genetically engineered/genetically modified (GE/GM) crops, but to emphasize decreased use of pesticides and increased use of no-till farming. Fedoroff et al. support "developing more ecologically sound farming practices"; a clear goal of alternative agricultural systems.

## Characteristics of Alternative/Organic Systems

For discussion of the relevant ethical position and technology of alternative agricultural systems, it is appropriate to begin by identifying the characteristics of each of the systems with different names, but similar objectives. There is not one alternative agriculture, there are several. Each of the five systems below is based on a holistic/biological conception of farming as opposed to a physical/chemical view (Vail, 1979). They share objectives and each title can be used as a synonym for the others. They are used independently, but their characteristics reveal their similarity. The most commonly used general terms are alternative- and agroecological/ecological-agriculture. Alternative agriculturalists regard these systems as sustainable compared to the modern agricultural system because they are designed to be low chemical input systems (but may require more labor and management)[3] that include long-term

[3] Management, deciding what to do and when and how to do it, has been simplified by modern agriculture. Farmers like GE crops because they eliminate the necessity of making difficult management decisions.

soil maintenance, avoidance of synthetic chemical pesticides and fertilizer, rotation of crops, and inclusion of animals. Participants advocate exclusion of nearly everything modern agriculture employs. All argue that modern agriculture should but does not employ technology, which is accessible, affordable, easy to use and maintain, and effective; in other words, technology that is appropriate to place and crop. Alternative systems strive to do each of these things. Although it is usually not mentioned, each of these systems reflects Schumacher's (1973) claim that "Small Is Beautiful."

1. Agroecological (ecological) systems are generally labor intensive as are all other systems. Agroecology is characterized by low high-efficiency in use of natural resources, limited dependence on synthetic fertilizers and pesticides, use of locally produced manure or compost, use of leguminous plants or trees to fertilize soil, and maintenance of a diverse cropping/livestock system that is sustainable. Advocates of agroecology have multiple objectives in addition to production. The system should achieve social (community maintenance), economic (profit), and ecological (sustainability of the environment, other creatures, soil, water, air, and so on) objectives.
2. Biodynamic farming generally means using standards that emphasize a farm's organic unity. Biodynamic farmers are encouraged to make their own fertilizer primarily by keeping animals. It advocates an ideal, diverse, integrated farm with crops and animals (Thompson, 2010, p. 213). Biodynamic farmers emphasize the uniqueness of each farm and conclude that uniform standards are not appropriate. Farmers are encouraged "to live fully in the organic spirit, including participation in activities intended to promote the biodynamic way of life" (Thompson).
3. Conservation agriculture advocates zero or minimum tillage, topsoil management to achieve preservation, crop rotations based on minimal soil disturbance, permanent soil cover, and biological pest control.
4. Organic agriculture seeks to maintain long-term soil health. It advocates use of crop rotation, green manure, composting, and biological pest control. Northbourne (1940) is credited with coining the use of "organic" as a modifier of agriculture (Darby et al., 2007). For a farm to be certified as organic, it must exclude or strictly limit use of synthetic fertilizers, pesticides, plant growth regulators, livestock antibiotics, food additives, and genetically modified crops (Buck and Scherr, 2011). One must wonder if and how the standards might be modified if some promised GM crops are developed. For example, cereal varieties may be modified to survive and even prosper under drought conditions or crop cultivars may be bred or engineered to mature more quickly independent of where they are grown (Worldwatch, 2011, p. 132).
5. Regenerative agriculture is essentially agroecological farming. It is rooted in knowledge of how to manage the complex dynamic among plants, animals, water, soil, and pests required to produce crops. The name implies the primary objective is to create a sustainable system (Buck and Scherr, 2011). Practitioners use and advocate use of crop residues and surface mulch for weed management, use of compost and green manure, growing legumes, and biological control of pests.

## The Farmers and Productivity

Each of these systems is described by users as a way to increase yield and sustain soil fertility with minimal or no dependence on outside chemicals and energy.

All reject use of soluble minerals from synthetic fertilizers and use of chemical pesticides. A primary objective is to increase economic returns to land labor and capital and satisfy household, community, and market needs. Those who choose to practice agriculture that conforms to the requirements of one or more of these definitions agree that it is an alternative, primarily organic agriculture. They are, in Thompson's (2010) view, modern agrarians. They are not simple-minded peaceniks who favor back-to-nature, live-on-10-acres, in rustic homes built with their own hands. They are not uneducated, nonvoting, antimodernity hippies who love their land and their guns. They are self-reliant farmers who strive to maintain soils, produce healthy food, and build communities. Some may be libertarians, but most accept regulations that restrict pollution and encourage matching the scale and character of a farm to ecological and cultural expectations. Inclusion of cultural expectations, and aesthetic standards, is unique to most alternative systems. Modern agrarian thought includes a moral obligation to community, neighbors, and the land. They are not simply producers of food. More than conventional farmers, agrarians considered themselves to be dependent on collaboration with nature. Their task is not to tame or transform nature, but to work with and learn from nature.

Each of the alternative agricultural systems can be highly productive per unit area (Bello, 2009, p. 139), but each is labor intensive. The seasonal requirement for abundant, hard-working, low-paid labor is a disadvantage of alternative systems. After all, those engaged in modern agriculture know, perhaps everyone knows, that they don't want to do hard farm labor and they assume no one else wants to either. Alternative modes of production may be highly productive per acre and very efficient in use of natural resources, but when labor is scarce and those who are available don't want to do farm work, alternative systems will lose in pure economic competition with modern agriculture. Buck and Scherr (2011, p. 23) suggest limiting the modern/alternative debate to whether alternative agriculture can meet the entire global food demand is incorrect. If one levels the playing field, and combines the (usually nonquantitative) potential achievements of these systems to prevent environmental degradation, secure farmer's livelihood, and decrease poverty with their productive capabilities, proponents claim they will be much more competitive. It must be noted that alternative/organic systems are not necessarily environmentally benign or neutral. Weeding, done by soil cultivation, uses polluting fossil fuel (unless one uses horses or mules), is likely to increase soil erosion, and disturb ground-dwelling creatures (Trewavas, 2001). In addition, and of more importance, using productivity/unit of land area as the sole criterion for rejecting alternative/organic agriculture, except as a niche market for rich ideologues, ignores the question of distribution (Röling, 2006). Sen (1981, pp. 162–166; see Chapter 3), in an incisive analysis of famine, concluded that starvation is characteristic of some people not having enough food to eat. It is not accompanied by there not being enough food to eat. Lack of food production and subsequent availability can cause famine and starvation, but it is one of many possible causes. Thus, whether or not alternative/organic agriculture or modern agriculture can feed the world is not simply a production question: it is or ought to be considered as a distribution question. Making the accusation and then reaching a decision based solely on productivity per unit area "is a technocratic argument that ignores the growing international evidence that institutional factors are the real bottleneck to making the most productive use of agricultural assets (Röling)." Bindraben and Rabbinge (2005) agree that distribution is the problem. "Food availability per person has increased worldwide in the last 40 years by an average of 30%." At the same time, food availability to people in African countries south of the Sahara has declined an average of 12%. Bindraben and Rabbinge argue that without restoring and maintaining the soil's nutrient balance and preventing soil degradation the production problem cannot be solved. They also claim that "currently available technologies are not sufficient to ensure sustainable development" in Africa. Technological innovations must be combined with favorable institutional and economic conditions, which can be, but have not been, created by national governments. Advanced technologies of the developed world (e.g., pesticides, GM crops, highly productive monoculture) may play a role in the future, but Bindraben and Rabbinge suggest that immediate progress can be achieved with production technologies used in alternative/organic agriculture. To wit:

- Using specific crop sequences that diminish pest presence
- Introducing legumes to improve soil nutrient soil balance
- Planting in specially shaped holes to catch rainwater
- Using plant ground covers for fertility and water retention
- Building ridges to retain rainwater and improve infiltration

Consistent with the principles of alternative agricultural systems, these techniques can be adjusted to fit local conditions and they use local resources.

It is true that for the foreseeable future, these systems cannot compete in terms of production volume or economic return with large-scale, capital, energy, and chemical-dependent (i.e., modern) agriculture, but to suggest they never will is to ignore their productive capability and their other achievements. For example, Pretty (1999) examined 286 projects in 57 developing countries, which represented 37 million hectares and found the average crop yield was 79% greater than previous production practices had achieved. In polycultures developed by small-acreage farmers in Mexico, productivity of harvestable products per unit area was higher than under single cropping with the same level of management. Yield advantages ranged from 20% to 60%, because polycultures reduced pest losses and made more efficient use of available resources (Altieri and Nicholls, 2008). A special report from the office of the United Nations High Commissioner for Human Rights says that "small-scale farmers can double food production within 10 years in critical regions by using ecological methods." The study calls for adoption of "agroecological" principles as the best way to boost food production and feed the poor[4] (De Schutter, 2011). Other work has shown that the yield of organic compared to modern/conventional systems was 68% lower for cereals and 73% lower for potatoes (Paarlberg, 2009). Trewavas (2001) suggests a conventional farm can match yields on an organic farm with 50–70% of the land. It is clear that more research is required to answer the production question. The challenge is not to be as productive as modern agriculture. If alternative/organic production systems

[4] Agroecology and the right to food. www.srfood.org and http://www2.ohchr.org/english/issues/food/annual.htm. Accessed May 2011.

are to achieve the goal of feeding the world, they will have to be at least 50% better than modern systems over the next 50 years. Otherwise they will remain ecologically sound alternatives that are only locally attractive.

The data, albeit limited and primarily from developing countries, support the claim that alternative systems can produce yields equivalent to or perhaps better than modern agricultural systems. Therefore, for many, the question of whether or not alternative systems can feed the world should be dismissed. It is resolved. They can (see Hamer and Anslow, 2008; Badgley et al., 2007). Data from carefully conducted, replicated studies that compare alternative and modern systems in places where the yields of modern systems are quite high are needed (e.g., Klepper et al., 1977). In short, dependent on whose work is believed, the answer to the question is affirmative or unknown. The debate about competing evidence has been limited, because neither side recognizes the benefits and achievements of the other. A challenge to the alternative movement is to recognize that the claims that have been and are still offered in support of alternative agriculture tend to be uncritical, anecdotal, and testimonial (Vail, 1979). This kind of argument while interesting is not credible within the scientific or agricultural communities. Thirty-three years ago, Oelhaf (1979) presented an honest interpretation of existing scientific and economic evidence. It was carefully critical of modern agriculture and strongly advocated continued research and emphasis on the positive externalities of alternative systems. His book does not specifically argue, although one must assume that he agrees with the argument, that if alternative systems are to achieve greater acceptance, the nature of their arguments must change. To be competitive and survive, they must also gain research support from universities, governmental research organizations, and states. Support will not be obtained with anecdotes.

## Transition and Advantages

The transition to an alternative agriculture, if it occurs, will be most likely if there is an affirmative answer to the production question. Such change also must have a supportive social, economic (Bello, 2009), and philosophical context, which recognizes and supports the human and environmental justification for an alternative agricultural system that produces adequate food but cooperates with and adapts to the natural world rather than dominating and harming it. Adequate production to meet future needs alone is not sufficient for the public to recognize and support alternative agriculture.

Greater public recognition of the claim for the benefits of alternative agriculture might result in a shift away from the current system of subsidizing production toward a system that provides incentives to family farmers to keep them on the land and thus support rural economies, assist with soil conservation, and the transition to sustainable farming practices. Government subsidies for production using modern agricultural techniques for some large acreage commodity crops (e.g., corn, soybean, cotton, wheat) and livestock in the USA and the European Union are still available. The fact that yields of alternative agricultural systems have been achieved with limited research funding, support from university scientists, and other research institutions is an impressive achievement (Darby et al., 2007). An indication of the lack of research attention to alternative systems is a recent CAST report (2010) on agricultural productivity for the future. It did not mention alternative/organic agriculture. Eight examples of ongoing scientific research and a ninth on new innovations, each of which promises improved worldwide productivity, were included.

In the USA, public support for all agricultural research has been steadily declining for several years. This is a challenge to all agricultural research and a particularly difficult one for supporters of alternative agricultural research. Of the $7.8 billion invested in agricultural research in the USA in 1986, 53% came from the public and 47% from the private sector. In 1992 research support from agribusiness, primarily seed and agrochemical companies, which are often the same company, increased 72% from 1979 to 1992 and was 1/3 greater than public support.

Although each alternative system has characteristics shared with other systems, they differ because the use and purpose of nature are contested (Hansen et al., 2006). They disagree about whether:

1. Nature is a partner in organic production. In contrast to culture, nature is the "parts of the world which have not been touched or influenced by humans." These places should be protected and preserved.
2. Nature has inherent biological value and there is a close relationship between it and agricultural production. Nature can be uncultivated or cultivated land. It is the force that makes crops and animals grow, and farming possible. Nature is a partner in farming, but it must be controlled. This view is similar to the modern agricultural view that nature must be tamed and controlled or agriculture will not be possible.
3. Nature is a place for relaxation of mind and body. It provides multiple sensory experiences and a place for recreational activity. This view focuses on the relationship between humans and the environment. Modern agriculturalists are familiar with and sympathetic to this view but clearly separate it from their production responsibility.

Hansen et al. suggest the question of who owns and therefore has a right to control nature is a central aspect of the debate between these different, yet similar, views. The distinction is drawn from within the alternative agricultural movement and it has no relevance to the ongoing debate about which agricultural system should be adopted: which is best. I do not suggest that proponents of modern agriculture do not consider the purpose of nature. The point is that nature's purposes are not part of the debate between the systems. A reasonable conclusion that can be drawn from Hanson et al.'s distinction is that the case to be made for alternative agriculture will be strengthened by developing a story about the virtues of alternative systems. The story must be credible, general, and inclusive of all perspectives and expectations of alternative systems. The story must meet consumer's expectations and societal demands concerning food quality. To date this challenge has not been met. Knowing and telling their story may be vital to the continued success of those who practice and promote these systems. The story should explain that sustainability is a goal that will be achieved by creating a set of agricultural practices that regard nature as a partner that has inherent value. Nature, the natural world, includes a place for activity and experience directed toward human development and maintenance of local communities, which, in turn, will maintain the social organization and culture of rural life. The alternative agriculture story should emphasize the positive aspects of alternative agriculture and compare them clearly to its disadvantages. The most frequently mentioned disadvantage is that these systems must become more productive because presently they are not able to feed North America or Europe and may never be able to feed the world. Therefore, those who are developing the story that alternative/organic agriculture can feed the world usually begin by addressing the question of yield. Hamer and Anslow (2008) provide limited evidence that yields on US farms that converted to organic production systems initially dropped 10% to 15%, but quickly recovered and became equal to or greater than farms in the same area that retained the modern system. They concluded that the UK can be fed with alternative systems if farmers adopt its methods and consumers are willing or can be convinced of the validity of the reasons proposed for changing their dietary practices. For farmers, change means accepting the methods and requirements of one of the systems described above. For consumers, change means eating food that is produced under alternative systems and, perhaps of most difficulty, reducing consumption of meat and dairy products by at least one-third and increasing consumption of vegetables. Hamer and Anslow offer the quite reasonable argument that the modern system uses about 10 calories of fossil energy to produce one calorie of food energy. Part of the story is the claim that alternative systems can reduce energy consumption by at least 25% or more for certain crops. Modern agriculturalists respond that the task of farming is to produce food, not energy. The story should also describe the advantages claimed for alternative systems:

- Reduced greenhouse gas emissions, which will reduce agriculture's effects on climate change.
- Reduced, or perhaps eliminated, pesticide use.
- Reduced effects of agriculture on the ecosystem.

They also make claims for which there is limited support and little, if any, scientific evidence. These include:

- Food produced by these systems is better for consumers, that is, it is more nutritionally adequate, less harmful and perhaps beneficial to human health.
- Alternative agricultural systems will create jobs. This is a dubious claim because few people will be willing to do the hard work required for little pay. There may be jobs, but no one will want them.

The better food argument is essentially dismissed by proponents of modern agriculture, who claim there is no credible evidence that it is true. Miller (2004) states that if consumers "want a safer, more nutritious, and more varied supply at reasonable cost" they must become better informed and "deny fringe antitechnology activists permission to speak for consumers." Miller challenges the claim of adequate production, safer food, and better nutrition, all of which may be false. He argues that our food is safer and more nutritious than it has been at any time in the past. He decries the use of the word organic. In his view all food is organic. The USDA makes no claim that organically produced food is safer or more nutritious than food produced by modern agriculture, because there is no scientific evidence that organic food is healthier or safer. Miller cites former US Secretary of Agriculture, D. Glickman, who said "the organic label is a marketing technique not a statement about food safety."

Miller disputes claims that yields of alternative systems are higher or even competitive with modern agriculture.

Alternative/organic agricultural products are now at least a $40 billion market in USA (Bello, p. 144), but only 0.04% of US farmland is organic. It is 4% in the European Union. Supermarkets now devote significant shelf space to organic brands. Many food corporations, including major multinationals, have their own organic products. Bello suggests alternative agriculture is in danger of becoming a profit center for corporations and modern large-scale agriculture and thereby lose some of its allure of health and safety; that is, its panache. However, based on present trends, consumer demand and scientific research, although still limited, are likely to increase (Ronald and Adamchak, 2008, p. 37).

Discussion of the virtues of alternative vs. modern agricultural systems often results in arguments about production and environmental benefits. The alternative/organic side claims greater food safety, better nutrition, compatibility with the environment, ecological sustainability, and proper ethical behavior (they protect communities and their associated values and are socially just). The other side says this is nonsense. But, even if these claims are true, opponents argue, the larger question remains, is any system of alternative agriculture sufficiently robust to meet present and future demand for food? Proponents claim, with limited experimental evidence, that because these systems are founded on sound ecological principles, they will alleviate environmental problems caused by modern agriculture and be able to feed the world. Proponents of modern, industrial agriculture generally do not dispute the claim of ecological sustainability, but argue we are too. However, they question the claim of social justice when it is common knowledge that the need for farm labor will increase and farm workers will still be underpaid and overworked. Are the labor relations just? Alternative agriculture commonly emphasizes ecological sustainability first and social justice second (Clarke et al., 2008, 2009). Clarke et al. also note that the ethics of organic agriculture deal with undeniably good things: caring for families and communities, healthy, tasty food, but not with a plan to achieve ecological sustainability, which they claim they have achieved, and increased production, which they claim they will achieve.

## Ethical Problems

The question that remains is: are there ethical problems with alternative systems? Is there an ethical dilemma here? The debates are dominated by statements of opposing views and rarely achieve resolution or agreement. They assume too much about the goodness of organic food and leave out or ignore the ethical dimension of alternative/organic agriculture. For example, as mentioned above, criticism of modern intensive farming is a privilege of those who are rich and well fed (*Economist*, 2011). Walsh (2009) offered a comparison of the cost of the same food items (e.g., bread, milk, apples, boneless chicken breast) produced using organic vs. conventional methods. Food produced with the organic system costs a bit more than twice as much, which demands a response to the ethical question—is a system that produces food only the rich can afford just? Is it right? Consumers tend to buy or favor organic produce when they have sufficient income, if they have a health scare (e.g., cancer), and perhaps when a baby arrives. These purchasing criteria do not imply any environmental concern, which weakens the effectiveness of a claim of virtue for diminished environmental effect. Who cares? The emphasis is not on what ought to be done, but rather on what is being done, why it is good, and why modern agriculture is bad. Good and bad are defined in the terms above, but do not include a fundamental ethical foundation.

Paarlberg (2009) notes that "all agriculture is damaging to the natural environment, without exception." Agriculture is the largest human interaction with the environment and, with few exceptions (hydroponics, vertical agriculture,[5] greenhouse crops) must be practiced out-of-doors. Paarlberg provides some data to show the green revolution—modern agriculture—increased yields in India and saved land that might have been plowed for crops. He goes on to claim that organic agriculture does not do less harm to the environment. Lower yields and a dependence on animal pasture or cover crops for fertilization require more land. He concludes, in a very short article with very little data, that on a per-unit-of-production basis, modern agriculture is superior to any alternative system in its treatment of people and the environment. The controversy continues.

Countries that employ intensive capital, energy, and chemically dependent agricultural systems for which continued yield growth is a primary objective, have met the needs and many of the wants of their citizens, but, in the view of many, they have made unsustainable demands on the ecosystem which was less valued (see Daly, 1996). Normatively, many thoughtful people believe that all societies want to adopt modern, Western agricultural methods, institutions, and their associated values, because they embody the best, most rational, most modern thinking, and they have succeeded. Western agricultural methods are regarded as essential to development. Belief in the universality of Western agriculture, its values, and culture suffers from three problems: it is false, it is immoral, and it is dangerous (Zimdahl, 2009).

Increasing production, a good thing, has been the primary, if not the only, objective of modern, Western agriculture. The agricultural community has understandably celebrated its productive achievements, but has given little or no thought or moral consideration to the effects of uneven distribution of its abundant production (see Sen, 1981). When the benefits of modern/intensive agriculture are spread unevenly, not gained by or shared with small-scale farmers or urban and rural workers, rural communities and the environment will suffer. This unintended and preventable consequence of the modern system's systems singular emphasis on increasing production raises important moral questions.

Alternative agricultural systems purport to promote and enhance biodiversity, maintain soil fertility, and gain precisely those things, the modern system does not. The unwarranted[6] but not entirely false assumption is that all modern farmers

[5] Vertical housing is a widely used solution to urban crowding. Vertical agriculture, an interesting, unproven idea to use multistory buildings as a platform for crop growth using hydroponics with supplemental lighting (Anonymous 2010).

[6] An assumption not shared by most academic agricultural scientists.

and all modern agriculture are brutal to the environment and produce bad food. An example denying the assumption that modern agriculture is harmful to soil health is the work of Peterson and Westfall (2004). Their work shows that over 12 years, an intensified cropping system had distinct advantages over the unsustainable wheat–fallow rotation of the central Great Plains region of the USA. Proper use of no-till soil cultivation resulted in an economically and environmentally sustainable system that increased soil organic carbon content by as much as 39% relative to the wheat–fallow system and increased yield and water use efficiency. However, this system is dependent on chemical pesticides that are not acceptable in alternative systems.

A basic premise of all alternative systems is minimal or no use of off-farm inputs, chemical pesticides, growth hormones, genetically modified organisms, and chemical fertilizers (see item 4 above under characteristics of systems in page 200). This raises questions about the scale of production and links between consumers and producers. Most importantly, it raises questions about and the justification for a food production system based on what are promoted as sound ecological principles. Thus, the discussion becomes philosophical as well as agronomic. It involves thought about what is right and wrong and good or bad. Ethics requires identification and discussion of foundational moral theories and scientific claims (Zimdahl, 2009). The discussion involves questioning assumptions about what is right and wrong, good and bad, and the extent and reliability of agronomic and ecological knowledge. Most people resist questioning their assumptions, they want to retain and use them. There are many assumptions in the alternative vs. modern agricultural debate. All assumptions about the productivity, sustainability, and moral justification for modern and alternative systems should be questioned. The questions and the ensuing discussion ought to proceed in a climate of openness, acceptance of varying views, and without preconceived ideas of what is right and wrong, good and bad. There is clearly an opportunity for agricultural scientists to initiate the discussion. Indeed, the willingness of any profession to engage in critical self-examination of its goals is a mark of its maturity.

## References

Altieri, M. and C. Nicholls. (2008). Scaling up agroecological approaches for food sovereignty in Latin America. *Development* 51(4):474.

Anonymous, (2010). Does it really stack up? *Economist* December:15–16.

Badgley, C., J. Moghtader, E. Quintero, E. Zakem, M.J. Chappell, K. Avilés-Vázquez, A. Samulon, and E. Perfecto. (2007). Can organic farming feed the world? *Renewable Agric. Food Syst.* 22(2):86–108. Also cited *in* T. Easton, (ed.) 2011. *Taking Sides: Clashing Views on Environmental Issues*. 14th Ed. Pp. 276–285.

Bello, W. (2009). *The Food Wars*. London, Verso.

Bindraben P. and R. Rabbinge. (2005). Development perspectives for agriculture in Africa—technology on the shelf is inadequate. *North-South Discussion Papers*. Wageningen, The Netherlands. 7 pp.

Buchanan, G., R.W. Herdt, L.G. Tweeten, and N.E. Borlaug. (2010). Agricultural productivity strategies for the future: addressing US and global challenges. *CAST Issue Paper*. Number 45:16.

Buck, L.B., and S.J. Scherr. (2011). *State of the World—Innovations That Nourish the Planet*, Pp. 15–26. New York, W.W. Norton & Company, Inc.

CAST ( 2010). Agricultural productivity strategies for the future: addressing US and global challenges. *CAST Issue Paper* No. 45, Ames, IA, CAST.

Clarke, N., P. Cloke, C. Barnett, and A. Malpass. (2009). The ethics of organic agriculture. *EurSafe News*. June. http://www.eursafe.news/org.html

Clarke, N., P. Cloke, C. Barnett, and A. Malpass. (2008). The space and ethics of organic food. *J. Rural Stud.* 24:219–230.

Conway, G. (1997). *The Doubly Green Revolution: Food for All in the 21st Century*. Ithaca, NY, Comstock Publishing Associates, a Division of Cornell University Press.

Darby, H., J. Dawson, K. Delate, W. Goldstein, J. Heckman, K. Leval, and S. Seiter. (2007). Letters to the editor in response to 'Going Organic'—Spring 2007. *Crops Soils*. Summer. p. 14.

Daly, H.E. (1996). *Beyond Growth—The Economics of Sustainable Development*. Boston, MA, Beacon Press.

De Schutter, O. (2011). Foreward(2011). *State of the World—Innovations that Nourish the Planet*. New York, W.W. Norton & Company, Inc. Pp. xvii–xix.

Economist. (2011). The 9-billion-people question: a special report on feeding the world. February 26:1, 3–16.

Economist. (2009). Whatever happened to the food crisis? July 4:57–58.

Federoff, N.V., D.S. Battisti, R.N. Beachy, P.J.M. Cooper, D.A. Fischoff, C.M. Hodges, V.C. Knauf, D. Lobell, B.J. Mazur, D. Molden, M.P. Reynolds, P.C. Ronald, M.W. Rosegrant, P.S. Sanchez, A. Vonshak, and J.K. Zhu. (2010). Radically rethinking agriculture for the 21st century. *Science* 327:833–834.

Hamer, E. and M. Anslow. (2008). 10 reasons why organic can feed the world. *The Ecologist* 38(2):43–46. Also cited *in* J.M. Colson, (ed.) 2012. *Taking sides: clashing views in food and nutrition*. 2nd Ed. Pp. 165–172.

Hansen, L., E. Now, and K. Højring. (2006). Nature and nature values in organic agriculture. An analysis of contested concepts and values among different actors in organic farming. *J. Agric. Environ. Ethics* 19:147–168.

Klepper, R., W. Lockeretz, B. Commoner, M. Gertler, S. Fast, D. O'Leary, and R. Blobaum. (1977). Economic performances and energy intensiveness organic and conventional farms in the corn belt: a preliminary comparison. *Am. J. Agric. Econ.* 59(1):1–12.

Miller, J.J. (2004). The organic myth: a food movement makes a pest of itself. *Natl. Rev.* 56:35–37. Also cited *in* T. Easton, (ed.) 2011. Taking Sides: Clashing Views on Environmental Issues. 2nd Ed. Pp. 35–37.

Northbourne, C.J., 5th Lord. (2003). *Look to the Land*. 2nd Rev., Spec. Ed., Hillsdale, NY, Sophia Perennis. First Ed. 1940. J.M. Dent & Sons.

Oelhaf, R.C. (1979). *Organic Agriculture: Economic and Ecological Comparisons with Conventional Methods*. Montclair, NJ, Allanheld, Osmun & Co.

Paarlberg, R. (2009). The ethics of modern agriculture. *EurSafe News* 11(2):3. An online journal.

Peterson, G.A. and D.G. Westfall. (2004). Managing precipitation use in sustainable dryland agroecosystems. *Ann. Appl. Biol.* 144:127–138.

Pollan, M. (2006). *The Omnivore's Dilemma—A Natural History of Four Meals*. New York, The Penguin Press.

Pretty, J. (1999). Can sustainable agriculture feed Africa? New evidence on progress, processes and impacts. *Environ. Dev. Sustain.* 1(3&4):253–274.

Röling, N. (2006). Organic agriculture and world food security. Zie en Onzin van Biologische Landbouw—An academic debate. Wageningen, Netherlands, March 23.

Ronald, P.C. and R.W. Adamchak. (2008). *Tomorrow's Table—Organic Farming, Genetics, and the Future of Food.* Oxford, UK, Oxford University Press.

Schumacher, E.F. (1973). *Small Is Beautiful—Economics as if People Mattered.* New York, Harper & Row.

Sen, A. (1981). *Poverty and Famines: An Essay on Entitlement and Deprivation.* Oxford, UK, Clarendon Press.

Thompson, P.B. (2010). *The Agrarian Vision: Sustainability and Environmental Ethics.* Lexington, KY, University Press of Kentucky.

Trewavas, A. (2001). Urban myths of organic farming. *Nature* 410:409–410.

Vail, D. (1979). The case for organic farming. Review of R.C. Oelhaf, Organic Agriculture, 1979. *Science* 205:180–181.

Walsh, B. (2009). America's food crisis and how to fix it. *TIME* August 31:31–37.

Woese, K., D. Lange, C. Boess, and K.W. Bögl. (1997). A comparison of organically and conventionally grown foods—results of a review of relevant literature. *J. Sci. Food Agric.* 74:281–293.

Worldwatch Institute, (2011). *State of the World—Innovations that Nourish the Planet.* New York, W.W. Norton & Company, Inc.

Zimdahl, R.L. (2009). Ethical merits of agricultural types—a few thoughts. *EurSafe News* 11(2):2. An online journal.

# 10 Animal Agriculture

*I am dragged along by a strange new force*
*Desire and reason are pulling in different directions*
*I see the right way and approve it, but follow the wrong.*[1]

***Ovid–Metamorphoses***

## Western Thought and the Line

Western thought has created a division and a hierarchy in the animal kingdom. There is a conceptual separation, a definite division, that leads to a moral separation between humans, those who deserve rights and moral consideration, and other animals, that do not. Humans were historically understood as different in kind, not different in degree, from other creatures. For example, humans have souls, but rats do not, nor do whales, pandas, bears, household pets, elephants, dolphins, and so on. Humans are therefore at the top of the hierarchy and, in a perfect world, always deserve moral consideration. For most people, the bottom includes creatures that live in or crawl on the ground. The creepy, slimy, yucky things: snails, slugs, scorpions, snakes. Where one draws a line of separation is an empirical, individual, but not a moral question. The line defines the division; it creates the hierarchy. Rachels (2004) points out that people who are skeptical of or reject the idea that we have moral responsibilities for animals have one of two things in mind when they ask where the line should be drawn. The first is, where should the line be drawn with respect to the kinds of animals for which we have moral responsibilities? Do my duties include only my dog, dolphins, and elephants, or am I obligated to include all the yucky creatures? The second question about where to draw the line concerns the kinds of duties the line defines. Does our duty end with doing no harm or does it include protecting from harm, assuring habitat for and even feeding animals?

Humans know they have bodies, minds, consciousness of themselves and others, and an understanding of themselves and their world. Nonhuman animals also have bodies and, in some cases, a level of consciousness, but do not have minds capable of self-awareness and symbolic abilities, such as the creation of language. Finally, plants and the creepy/crawly things have only a body. They are biological beings, but lack minds and consciousness of themselves or their world.

The central belief, the dogma, of humans is that they are singular and superior to all other life forms.[2] What we know or assume we know about ourselves and the

[1] Medea has fallen in love with Jason—she was literally struck by Eros' arrow—but her love for Jason conflicts with her duty to her father.

[2] Personal communication: G.L. Comstock, Professor of Philosophy, North Carolina State University, Raleigh, NC.

**Agriculture's Ethical Horizon. DOI: 10.1016/B978-0-12-416043-9.00010-6**

animals assures us the dogma is correct. We are of a different kind and the difference makes it easy to assume that we have rights and animals do not. Animals do not deserve moral consideration and therefore can be treated instrumentally. An important source of this belief is the Judeo–Christian idea that humans have been given dominion over animals, indeed over the earth. But the Genesis account includes an interesting, relevant distinction. Humans are called to name the animals, that is they are the masters of the animal kingdom, but that also means they are responsible for them because they are the namers, that is those who have conceptual and moral power over animals. To name something is to give them their identity and that involves relating to them by the very act of naming. Naming is not an objective, conceptually cold relationship—think of what it means to name a child. To name a child is a way of establishing a living, responsible identity and connection with the child. Naming children or animals creates a moral responsibility.[3]

The Christian tradition, but not all religious traditions, holds that man, but not other creatures, is made in God's image. Animals do not have souls and this affirms that they are lesser creatures and humans assume they can do anything they want with or to them. There is a variety of interpretations of the tradition that reach this conclusion. However, if the Genesis story is read carefully it denies the most common interpretation that gives humans license to do whatever they like to nonhuman animals; to use them for human purposes (Nussbaum, 2006): food, entertainment, hunting, beasts of burden, scientific experiments, a source of power for farmers. They have no moral standing. Moral concern, if it exists, begins and ends with the convenient, but incorrect view that duties to animals are simply indirect duties to humans. Animals, independent of where they are in the hierarchy, have no independent rights. Nussbaum argues that while we may have sympathy for how animals are treated, it is corrupted by our interest in protecting the comforts of our life that include using animals as objects for our gain and pleasure. We put humans at the top of the hierarchy because we are humans (Posner, 2004[4]). Ethical reasoning does not affect our placement.

## A Person

In US, only persons have moral status, which bestows rights to have those things that make life worthwhile: life, freedom, safety, freedom from torture. But these turn on whether the law classifies a living thing as a person or just a thing. Examples of the denial of personhood include: Jews belonged to Pharaoh, Syrians belonged to Nero, Nazis killed Jews, George Washington and Thomas Jefferson owned slaves, in Saudi Arabia women cannot vote or drive a car. *Homo sapiens* is a biological, not a moral, category. Personhood, a moral and legal category, is used incorrectly to justify separating man and the beasts.

[3] I am indebted to my colleague and friend, Dr. J.W. Boyd, Professor Emeritus of Philosophy for this insight.

[4] Interested readers are referred to a debate between Posner, a legal scholar and Singer, a philosopher in *Animal Rights—Current Debates and New Directions*. C. Sunstein and M. Nussbaum (eds.). Oxford University Press. 2004.

There are criteria for personhood, including the following:

1. In the philosophical/academic context, a person is the subject of a life and is aware of itself. A person has a concept of self. The legal definition is:

   *"Any being whom the law regards as capable of rights and duties. Any being that is so capable is a person whether a human being or not, and no being that is not so capable is a person, even though he be a man. Persons are the substances of which rights and duties are the attributes. It is only in this respect that persons possess judicial significance, and this is the exclusive point of view from which personality receives legal recognition."*

   *(Salmond, 1947[5])*

2. A person can experience pleasure (happiness) and pain (suffering) and is capable of showing feelings, for example, anger, happiness.
3. A person is sentient—has a capacity for feeling or perceiving—consciousness.
4. A person has language—the ability to communicate.
5. A person has the ability to reason.
6. A person can anticipate the future.

The common view is that animals are lesser because they do not possess any or only some of the above six characteristics[6] and therefore they ought to be treated as lesser beings. Legal scholars and philosophers agree that being human is sufficient for acquisition of legal rights (Wise, 2001). Arguments that begin with the common view, claim there is a difference between scientific facts and values and only the facts are meaningful. It is logical positivism: there is a difference between empirical (scientific) facts and value judgments and only the former are meaningful. It is consistent with a strong tendency within the scientific community to deny the meaningfulness and relevance of value questions. Science deals with facts. Ethics deals with what ought to be, with values. It is fallacious to attempt to move from facts to values. One should not attempt to, indeed one cannot, get values from facts. Facts do not provide reasons for action. Facts help us make sensible decisions but no amount of facts can make up my mind about what I ought to do (Singer, 1981, p. 75). Most, but not all humans (e.g., a baby born without a brain, a brain-dead adult kept alive by medical technology, a victim of a severe stroke who cannot move or communicate and is in a deep, apparently permanent, coma) possess each of the six characteristics and are therefore regarded as persons. Being a person and having the right to moral consideration does not mean that all members of the animal kingdom should have equal rights or that animals should be (must be) treated as humans are. No one suggests that dogs should have the right to vote or be entitled to own property. Cats have a right[7] to be fed and sheltered but do not need a driver's license. The nature

[5]This information was found in the 10th edition of *Black's Law Dictionary*. 2004. (Editor-in-Chief B.A. Garner), which quoted p. 318 of the 10th ed. of J. Salmond. (1947). *Jurisprudence*, edited by G.L. Williams.

[6]Research shows that elephants possess characteristics 3 to 6 above. Woodard, C. 2011. The Intelligence of beasts. *The Chronicle Review*. June 26. See http://chronicle.com/article/The-Intelligence-of-beasts/127969/?sid=pm&utm_source=pm&utm_medium=en. Accessed July 3, 2011.

[7]As mentioned in Chapter 8, no nation has affirmed a human right to food as national policy.

of animals does not demand that they have religion, property, voting privileges, or schools. Their treatment should conform to their nature. It should be "based on the insight that what we do to them, matters to them" (Rollin, 1995, pp. 17–18). Thus, the common view is: animals do not have or deserve the rights humans enjoy. For many it is a fact that animals are lesser (i.e., less important, not human) beings but it does not logically follow that they *ought* to be treated as lesser.

The utilitarian view is that it does not matter whether an animal can anticipate the future, has consciousness, can reason, has language, or is aware of itself as the subject of a life. What matters is whether an animal can experience happiness or suffer pain. Rachels and Rachels (2010, p. 105) following Bentham argue that "whether an animal is human or nonhuman is just as irrelevant as whether she is black or white." The important question is not whether they are lesser, can speak, reason, or anticipate the future. It is: Can they experience misery? Can they suffer?[8] If they can suffer, and there is no doubt they can, it confers moral status and humans have a duty to consider their rights when we decide what to do with or to them. For the utilitarian, suffering is what counts. It trumps inherent value. There is no inherent value that counts more than suffering.

There is no doubt that Americans love some animals. Forty-six million American families own at least one dog, and 38 million keep cats. If one adds the 13 million households with freshwater aquariums, the total expenditure on these creatures is about $40 billion per year (including $17 billion for food and $12 billion for veterinarians) (Kolbert, 2009). However, our pets are only about 2% of the animals with which we interact (Wolfson and Sullivan, 2004). The other 98% are the animals we eat, and what is known about how they are treated seems to deny the claim that we love animals (Myers, 2005). One wonders how people who are so concerned about the welfare of their dogs, cats, and horses and delight in the clear, thoughtful look on the face of a cow can be unconcerned about the source of the plastic-wrapped meat displayed in the supermarket. Are they ignorant of the conditions under which these animals are raised, or do they just not care? The American Meat Institute, a national trade association that represents companies that process 95% of red meat and 70% of the turkey in US, claims that "caring for animals is an American Value." McDonalds advertises that animal welfare is included in their purchasing decisions. We love our pets, but the evidence is slim that we even care about the other 98%.

**Highlight 10.1**

The Chinese have eaten dogs and cats for many years, a repugnant practice in the Western world. Things are changing. Public pressure is building to persuade the National People's Congress where the Communist party decides what laws are to be drafted and passed to ban eating dog and cat meat. A proposed animal rights law was circulated in September 2010 by animal rights activists. It would

[8] To suffer is to feel or endure physical pain or psychic/mental distress/anxiety.

be the first of its kind in China where animal welfare is not a priority, and is for all practical purposes an unknown concept among farmers and consumers. If the law were passed there would be inevitable conflict with ancient cultural tradition where the belief is that eating dog meat helps keep the body warm in winter. Cat meat, with it sweet taste, is popular in southern China. Modern restaurants that serve dog meat have been and are still popular in many parts of China, especially among ethnic Koreans in the northeast.

Dogs were once banned in urban areas. In recent years, the government has recognized the increasing demand for pet dogs and loosened or eliminated the ban. There is an enforced one-dog policy which decrees the dog must be less than 35 cm high (14 in.). The law, if enacted, would prescribe fines or imprisonment for illegal consumption or sale of dog or cat meat. Opponents argue that eating dog meat is a time-honored tradition. A primary objection is that this is a Western cultural norm that should not be adopted by China. (Anonymous 2010)

In 2002, Americans ate 124.8 kg (275 lb) of meat per person, up from 112 in 1990.[9] The French were next with 101.4 kg, whereas Indians ate only 5.2. Fifty percent of the total grain consumed in US is fed to the animals we like to eat, whereas in India it is 4.7%.[10] Americans and others eat meat because we like it, and know it is a good source of protein. Can meat consumption be morally, economically, or environmentally justified? There are several arguments in favor of and opposed to animal agriculture and meat consumption. The arguments frequently present exactly opposing views on several points.

## Arguments in Support of Animal Agriculture

1. Animals are essential to a sustainable agriculture (see Chapter 7). This is especially true for the world's poor in developing countries. Smallholder-farming systems now feed most of the world's poor (Herrero et al., 2011). The growing number of people moving out of poverty will create a demand for more animal foods: meat, milk, and eggs, which has been true for all societies as they developed. It is projected that demand will double over the next two decades (Herrero et al.). It is the sustainable mixed crop and livestock production systems that have been and continue to be the backbone of agriculture in developing countries that will satisfy the demand. Proponents of alternative/organic agricultural systems advocate (if not mandate) inclusion of animals that contribute to fertility through their manure. They claim a diverse farm that integrates crops and animals is the surest route to sustainability.

[9] www.guardian.co.uk/environment/datablog/2009/Sep/02/meat-consumption-per-capita-climate-change. Accessed June 8, 2011.

[10] http://earthtrends.wri.org/searchable_db/index.php?action=select_countries&theme=8&variable_ID = 348. Accessed June 8, 2011.

2. Animals, when properly managed, are not environmentally destructive. Livestock use land that cannot be used for food production because it is too steep, dry, cold, saline, acidic, and so on. Those who oppose conversion from pasture or range land to condos or to "development" (housing, malls, golf courses, and so on) argue that ranching preserves wildlife and maintain the integrity and beauty of the land.

   > *A thing is right when it tends to preserve the integrity, stability and beauty of the biotic community. It is wrong when it tends otherwise.*
   >
   > *Leopold, 1947, p. 262*

   Leopold's is an environmental, aesthetic, ethical argument. He prescribes what we should do and why. We Americans value the environmental/aesthetic accomplishment roles of ranching. If given a choice, we might prefer cattle over condos.
3. Cattle production has become more efficient in the sense that improving productivity per animal has reduced the environmental effects of each animal and of the cattle industry. Capper (2011) reported that in 1977, it took 602 days to raise an animal from birth to slaughter; in 2007, only 482 days -; three-fourths of the time to produce 30% more beef. Marks of efficiency including a defined weight of beef in 2007 required 70% of the animals, 81% of the feed, 88% of the water, and 67% of the land required in 1977. The same amount was produced (Capper, 2011) in two-thirds of the time using less land, water, and feed and producing less waste.
4. Ruminant animals convert what we cannot eat to what we can eat. They are not direct competitors for food, but they do compete because the land used to grow feed for animals could be used to grow food for humans (see point 1 in *Arguments against*). The grain fed to animals raised in confinement is not suitable for human consumption. Monogastric animals (e.g., poultry and pigs) are similar to humans and can compete with humans for nutrition from soybean, cereal grains, and other crops. Ruminants (e.g., cattle, sheep, buffalo, goats) are able to prosper on high-fiber diets and nonprotein nitrogen that humans cannot digest. In this sense, they are not competitors.
5. Animals produce abundant by-products, including fiber (e.g., wool), milk, blood (consumed in some cultures), hides (leather), lubricants, fuel, and waste products for fertilizer. Animal manure contributes 14% of nitrogen, 25% of phosphorus, and 40% of the potassium used to produce agricultural crops in developing countries (Herrero et al., 2011).
6. Animals, in some cultures, are a source of wealth. They are assets similar to land or a home. They serve as a place to store wealth (similar to a bank).
7. Animals are a way to distribute agricultural risk. They are economic stabilizers that can be kept in good times and slaughtered or sold for food in bad times. They are one of the few means for the poor to generate income and begin to escape from poverty (von Kaufmann and Fitzhugh, 2004).
8. Animals have many nonfood uses, including hunting, pets, recreation (riding, racing, rodeo), religious sacrifice, exhibition, biological pest control, inclusion in cultural practices (e.g., religious sacrifice in India), traction or draft power for agriculture, and travel.

## Arguments Against Animal Agriculture

1. Land should be used to feed people. The percent of total grain produced consumed by livestock (USA, 50%; European Union, 52%; Brazil, 59%; Japan, 44%; China, 29%; India and Sub-Saharan Africa, 2–4%) varies around the world. However, in opposition to the

second argument above in support of animal agriculture, it is reasonable to claim that much of the land used for grazing or to produce crops to be fed to animals could, without significant modification, be used to produce food for people. As demand for food from animals increases, demand for grain to feed animals will increase, which may account for more than 40% of global cereal use by 2050 (Herrero et al., 2011). Where will this grain come from, given that nearly all of present production is consumed? Brown (1995) says no place can meet the demand because of loss of crop land to urbanization, diminishing water supplies, and increased demand for animal food. Smil (1985) and Prosterman et al. (1996) agree with Brown's pessimistic view, with the caveat that governmental policy changes could make a difference.

2. Animals are not efficient convertors of feed to food and they use a lot of water. To produce 1 lb of feedlot beef requires 7 lbs of feed grain and roughly 8.4 lbs of water or a bit over 6,000 gals to produce a 1,200 lb steer that may yield 720 lbs of beef in 420 days (a reasonable estimate for US beef production) (Guyer, 1977; Gadberry, 2010). The water consumption estimate depends on desired finish weight, days to slaughter, and temperature. Water consumption is more than twice as high in July/August as in December/January. A common estimate is 1,800 gal to slaughter, which may be true, but does not include the variables mentioned above. If only August consumption is used (23 gal/day), 10,074 lbs of water will be used in 420 days to produce 800 lbs of beef, which equals 12.6 gal/lb. Thus, beef production can look good or bad depending on the data chosen. In any case, cattle consume a lot of water.

   Thompson (2010, p. 86) challenges people to ask whether or not societies (not just Americans) will continue to feed large quantities of grain to animals to maintain diets rich in animal protein for the rich of the world? About 30% of global water now used in agriculture, the planet's major water user, is now used for livestock.
3. Some animals are environmentally destructive (overgrazing and desertification). Will societies dispassionately evaluate meat consumption in terms of how the way agriculture is practiced affects the planet and the future of our species (p. 86)? All trends indicate the most likely answer is, No, they will not.
4. There is an argument in opposition to item 3 (above) in support of animal agriculture. In 1950, there were 16.7 million head of beef cattle in US. In 1980, there were 37.1 million. By 2010, the number had dropped to 31.4 million head. Therefore, while it is undoubtedly true that production has become more efficient and the environmental effects of a single animal have been reduced, because the total number of cattle has almost doubled since 1950, one must question if the net environmental effect has indeed been reduced.
5. Animals emit methane, a greenhouse gas and a major contributor to global warming. A cow emits enough methane to light a 75-watt light bulb for 1 year. Five percent to twenty percent of global methane comes from cows. Rice paddies, landfills, and wetlands also contribute to atmospheric methane.
6. Eating too much red meat is not healthy. The medical community and many people are aware of the potential detrimental health effects of excessive meat consumption. Will individuals eventually act on what is known?
7. Cattle ranching in the tropics has led to deforestation. Extensive cattle raising has been responsible for 65–80% of the deforestation of the Amazon basin (Herrero et al., 2011). Between 400,000 and 600,000 hectares (0.988–1,482 million acres) of forest per year have been cleared to grow crops (e.g., soybeans) to feed to livestock.
8. Animal pollution includes undesirable odors, most notably from confined animal feeding operations (CAFOs) that produce huge quantities of waste (manure), nitrates, ammonia, and methane. The USDA (2002) estimates more than 335 million tons of dry matter

waste (after water is removed) are produced annually by the 95.5 million cows and calves and 60.4 million hogs and pigs on US farms. CAFOs annually produce 100 times more manure than the amount of treated human sewage sludge in US municipal wastewater plants.[11] Thirty-five thousand miles of American waterways have been contaminated by animal excrement (Kolbert, 2009). Growth of pollution from animal manure has brought to the fore issues concerning livestock production practices. The problems have led to new EPA regulations (Chassy et al., 2001). Buchanan et al., 2010 note that such regulations can significantly increase capital requirements and costs in livestock production systems. The authors suggest markets usually provide "desirable levels of environmental protection and animal welfare." If markets do not function in the manner the authors suggest, then a public rule *may* be appropriate. Buchanan et al. clearly support modern livestock production systems and object to regulation because it increases costs. Environmental benefits of regulations and the rights of animals are not mentioned.

When raised in CAFOs, animals cannot avoid excreting their waste, which is stored in large holding ponds. There have been several well-documented instances where the banks of a pond have broken and the excrement has moved to a nearby waterway where microbes, nitrates, and drug-resistant bacteria are added to the water. This, of course, is not the animal's fault. It is the fault of those who choose or are compelled by market factors and economically rational managers to raise animals in large confinement facilities.

**9.** Raising animals just to kill and eat them violates their basic rights as sentient creatures. This is the view of People for the Ethical Treatment of Animals (PETA, which has more than 1.8 million members) and the Animal Legal Defense Fund (ALDF), which sells posters and T-shirts that proclaim "We May Be The Only Lawyers On Earth Whose Clients Are All Innocent." A goal of the ALDF is to "spread the word that all animals deserve basic legal rights!" They want to show Congress the groundswell of support for legislation that protects animals and recognizes that animals ought to be entitled to basic legal rights. If animals are compelled to live their short lives in what Patterson (2002) calls an Eternal Treblinka, PETA and the ALDF believe we have a responsibility, a moral obligation to, at the very least, pay attention, if not change some practices. For example:

- Broiler chickens spend their lives in windowless buildings with thousands of other birds and generations of accumulated waste.
- Egg-laying hens spend their lives in small cages with several other hens. The argument is that for their own protection, their beaks are cut flat. They cannot spread their wings or do what normal chickens do. When they no longer add eggs to the conveyor belt, they are killed for their meat.
- Piglets have their tails cut off soon after birth. Males are commonly castrated without anesthetic.
- Slaughterhouses try to kill animals before they are butchered. Some are not killed and are butchered while still alive.

Animals are essential to the recreational sport of hunting. The meat is usually eaten and heads may be kept as trophies. One could defeat the rights argument if it could be shown that hunted animals die painlessly, have a good life, and live a full life. Similarly, the argument could be defeated if it could be shown that animals raised for slaughter do not suffer (branding, early weaning, inoculations, pigs' tails cut off without anesthetic, CAFO life), and were killed instantly without pain. Slaughterhouses now use a captive bolt, a forceful,

[11] http://www.sustainabletable.org/issues/wastes/. Accessed June 13, 2011.

stunning strike on the forehead, which induces unconsciousness. Proper stunning prevents pain and suffering during bleeding and butchering. One cannot know if there is no pain. Grandin (see http://www.grandin.com) has created and perfected techniques to gather and guide animals to slaughter with minimal stress, but they are still slaughtered. The way animals raised for human consumption are treated raises questions about respect for or even acknowledgment of their rights. They are raw material for a process that is quick and efficient for transforming them into plastic-wrapped cuts of meat displayed in the supermarket. Their moral worth may be less than the barriers that enclose them.[12] Regular consideration of their rights by those who raise animals has not been demonstrated conclusively. Animal husbandry emphasized animal care, not just production. There was reciprocity between the animal and the farmer. Husbandry had a personal dimension: animals had names. There was an emotional bond. Raising them was not a business that emphasized efficiency and profit. All that was lost when husbandry was replaced by science and business rationality. CAFOs and industrial agriculture are impersonal. There are no emotional consequences of killing. However, ignoring moral consideration, the rights of animals, does not seem to have affected the psychological well-being of those who raise them.

PETA's argument was brought home with great force when I saw a bumper sticker that read PETA—People for the Eating of Tasty Animals. The opposing arguments are vigorously debated in some quarters. Generally, in Departments of Animal Science, the issue has been resolved. The views of those opposed to animal agriculture are known and dismissed as invalid, poor arguments accompanied by weak or no supporting evidence. They are just someone's opinion. The challenge is—*Tu quo que*—You do it too! Unless you are a strict vegetarian, you eat animal products and you like them.

There is clear evidence that animal scientists and society are changing their view of organizations like PETA and of the legitimacy of their quest for recognition of our moral obligation to recognize and act on the rights of animals. For example,

- All US states have laws prohibiting cruelty to animals.
- There is increasing media attention to the rights of animals.
- There are laws governing the use of animals in research and laws at state and municipal level about horse welfare. Veterinary schools have stopped cruel practices (e.g., multiple, unneeded surgeries on one animal) (Rollin, 1989).
- It is common that calf roping at rodeos has been eliminated.
- Significant shelf space in supermarkets is now devoted to cage-free eggs, open-range beef, hormone-free animals, and so forth.
- Cockfighting and dogfights have been outlawed.
- Several EU nations have banned raising animals in CAFOs.

It took the time and efforts of a lot of people, but Americans and others have accepted that they have a duty to protect the environment and, by implication, its creatures, some known, but most unknown. It is a general concern. Animal rights, in contrast, focus on individual rights and, while a great majority of Americans think they are important, they have not achieved a similar level of public demand for legislation or action against those who abuse animals, but produce the inexpensive meat we demand and enjoy.

**10.** Exploitation of dairy cows is a related issue (Knaus, 2009). Milk yield has increased from about 2,500 to 9,000 kg per year from 1950 to 2005, which affects vitality, fertility,

[12] See McWilliams, J., 2011. http://www.theatlantic.com/life/archive/2011/08/the-dangerous-psychology-of-factory-farming/244063/

and longevity; cows live fewer years, calves the first time earlier, and, on average, have less than four lactations. Use of concentrated grain feed increases milk production, but is not in accordance with the evolutionary requirements of the bovine digestive and metabolic system. Increased milk production increases profit for diaries, but can be seen as a violation of the cow's basic right to a life span not determined solely by profit. Cows should also not suffer decreased health and fitness because of increased milk production or digestive disorders through massive feeding of concentrates.

**11.** The way animals are raised, especially in CAFOs, can result in bacterial resistance to antibiotics. When hundreds, perhaps thousands, of cattle, hogs, or chickens are raised in confinement, the best way, perhaps the only way, to keep them alive is to administer regular doses of pharmaceuticals (antibiotics). Without antibiotics, bacteria could quickly kill all in one confined structure. The loss would be borne by the individual operator, not by the company that supplied the animals/birds. Constant administration of antibiotics is the only way to maintain the scale and intensity of CAFO production.

Large numbers of animals that live in unsanitary (filthy may be a better word) confinement create an ideal incubator for disease. Life-time administration of antibiotics also provides ideal conditions for development of drug-resistant organisms. If inevitable drug-resistance was confined to CAFOs, it would only concern those dependent on the operation for their income. It would be their problem. But it is not just their problem, because the antibiotics used to protect the animals are, in many cases, the same ones used to treat human bacterial diseases. Public health officials are concerned that widespread antibiotic resistance is a human health disaster waiting to happen. Pollan (2007) suggests it has already happened. MRSA (methicillin resistant *Staphylococcous aureus)* is an antibiotic resistant strain of *Staphylococcus* bacteria, which he claims kills (19,000 in 2005, *Economist*, 2007) more Americans than AIDS (17,000). Some have called resistant bacteria super bugs, but they are not. They are resistant to formerly effective antibiotics that have been used to combat bacterial infections in humans. These infections can become difficult to control and can lead, without more effective antibiotics, to epidemics.

Managers of CAFOs have been adding antibiotics to livestock feed since 1946, when research showed that they contributed to faster growth and more efficient weight gain. Approximately 10 million pounds of antibiotics are fed to hogs each year; 11 million to poultry; and 4 million to cattle (Leutwyler, 2001) to promote growth and compensate for disease related to the unsanitary conditions in which they are raised. In the late 1990s, as much as 70% of all antibiotics used in US was fed to healthy farm animals.[13] In 2010, McKenna claimed that farm animals receive 80% (28.8 million pounds) and humans received 3.3 million pounds.[14] As a consequence, microbes that cause human disease are becoming resistant to antibiotics. The amount of antibiotics fed to healthy animals is nearly 10 times greater than the amount given to people. The connection between

[13] http://www.sustainabletable.org/issues/antibiotics

[14] Union of Concerned Scientists—Hogging it: Estimates of antimicrobial abuse in livestock (2001). http://www.ucsusa.org/food_and_agriculture/science_and_impacts/impacts_industrial_agri. Accessed August 30, 2011. And—http://www.wired.com/wiredscience/2010/12/news-update-farm-animals-get-80-of-antibiotics-sold-in-us/. Accessed August 15, 2011.

drug use in animals, the spread of resistance, and human risk has not been studied. The suspicion is that infectious diseases in animals and humans could become untreatable.

### Highlight 10.2

The Pew Commission report (2010) summarizes the arguments against CAFOs and other aspects of industrial farm animal production in the US. These include: a growing public health threat, the threat of antimicrobial resistance, harm to workers and neighbors, air and water pollution, high levels of respiratory problems among workers, possible transmission by workers of animal-borne diseases to a wider population and air, and water pollution from the enormous amounts of waste. Each of these is dealt with in detail in the full report. The report makes the important point that the transformation of farm animal production from extensive, decentralized, family farms that many remember and value, to the presently dominant concentrated, economically efficient, industrialized systems has been unnoticed by most Americans. As long as carefully selected, plastic-wrapped meat is available in the supermarket, the social, economic, environmental, and public health problems created by industrial farm animal production remain unknown and therefore not a matter of public concern. This illustrates that as long as food is abundant, agriculture's negative effects will be ignored.

## Animal Biotechnology

For many people, anything that involves biotechnology is suspect, if not already guilty and deserving of legislative banning or significant legal constraints. In this view, biotechnology belongs in the array of the arguments against animal agriculture, but not all agree. Chapters 3 and 8 suggest that biotechnology has changed the debate about the criteria used to determine the acceptability of any agricultural technology by introducing a new question: Do we need it? The question could also be moral, rather than just scientific or economic, by asking Should we do it? The dilemma is one that plagues biotechnology. Will animal biotechnology create a monster for which no one accepts responsibility or is it scientific technology with vast promise? Developments and arguments to date are not dominated by a clear demonstration of exclusively negative or positive effects. Herbicide tolerant and insect resistant crops, the primary technological achievements, have had positive worldwide effects on crop yields and have reduced pesticide use. Transgenic cotton grown in China improved yields 10–30%, reduced pesticide use 50–80%, and human insecticide poisoning nearly 75% (Chassy et al., 2005; Pray et al., 2002). The promises of continued scientific research includes increased crop yield, resistance to drought, salinity, and heat or cold, each of which limits crop growth, expansion of the geographic limits of crop growth, and improved efficiency of fertilizer utilization.

Promised accomplishments specific to animals include optimization of nutrients for faster growth and increased production, improved efficiency of feed utilization and animal health (disease resistance), and animal production of products for industrial or consumer use (e.g., fibers) (Wall et al., 2009). Transgenic livestock and recombinant DNA technology have the potential to play a critical role in the production of better vaccines to combat diseases, production of pharmaceutical products for human therapeutic use, and development of animal models for research on human diseases (pigs as models for cardiovascular disease; Keefer et al., 2007; Jackwood et al., 2008).

Another promising research area is cloning and transgenesis. Cloning is not genetic modification in the sense that genes that could not have been acquired naturally are incorporated into the genome by artificial means. It is more properly considered plant breeding by new more rapid and predictable techniques that are superior to and faster than traditional animal breeding ever could be. Cloning is a reproductive breeding technique that makes genetic copies of an individual and potentially decreases genetic diversity in a population. Transgenesis, as its name implies, is incorporation of genetic information (genes) at precise location in a genome. It differs from cloning in that the incorporated genetic information can come from almost any source. Species barriers that previously could not be crossed now can be. Cloning can be used to make rapid genetic gains, and eventually to create new genetic variation, conventional breeding is required. Transgenic technology can be used to address a variety of problems and introduce new characteristics. To date, it has not been applied to manipulate complex traits controlled by multiple genes (Jackwood et al., 2008). The previous possibility of crossing species barriers permits a breeder/geneticist to use any desirable trait that can be found. It is worthy of note that in 2006 (Phipps et al., 2006) and to date, while suspicion persists, there have been no proven detrimental human health effects from meat, milk, and eggs derived from animals fed genetically engineered crops. In spite of this, debate over regulation of genetically engineered foods continues. Present federal guidelines do not require labeling plant or animal products derived from genetic modification. Those who think the present regulatory system is adequate argue that increased requirements will involve unpredictable time and inhibit investment, development, and the opportunity to obtain the advantages of genetically engineered animals (Van Eenennaam et al., 2011). They claim that the current regulatory approach for labeling would not provide new information about safety, suggests to consumers the food is not safe, and increases cost primarily due to the necessary segregation from production to marketing. Those in favor of labeling argue that they have the right to know what they are eating. They have a right to choose.

The Council for Agricultural Science and Technology (CAST) has published several brief papers on different aspects of animal and plant biotechnology. Each is a balanced discussion of a particular aspect of biotechnology/genetic modification. Each also concludes that present and promised future benefits outweigh known or predictable disadvantages. Only one of the CAST papers I reviewed addresses the ethical aspects of animal biotechnology (Thompson et al., 2010), although the word is used in one other. Thompson et al. do not charge scientists with intentionally violating specific religious or ethical precepts. They do point out a basic ethical blind-spot: humans place themselves and their interests over and above the line that separates creatures that deserve moral consideration from those that do not. "The specific scientific interventions may be less characteristic of this ethical failing than is an overall attitude or manner of conduct regarding the development and governance of the technology" (Thompson et al.). An important aspect of the problem is that ethical review, if it occurs at all, occurs after a technology or product is finished and ready to use. The dominant questions are does it work as intended and is it safe for users? "The discussion does not focus on whether cloning, genetic engineering, or other biotechnologies are appropriate for the animals, the environment, or even whether the resulting products are socially or ethically appropriate" (Thompson et al.). Economic effects are considered, because no company wants to bring a product or technology to market that is too expensive, not profitable, or of minimal benefit to consumers. If social, economic, or ethical problems require further discussion to achieve acceptability or regulatory approval, proponents advocate education (of the public) as the way to achieve acceptability. A few examples:

> *"The authors suggest that education regarding the advantages and challenges associated with this new technology is the key to public understanding" (Keefer et al., 2007).*
>
> *"Inherent in the successful adoption of new and emerging biotechnologies is the need to increase public understanding ..." (Etherton et al., 2003).*
>
> *"Education addressing the realistic advantages and challenges of continued development and commercialization of biotech crops, as well as nonbiotech crops, will be key to public understanding and discourse related to future policy toward biotech crops" (Gealy et al., 2007).*

Those engaged in education recognize its value and efficacy. Education of citizens is regarded as fundamental to proper functioning of a democratic society. However, I fear that what is really advocated by those who favor and promote a technology or new product is neither education, nor a dialogue intended to encourage a broader, thoughtful analysis of an ethical claim. The scenario is: facts and evidence in support of a position or technology are presented within the parameters of scientific verification and falsification. Broader debate is not envisioned or encouraged. The issues to be discussed are limited to whether or not benefits to humans outweigh known or predictable disadvantages. The assumption is that only humans deserve moral consideration. Anyone who raises the basic issue of whether other forms of conscious life (those below the line) deserve to be included is met with suspicion and is assumed to need education. The oft-stated response is: "if critics would pay attention to the scientific facts they would understand the importance of the real issues." This fails to understand that science is based on a particular mind-set that assumes certain value claims. For example, science and scientists value objectivity over subjectivity, which is often misunderstood as being a "value-free" position. Subjectivity is the presumed realm of values. The scientific approach is but one of any number of value claims and ethical issues. The foundations of science stand within the arena of values and ethics:

*Science is value-free and free of the necessity of considering the ethical implications of scientific research.*
*humans are valued over other forms of life,*
*only humans deserve moral consideration and*
*objectivity is valued over subjectivity.*

Claims that ethical arguments and moral questions stand outside the provenance of science betray a naiveté regarding the assumptions on which science rests (Rollin, 2006).

Opponents of animal biotechnology (or some other technology) deserve a hearing, as they raise issues that ought to be addressed and listened to by those in the scientific community. Those who take a different position on the proper value orientation of animal and plant biotechnology should not be dismissed because, it is claimed the scientific debate should not be muddled by ethical arguments or moral claims, which, of course, is a position with a moral foundation, albeit an unrecognized one.

## A New Technology

Given the scientific capability to isolate stem and embryonic cells and grow them *in vitro*, a few scientists are studying whether or not it is possible to grow meat outside of an animal (Bartholet, 2011; Specter, 2011). The ultimate goal is to help rid the world of the present inefficient production of animals for food (Jones, 2010) and the often appalling conditions under which many are raised that ignores their suffering.

Can meat be grown in a laboratory? If it is possible to produce meat that looks, smells, and tastes like meat from cattle, pigs, sheep, or any other meat-producing animal, arguments opposed to animal agriculture would lose their force. They would be superfluous. Meat produced in the laboratory could, in theory, eliminate debate about the Animal Legal Defense Fund's (ALDF) claim that animals raised for the meat, dairy, and egg industries are among the most abused in the US, in numbers that are so staggering as to be almost incomprehensible. In theory, the industries would go out of business. It would no longer be necessary to slaughter animals for meat, to ask if the animals raised solely for meat, milk, or eggs died painlessly, had a good life, and lived a full life. It would no longer be necessary to ask if animals raised for slaughter were killed instantly without pain, or suffered from branding, early weaning, inoculations, surgery without anesthetic, and about CAFO life. There would be no reason for CAFOs.

The central belief of humans that they are singular and superior to all other life forms would still exist. What we know or assume we know about ourselves would assure us the dogma is correct. The belief that we are of a different kind and the belief that difference gives us rights animals do not have would persist. Many would still argue that animals do not deserve moral consideration and can be treated instrumentally. But if we could have meat produced without inflicting pain, would we do it?

Producing meat *in vitro* would diminish but not eliminate the need for the arguments above in favor of and opposed to animal agriculture. If the research is successful and the price is right, the subsequent debate will focus on whether or not these products are socially acceptable. Will people eat miracle meat? Four independent studies were cited by Bartholet.[15]

It has been a truism of development that as societies develop, people want to consume more meat. Meat supplies a variety of nutrients including iron, zinc, and vitamin $B_{12}$. We like it and it is good for us. But as vegetarians have shown, it is possible to survive without it. There is however, the fundamentally utilitarian ethical position which holds that if we could produce meat without animal suffering, we ought to do it. One must grant that removing animals from an ecologically sound, sustainable farming system is not a good idea. CAFOs and farming systems that are chemical, capital, and energy-based are neither sustainable nor ecologically sound.

## A Final Word

At this point, it would be presumptuous of me and beyond my philosophical skill to attempt to delve further into or critique the rich discussion of existing ethical arguments for animal rights. They have been presented, critiqued, and defended by several authors, each of whom has provided substantial justification for why animals deserve moral consideration. A partial list of publications follows: Cavalieri (2001), Foer (2009), Midgley (1984), Sapontzis (1987), Sherlock and Morrey (Ed.) (2002) (eight articles, pp. 247–358), Regan (1987, 2004), Regan and Singer (1989), Rifkin (1993), Rollin (1989, 2006, 2011), and Singer (1981, 1985, 1986, 2002, 2004).

We are, as Ovid's Medea was, dragged along by a strange force. Our desire and reason pull in different directions. We may see the right way, what we ought to do, and approve it, but we follow the wrong.

## References

Anonymous (2010). Off the Menu. *The Economist.* February 27, p. 51.

Bartholet, J. (2011). Inside the meat lab. *Scientific American.* 7(6):65–69.

Brown, L. (1995). *Who Will Feed China? Wake Up Call for a Small Planet.* The Worldwatch Environmental Alert Series. New York, W.W. Norton & Co.

Buchanan, G., R.W. Herdt, L.G. Tweeten. (2010). *Agricultural productivity strategies for the future: Addressing US and global challenges.* Issue Paper 45. Council for Agricultural Science and Technology. Ames, IA, CAST.

[15]Haagsman, H.P., K.J. Hellingwerf, and B.A.J. Roelen, 2009. Production of animal protein by cell systems. Univ. of Utrecht.; Thornton, P.K. 2010. Livestock production: Recent trends, future prospects. Trans. R. Soc. B, 365 (1554):2853–2867.; Jones, N. 2010. Food: A taste of things to come. *Nature* 468:752–753.; Bhat, B.F., and Z. Bhat. 2011. Animal-free meat biofabrication. *Am. J. Food Tech.* 6(6):441–459.

Capper, J.L. (2011). Replacing rose-tinted spectacles with a high-powered microscope: the historical versus modern carbon footprint of animal agriculture. *Animal Frontiers* 1(1):26–32.

Cavalieri, P. (2001). *The Animal Question: Why Nonhuman Animals Deserve Human Rights.* Oxford, UK, Oxford University Press.

Chassy, B.M., S.H. Abramson, A. Bridges, W.E. Dyer, M.A. Faust, S.K. Harlander, S.L. Hefle, I.C. Munro, and M.E. Rice. (2001). *Evaluation of the US regulatory process for crops developed through biotechnology.* Issue Paper 19. 14 pp. Council for Agricultural Science and Technology. Ames, IA, CAST.

Chassy, B.M., W.A. Parrott, and R. Roush. (2005). *Crop biotechnology and the future of food: A scientific assessment.* CAST commentary. QTA 2005-2. 6 pp. Council for Agricultural Science and Technology. Ames, IA, CAST.

Economist. (2007). Riding piggyback. December 1:94.

Etherton, T.D., D.E. Bauman, C.W. Beattie, R.D. Bremel, G.L. Cromwell, V. Kapur, G. Varner, M.B. Wheeler, and Wiedmann. (2003). *Biotechnology in Animal Agriculture: An overview.* Issue paper 23, Ames, IA, Council for Agricultural Science and Technology (CAST).

Foer, J.S. (2009). *Eating Animals.* New York, Little, Brown & Co.

Gadberry, M.S. (2010). *Water for Beef Cattle.* Little Rock, AR, University of Arkansas, Cooperative Extension Service. FSA 3021.

Gealy, D.R., K.J. Bradford, L. Hall, R. Hellmich, A. Raybould, J. Wolt, and D. Zilberman. (2007). *Implications of gene flow in the scale-up to and commercial use of biotechnology-derived crops: Economic and policy considerations.* Issuer Paper 37, Ames, IA, Council for Agricultural Science and Technology (CAST).

Guyer, P.Q. (1977). *G77-372 Water requirements for beef cattle. Historical materials from the University of Nebraska-Lincoln Extension.* 2 pp.

Herrero, M. with S. Macmillan, N. Johnson, P. Ericksen, A. Duncan, D. Grace, and P.K. Thornton. (2011). Improving food production from livestock. Pp. 155–163 *in State of the World: Innovations that Nourish the Planet.* New York, W.W. Norton & Company.

Jackwood, M.W., L. Hickle, S. Kapil, and R. Silva. (2008). *Vaccine development using recombinant DNA technology.* Issue Paper 38. Ames, IA, Council for Agricultural Science and Technology (CAST).

Jones, N. (2010). A taste of things to come. *Nature* 468:752–753.

Keefer, C. L., J. Pommer, and J.M. Robl. (2007). *The role of transgenic livestock in the treatment of human disease.* Issue Paper 35. Ames, IA, Council for Agricultural Science and Technology (CAST).

Knaus, W. (2009). Dairy cows trapped between performance demands and adaptability. *J. Science of Food and Agriculture* 89:1107–1114.

Kolbert, E. (2009). Flesh or your flesh; should you eat meat? *The New Yorker* 85(36):74–78.

Leopold, A. (1966). *A Sand County Almanac.* New York, NY, Ballantine Books. Original Pub. 1949. Oxford University Press.

Leutwyler, K. (2001). Most US antibiotics fed to healthy livestock. *Science* January 10.

Midgley, M. (1984). Animals and Why They Matter. Athens, GA, University of Georgia Press.

Myers, B.R. (2005). If pigs could swim. *The Atlantic Monthly* Sept. 2005:134, 136–139.

Nussbaum, M.C. (2006). The moral status of animals. *The Chronicle Review*, section B. February 3. Pp. B6–B8.

Patterson, C. (2002). Eternal Treblinka—Our Treatment of Animals and the Holocaust. New York, Lantern Books.

Pew Commission on Industrial Farm animal production. (2010). *Putting meat on the table: Industrial farm animal production in America.* A project of the Pew charitable trusts and Johns Hopkins Bloomberg school of Public Health.

Phipps, R.H., R. Einspanier, and M.J. Faust. (2006). *Safety of meat, milk, and eggs from animals fed crops derived from modern biotechnology*. Issue Paper 34. Ames, IA, Council for Agricultural Science and Technology (CAST).

Pollan, M. (2007). Our decrepit food factories. Pp. 25 in The Way We Live Now section of the *New York Times Magazine*. Dec 16.

Posner, R. (2004). Animal rights—legal, philosophical, and pragmatic perspectives. Pp. 51–77 *in* C. Sunstein, M. Nussbaum (eds.). *Animal Rights—Current Debates and New Directions*. Oxford, UK, Oxford University Press.

Pray, C.E., J.K. Huang, R.F. Hu, and S. Rozelle. (2002). Five years of Bt cotton in China—the benefits continue. *Plant J.* 31:423–430.

Prosterman, R.L., T. Halsted, and L. Ping. (1996). Can China feed itself? *Scientific American* November:90–96.

Rachels, J. (2004). Drawing lines. Pp. 162–174 *in* C. Sunstein, M. Nussbaum (eds.). *Animal Rights—Current Debates and New Directions*. Oxford, UK, Oxford University Press.

Rachels, J. and S. Rachels. (2010). *The Elements of Moral Philosophy*, 6th Ed. New York, McGraw Hill.

Regan, T. (1987). *The Struggle for Animal Rights*. Clarks Summitt, PA.

Regan, T., P. Singer (eds.). (1989). *Animal Rights and Human Obligations*, 2nd Ed. Englewood Cliffs, NJ, Prentice-Hall.

Regan, T. (2004). *The Case for Animal Rights*. Berkeley, CA, University of California Press.

Rifkin, J. (1993). *Beyond Beef: The Rise and Fall of the Cattle Culture*. New York, Penguin Books.

Rollin, B.E. (2006). *Animal Rights and Human Morality*. Amherst, NY, Prometheus Books.

Rollin, B.E. (2011). *Putting the horse before Descartes—My Life's Work on Behalf of Animals*. Philadelphia, Temple University Press.

Rollin, B.E. (1989). *The Unheeded Cry—Animal Conscious Animal Pain, and Science*. Oxford, UK, Oxford University Press.

Rollin, B.E. (1995). *Farm Animal Welfare—Social, Bioethical, and Research Issues*. Ames, IA, Iowa State University Press.

Sapontzis, S.F. (1987). *Morals, Reason and Animals*. Philadelphia, Temple University Press.

Singer, P. (2002). Animal Liberation. New York, Harper Collins Publishers.

Singer, P. (2004). Ethics beyond species and beyond instincts. Pp. 78–92 *in* C. Sunstein, M. Nussbaum. (eds.). *Animal Rights—Current Debates and New Directions*. Oxford, UK, Oxford University Press.

Singer, P. (1985). *In Defense of Animals*. New York, Harper and Row.

Singer, P. (1981). *The Expanding Circle: Ethics, Evolution and Moral Progress*. Princeton, NJ, Princeton University Press.

Smil, V. (1985). China's Food. *Scientific American* 253(6):116–124.

Specter, M. (2011). Test-Tube burgers. *The New Yorker* May 23. Pp. 32–37.

Thompson, P.B., F.W. Bazer, E.F.Einsiedel, and M. Foster Riley. (2010). *Ethical implications of animal biotechnology: Considerations for animal welfare decision-making*. Issue Paper 46. Ames, IA, Council for Agricultural Science and Technology (CAST).

Thompson, P.B. (2010). *Agrarian Vision—Sustainability and Environmental Ethics*. Lexington, KY, University of Kentucky Press.

USDA. (2002). *Census of agriculture 2002: Cattle and Calves—and—2002 Hogs and Pigs.*

Van Eenennaam, A.L., E.M. Hallerman, and W.M. Muir. (2011). *The science and regulation of food from genetically engineered animals*. Commentary QTA2011-2. Ames, IA, Council for Agricultural Science and Technology (CAST).

von Kaufmann, R.R. and H. Fitzhugh. (2004). The importance of livestock fo rthe world's poor. Pp. 137–159 *in* C.G. Scanes, J.A. Miranowski (eds.). *Perspectives in World Food and Agriculture* Vol 1: Ames, IA, Iowa State University Press.

Wall R., G. Laible, E. Maga, G. Seidel, Jr., B. Whitelaw. (2009). *Agricultural productivity and genetic diversity: cloned transgenic animals.* Issue paper 43. Ames, IA, Council for Agricultural Science and Technology (CAST).

Wise, S.M. (2001). Why animals deserve legal rights. *The Chronicle of Higher Education* Feb. 2:B13.

Wolfson, D.J. and M. Sullivan. (2004). Foxes in the hen house: animals, agribusiness, and the law: a modern American fable. Pp. 205–233 *in* C. Sunstein, M. Nussbaum (eds.). *Animal Rights—Current Debates and New Directions.* Oxford, UK, Oxford University Press.

# 11 A Glimpse Ahead

*We shall not cease from exploration and the end of all our exploration will be to arrive where we started and know the place for the first time.*

***T.S. Eliot, 1942.* Little Gidding**

*There are more things in heaven and earth, Horatio, than are dreamt of in your philosophy.*

***Shakespeare,* Hamlet*, Act 1, Scene 5***

When I was a young boy, everyone had read, or at least said they had read, J. D. Salinger's novel *The Catcher in the Rye* (1951). Salinger introduces Holden Caulfield, a young man in search of himself, as all young men seem to be. Holden wants to be somebody who engages in some form of service, but he is having trouble, as young boys do, in defining what that service might be.[1] At one point in the story (pp. 224–225), Holden is conversing with his sister, Phoebe, who asks him what he would like to become.

> *"You know what I'd like to be? I mean if I had my goddam choice?…"*
>
> *Phoebe responds, "What? Stop* swearing."
>
> *"You know that song, 'If a body catch a body comin' through the rye'! I'd like—" (Phoebe breaks in.) "It's 'If a body* meet *a body coming through the rye'! It's a poem. By Robert Burns."*
>
> *"I know it's a poem by Robert Burns."*
>
> *Holden regains his composure from this unwanted correction offered by his younger sister and continues.*
>
> *"I thought it was 'If a body catch a body.'"*
>
> *… "Anyway, I keep picturing all these little kids playing some game in this big field of rye and all. Thousands of little kids, and nobody's around—nobody big, I mean—except me. And I'm standing on the edge of some crazy cliff. What I have to do, I have to catch everybody if they start to go over the cliff—I mean if they're running and they don't look where they're going I have to come out from somewhere and* catch *them. That's all I'd do all day. I'd just be the catcher in the rye and all. I know it's crazy, but that's the only thing I'd really like to be. I know it's crazy."*

I suspect many of those engaged in agriculture, whether they are farmers, ranchers, equipment dealers, grain dealers, farm supply dealers, university researchers,

[1] My source of this view of Salinger's story is B.C. Birch and L.L. Rasmussen, 1978. The Predicament of the Prosperous. Philadelphia, PA. The Westminster Press, pp. 97–98.

**Agriculture's Ethical Horizon. DOI: 10.1016/B978-0-12-416043-9.00011-8**

and so forth, see themselves and their profession as having achieved what Holden Caulfield wanted to become. They are the only big ones around and they are quite literally catching the helpless of the world who do not or cannot produce their own food. They are the catchers in the rye, saving those who are about to fall off the cliff of starvation. However, as Phoebe Caulfield points out, they have misread the lines. Burns' poem does not say, "if a body catch a body." It says, "If a body meet a body."

All in agriculture (farmers, ranchers, suppliers, marketers, and, yes, even professors) need to "meet" head on the deeper ethical challenges. Our planet now (2011) has 7 billion people and perhaps 9 billion by 2045, when population may stabilize. However, over the next few decades, hundreds of millions of people who are not fed well now and have poor agricultural practices will suffer. It can be argued that the rich and prosperous have some duty to help the less fortunate. In general, they are helped, but not enough. If all in agriculture and all in the developed world view themselves as catchers, it is likely they will see the solution of the problem of hunger as more production so they can "catch" those who may starve. But if they are to "meet" the body of challenge(s) then they must find the root causes of hunger, which may include lack of food, but as Sen (1981) found, is more likely to be caused by inadequate or absent food distribution, inappropriate or inadequate agricultural technology, oppressive political systems, and omnipresent poverty. These exist in several nations outside US, but are also here. All engaged in agriculture ought to confront the large body of difficult questions that lie ahead and thereby arrive where they began, and know it well. If they do not they may be able to "catch" only a few until the hard questions are met addressed and answered.

*Is more production in developed countries really a solution?*
*Are any existing agricultural systems sustainable?*
*Are present agricultural systems ecologically responsible?*
*Do we, does anyone, know how to successfully transfer appropriate technology?*
*Is US foreign aid organized to help those in need?*
*What will be the effects of climate change on food production and on the poor?*

Living with such hard questions is never comfortable. If we are to meet those in need, we will necessarily proceed with the nagging uncertainty that we cannot know for certain that the right course of action is being followed. We will always see through a glass, darkly and will only know the wisdom of our actions later, after we are compelled to begin. It is certain that the glass will be much darker and the future much less certain if we proceed without considering what ought to be done as well as what can be done. Knowing what can be done is not sufficient. The difficult ethical questions must be addressed to create a firm ethical foundation as we go forth to meet the people and act to solve well-defined problems that lack well-defined solutions.

It is common to believe that Americans have a God-given right to the American dream, in its purest form. Economic growth will continue and, in fact, must continue if we are to have the prosperity and continued consumption we have earned and deserve. Many see this attitude as the epitome of American arrogance and misunderstanding of the world. Yet, we continue to charge ahead without a firm ethical foundation and without even taking the time to consider that foundation. We are like Dickens' Mr. McCawber[2] in *David Copperfield.* We are sure that "something will turn up" to enable us to go on. We will somehow continually expand the horizon of accomplishment in agriculture. We will somehow feed more people a better diet because agriculture's progress, we assume, is limited only by our skill and scientific knowledge. In fact, with the advent of biotechnology, many believe that agriculture's horizon is, once again, unlimited. Many of the strongest defenders of modern agriculture, its wonderful technology, and its undeniable achievements live in what could be characterized as an echo chamber of their own opinions. They grant credence to good science that supports their opinions and ignore all other information or put it in the category of bad science, whereupon it is dismissed. The dominant world view is reminiscent of Dr. Pangloss, Candide's mentor and a philosopher (Voltaire, 1759). Pangloss is responsible for the novel's most famous idea: that all is for the best in this "best of all possible worlds." The optimistic sentiment is the main target of Voltaire's satire.

Agricultural scientists, the larger scientific community, and the general public now recognize a large set of problems that have been created, at least partially, by agriculture: contamination of water, food and feed by pesticides, eroded soil, "mining" soils of their natural fertility, fertilizer pollution, pesticide harm to people and other living things, atmospheric contamination by ammonia and methane and their relation to ozone depletion, global warming, overuse of nonrenewable resources, loss of wildlife habitat, and groundwater mining (Pretty, 1995; Kirschenmann, 2000b). It is not unreasonable to claim that the agricultural community has been late in acknowledging and addressing these problems. Kirschenmann, citing Baskin (1997), identified six agricultural problems not all of which were caused by the practice of agriculture, but agriculture is intimately involved in and affected by all.

## Six Important Issues/Problems/Matters of Concern

### Highlight 11.1

During the next 50 years, agriculture "has the potential to have massive, irreversible environmental impacts," producing sources of global change that "may rival climate change on environmental and social impacts." If past global effects of agriculture on human population and consumption continue, 1 billion hectares of ecosystems would be converted to agriculture by 2050, accompanied by at least a doubling of nitrogen and phosphorus driven eutrophication of terrestrial, fresh water, and near-shore marine ecosystems, and comparable increases in pesticide use.

*Science* April 13, 2000 *and Alternative Agriculture News.* (2001) 19(5):1.

[2] "I have known him to come home to supper with a flood of tears, and a declaration that nothing was now left but a jail; and go to bed making a calculation of the expense of putting bow-windows to the house, 'in case anything turned up,' which was his favorite expression." D. Copperfield, Chapter 11.

## *Agricultural Production/Soil Erosion/and Desertification*

### *Production*

A major challenge for the agricultural community is to design sustainable production systems that produce sufficient, high-quality food without causing further harm to the ecological systems on which production depends.

Many citizens acknowledge and agree with the negative view of agriculture's environmental effects. For example, McNeill (2000, p. 358) says "In any case, human history since the dawn of agriculture is replete with unsustainable societies, some of which vanished, but some of which changed their ways and survived. They changed not to sustainability but to some new and different kind of unsustainability." McNeill suggests that ecological buffers—available new agricultural land, unused water, unpolluted spaces—that made it possible for earlier societies to make it through difficult times are now gone. Every technological advance designed to increase agricultural production created some negative ecological and social effects.

However, other authors argue forcefully that the negative views of agriculture are wrong. Bailey (1995, p. 12) states that the environmental movement that began with Earth Day 1970 has "scored some major successes" but has "been spectacularly wrong" about many things. Global famines predicted in the 1970s (see Paddock and Paddock, 1967) have not happened, all forests have not been cut down, global warming is not as serious a problem as predicted (a brief discussion of global warming follows), and far less damage has been caused by pesticides than Carson predicted in 1962. Bailey cites several things that support his optimistic view:

- Global life expectancy more than doubled in the twentieth century.
- Despite a tripling of world population, global health and productivity have exploded.
- The world's population growth rate has steadily declined.
- Problems typically associated with overpopulation (hunger, overcrowding, poor living conditions) are more properly identified as problems of poverty.
- Global per capita food availability rose by almost a third from the 1930s to the 1980s.
- Worldwide per capita food availability has kept pace with population growth.
- Where natural resource supplies have dwindled, they are more properly related to poor government policies.

Bailey's (1995) book was followed by the much more successful work by Lomborg (2001). He argued that all (not just some) of the literature and science of environmental pessimism has been written by dissembling environmentalists whose aim is to panic citizens and legislators into inappropriate action to save a planet that is not in danger. The environment, in Lomborg's view, is not bad and getting worse, it is good and getting better. "On practically every count, humankind is now *better (*italics in original) nourished. The Green Revolution has been victorious." Production has tripled in developing countries, calorie intake per capita has increased, and the proportion of starving people in the world has decreased. (Due to population growth this is correct, but the total number has remained about the same.) In short, the negative environmentalists have been totally wrong.

Thus, Bailey and Lomborg see progress wherever they look and discount the fears of the pessimists. Many find their views refreshing.[3] Others argue with good evidence that pessimism is warranted.[4]

## Soil Erosion

Under agricultural conditions, it takes about 500 years or more to create an inch of topsoil, which can be lost in minutes. For all practical purposes, topsoil is a nonrenewable resource. World agriculture contributes to loss of 24 billion tons of soil each year (Baskin, 1997). In 1982, the USDA[5] estimated that 3.1 billion tons of US topsoil were lost annually from wind, sheet, and rill erosion on cropland and conservation reserve land. The situation has improved, but not enough. The US average for sheet and rill erosion was 2.9 tons/acre/year in 1987, 2.2 in 1992, and 1.9 in 1995 and 1997. The USDA[6] also reported that erosion rates in some areas in the 1970s were above soil's estimated natural renewal rate (5 tons per acre) on 33% of corn, 34% of cotton, 39% of sorghum, and 44% of soybean acreage. Cropland soil erosion varies from an average of 10 metric tons per hectare in US to 40 in China, and as high as 5,600 in parts of India (Pimentel and Wilson, 2004). The 2011 data show US soil loss is 10 times faster, but China and India are losing soil 30–40 times faster than the natural replenishment rate. Soil erosion is in the US costs about $37.6 billion/year in lost production. Worldwide damage from soil erosion is estimated to be $400 billion/year.

As a result of erosion over the past 40 years, 30% of the world's arable land has become unproductive.[7]

Soil is agriculture's, indeed the world's, ultimate resource. Modern agriculture is dependent on maintaining soil as its productive base but is failing to do so. Since widespread farming began in the eighteenth century in US, it is estimated that 30% of all farmland has been abandoned because of soil erosion, salinization, or water logging (Pimentel, 1995). As much as one-third of all US topsoil has been lost and most US land is eroding at a rate above the regeneration rate (Pimentel). Soil tillage, a mainstay of modern agriculture, is estimated to erode soil at 1–2 times the rate of formation (Myers, 2009).

## Desertification

Baskin (1997) suggests that 70% of the world's drylands are now threatened by desertification and no one knows how to reverse the process once it has begun. Douglass (1994) estimates that desertification is removing at least 50 million productive acres in the world's arid and semi-arid regions. Today the world will lose

[3] A particularly favorable review appeared in The Economist, September 8, 2002, pp. 89–90.

[4] For example see: Bell, R.C., 2002. Media Sheep. World-Watch. March/April, pp. 11–13. Rennie, J., 2003. Misleading Math about the Earth. Scientific American.com http://www.sciam.com/print_version.cfm?articleID=000F3D47-C6D2-1CEB. Accessed April 3, 2003.

[5] http://www.nrcs.usda.gov/technical/land/meta/m5852.html, Accessed February 2005.

[6] http://www.nrcs.usda.gov/technical/nri/1977/summary_report/table10.html. Accessed January 7, 2005.

[7] Preceding data from: www.news.cornell.edu/stories/march06/soil.erosion.threat.ssl.html. Accessed July 19, 2011.

another 72 square miles to encroaching deserts, an area equal to the size of West Virginia over a year (Orr, 1994, p. 7).

## *Depletion of Water Resources*

About 1% of the earth's water is all that is available for human consumption. In US, 70% is used in agriculture; globally 60+ % is used for agriculture, primarily to irrigate crops that provide nearly more than one-third of the world's food. In western US states, 85% of available water is used to irrigate crops. The US Geological Survey estimates that agricultural withdrawal (e.g., Ogallala aquifer) averages 34% of total withdrawals. In the Rocky Mountain region agricultural withdrawal, primarily for irrigation, is almost 90% of total available water.[8] Irrigation, a proven technique to increase and assure yield, has allowed production of high value crops in areas where only low-yield, dryland agriculture was possible. In many of the world's irrigated areas (e.g., the southern Ogallala aquifer[9] under the high plains of the western US where withdrawal is three times greater than recharge, India's irrigated areas, China), water is being used at a rate faster than the source is being replenished. Therefore, water is being mined with abundant short-term gain leading inevitably to long-term failure. Doubling agricultural production will require at least 2,000–3,000 $km^3$ (yes, kilometers) of irrigation water each year, which more than triples current demand. Myers (2009) estimates it is equal to the flow of 110 more Colorado rivers. Water scarcity is the biggest threat to world food production. A blue agricultural revolution may be as or more essential than another green revolution (Postel, 1999). People can't do anything to change the amount of water on the planet, but can and do change its location and quality. As the worldwide demand for fresh water increases and the supply of good quality water diminishes, it is becoming more a manipulated commodity than a free good and its inequitable distribution has enormous political ramifications. For example, India has 2.2% of the world's arable land, 4% of its fresh water, and 17% of its population. It would require 2.5 billion gallons of water to support 4.7 billion people with the UN daily minimum water requirement. Worldwatch (Anonymous, 2004) estimated that is equal to the water used to irrigate the world's golf courses. National Geographic (2010) estimated 2 billion gallons were used daily to irrigate US golf courses. Which, if correct, pales right pales in comparison to the ethical question, which is not foremost among US citizens. Several examples illustrate water's moral dimension (National Geographic).

- Americans use about 100 gal of water daily at home.
- Millions of the world's poor subsist on less than 5 gal.

[8] Personal communication, Dr. Reagan Waskom, Director, Colorado Water Institute, Colorado State University, Fort Collins, CO.

[9] US agriculture irrigates 56 million acres, 14 million acres (25%) are irrigated by the Ogallala aquifer. It is used in Nebraska, eastern CO and NM, and western KS, OK and TX. In 1990, it was estimated the aquifer held 3.2 billion acre feet. Eight percent (270 million acre feet) had been withdrawn by 2007. The aquifer recharge balances withdrawals in parts of NE but consumptive withdrawals exceed recharge in other states. Current annual consumptive withdrawals are 19 million acre feet. Personal communication, Dr. Reagan Waskom, Director, Colorado Water Institute, Colorado State University, Fort Collins, CO.

- Women in the world's poor countries walk an average of 3.7 miles each day to get 5 gal of water.
- Forty six percent of the earth's people do not have water piped to their home.
- One of eight (12%) of the world's people lack access to clean water.
- 3.3 million people die annually from water-borne diseases

**Highlight 11.2**

Coleridge had it right when, in 1798, he published the first version of the *Rime of the Ancient Mariner*:

*As idle as a painted ship*
*Upon a painted ocean.*
*Water, water, everywhere,*
*And all the boards did shrink;*
*Water, water everywhere,*
*nor any drop to drink.*

That is a reasonable description of water supply for many people in the world. Water is the most common substance on earth, but 97.4% of it is in oceans. Of the 2.6% that is fresh water, almost 2% is in polar ice and glaciers. All that is available to us in rivers, streams, lakes, and groundwater is about 0.32% of all the water on earth. That is all we have to drink, bathe in, swim in, irrigate crops with, and do all the other things we do with water. It takes about 2 liters per person per day to keep us hydrated. That is about the volume of 5.6 cans of soda pop. For a life acceptable to most people in the world's developed countries, each person requires about 22 gallons a day, which equals 7900+ gallons per year. That means the planet's water supply could support a population of 20 to 25 billion people or 3.5 times the present population.

No matter what we do, we cannot affect the total amount of water on earth but we can and do affect its quality. Postel (2000) asserts that because water is essential to the lives of humans and all other creatures, every decision made about water is an ethical decision. There is a finite supply of usable water that can support life. We value a continued healthy life, but Postel makes the moral claim that no living creature has a greater right to life than any other living creature. Environmental preservation and sustainability, for all creatures, are dependent on water.

In the United States, about 85% of the fresh water used is used in agriculture and most of that (at least 80%) irrigates crops. Most US irrigation is in the 17 Western states, on 12% of the US crop acreage that produces 27% of the US crop value. Worldwide, about 18% of crop land is irrigated, and that land produces about one-third of the world's crops.

The following are some examples of the dimensions of the water problem.

- An unrealized (in 2011) Colorado plan illustrates the urban rural water dilemma. Suburbs of Denver formulated a plan to bring water from northeastern Colorado to thirsty front range cities through a 140 mile pipeline that will cost more than $1 billion. Farmers were to be paid to fallow 20% of their land each year and sell the water to the front range cities.

- An estimated 2 million children die each year (6,000 each day) from diseases linked to bad water. Most of these children live in Africa and Asia, but some live in the United States and Europe.
- The world's golf courses use 2.5 billion gallons for irrigation each day. The same amount of water would support 4.7 billion people with the UN daily minimum intake (World Watch March/April 2004, p. 36).
- My household water costs $34.50 per month for up to 5000 gallons. I irrigate pasture for sheep. My allocation is 4 acre feet/share at $110/share (1 acre foot = 325,000 gal). I pay the bargain rate of $0.42 for 5000 gals of irrigation water. Almost 60% of the world's fresh water withdrawals are for irrigating agricultural crops. In 2000, this amounted to 137,000 million gallons per day (153,000 acre feet per year). The amount of fresh water stored behind dams has quadrupled since 1985, and agricultural use has exceeded long-term supplies by 5% to 25%. By 2030, most estimates project that farmers will need 45% more water and probably won't get it due to demand and power of urban users. What we choose to eat and how efficiently it is produced matter. It takes 1150 to 2000 liters of water to produce 1 kg of wheat and about 16,000 liters to produce 1 kg of beef.
- The wetlands area (150,000 acres) of the Colorado River delta receives about 0.1% of the water that once flowed through it. The same area could be covered to a depth of 2 feet with water drawn from the river by the city of Las Vegas, NV, which uses much of the water to irrigate more than 60 golf courses (World Watch March/April 2004, p. 36).
- The human demand for water has been particularly devastating to wetlands. Globally the world has lost half of its wetlands, most in the last 50 years. One-fifth of the world's fresh water fish are endangered, vulnerable, or extinct (see Greenbiz.com Feb. 5, 2003).
- Wealthy citizens of the world spend US $14 billion on ocean cruises each year. According to the World Watch Institute (State of the World 2004, p. 10), US $10 billion annually could provide clean water to the estimated 1.1 billion people who lack it, in a world that spends about $240 million a day on tobacco products.
- The Glen Canyon dam created Lake Powell, which was designed to hold 24.3 million acre feet of water. In 1999, the lake was full, forcing water releases. In April 2004, the lake had only 10.2 million acre feet (42% of capacity), a level last seen in 1971. Given the continuing drought in the Western United States, experts predicted in 2004 that the lake could be dry by 2007. However, in July 2011, Lake Powell was 45.8 feet below its full level and 72.9% of its full capacity (24,322,000 af). In 2004, Lake Mead behind Boulder Dam was at 59% of capacity. Partially due to the upstream Lake Powell, scientists predicted in 2008 that there is a 50% probability that Lake Mead would be completely dry by 2021, because of climate change and unsustainable overuse of Colorado River water. The conclusion was the lake was at or beyond the sustainable limit of the Colorado system. The alternative to reasoned solutions to the coming water crisis is major societal and economic disruption in the desert Southwest; something that will affect all who live in the region.

  In 2010, the prediction was that if Lake Mead's water level drops below 1075 feet, it will automatically trigger emergency measures, including rationing, agreed on by the seven states that depend on Lake Mead's water. Ironically, the proposed rationing does include California, whose water demands get first priority.

- Half of all the world's hospital beds are occupied by people with water-borne diseases.
- In India, only 30% of the population has access to clean water. India has 2.2% of the world's people, 4% of its fresh water, but 17% of its population.
- Over-pumping of ground water is causing water tables to decline in important agricultural regions of Asia, North Africa, the Middle East, and the United States. The quality of groundwater is also declining (State of the World—2004; Worldwatch Inst., p. 17). The Ogallala aquifer, the nation's biggest source of underground water, is being drawn down eight times faster than the rate of replenishment. Total decline of the Ogallala reservoir since it has been monitored is about 200 million acre feet. It provides about 25% of the irrigation water used in the United States (14 million acres in Ogallala, 56 million total). Egan (2006) estimates that the rate of decline in Ogallala is about 1.1 million acre-feet a day. This is for withdrawal without recharge. The USGS estimates annual withdrawal is 19 million acre feet, still a large amount.
- Annual water withdrawals per person in cubic meters. Withdrawal is not equal to consumptive use. The thermoelectric (power generation) sector withdraws more water than agriculture but consumes very little.

| | |
|---|---|
| US | = 1,688 |
| Australia | = 945 |
| Germany | = 712 |
| China | = 431 |
| UK | = 201 |

More than a third of the world's people may soon live in areas that are water stressed. One can only conclude that water will be one of the primary factors that limits future world population growth and economic development. It is right to begin to consider if the proper goal is water for the wealthy nations or fresh water for the two-thirds of the world population that faces daily water stress?

Postel, S. 2000. Troubled waters. *Utne Reader*. July/Aug. P. 63.

No irrigation-dependent society, with the possible exception of Egypt, has survived, all have failed due to water logging or salinization of the soil, or both. "The overriding lesson of history is that most irrigation-based civilizations fail. As we enter the third millennium A.D., the question is: Will ours be any different?" (Postel, 1999, p. 12) It is accepted that these failures have been caused by poor irrigation practices, but salinization and water-logging of soil still occur. The question is an agricultural not a moral issue. However, we must also ask if feeding the rich by using water to maintain growth of expensive crops, irrigating golf courses, and consumptive home use, should rank above the very survival of others (see Singer, 2009). Those are moral questions.

## *Climate Change*

When addressing climate change/global warming, a major issue for most agricultural scientists, we humans have an advantage over other species. We are able to think ahead—anticipate the future—and prepare for change. We don't always use our advantage. It is likely that the earth will be at least 3°C warmer at the end of this century than at the beginning of the industrial revolution. The accepted projections by crop ecologists are that for every 1°C (1.8°F) rise in average temperature, wheat, rice, and corn yields are likely to decrease 10%. Days above 30°C can decrease yield at least 1%; days above 32°C may be twice as harmful. This is especially important in much of the corn growing region of Africa. If drought is added, the effects multiply. The Intergovernmental Panel on climate change forecast that the models are not keeping pace with the change. Greenland is losing about 52 cubic miles (miles is correct) each year and melting is increasing. Food security will become a major issue for the rich and the poor. The rich in industrial nations will be able to deal with global warming and higher food prices. The poor cannot.

Food production is a major issue, but there are also important environmental issues: Yellowstone Park is experiencing more severe fires,[10] partially due to climate change, which could shift forests to the North and by the end of this century, Yellowstone could be dominated by scrub and grasslands. Few doubt the scientific basis of the projections. Many fear the presently inadequate policy response. Skepticism that the political response will be adequate dominates.

Climate change may shift agro-ecological zones away from the equator toward the poles (Zilberman et al., 2004). The International Food Policy Research Institute (IFPRI) found that nearly all the results of climate change studies suggested yields of the world's primary cereal crops (wheat, rice, corn) are likely to be lower in 2050 than they were in 2000 (Economist, 2011). Half the studies predicted reductions between 9% and 18%. Wheat was the most vulnerable crop. However, although the planet has warmed during the past 30 years, temperatures in the mid-US, where up to 50% of corn and soybeans are produced, have not warmed. No one knows why and no one knows if it will last. Severe drought in 2011 in the southern US reduced the winter wheat crop. The results of a study by Deschênes and Greenstone (2007) were sharply different than most others and critically disputed by others (Fisher et al., 2007[11]). They conclude the effects of climate change will be insignificant or positive and project that it will increase annual profits on US agricultural land by 4% or $1.3 billion (in 2002 dollars). Similarly, some argue that because plants require carbon dioxide for photosynthesis, increasing atmospheric levels will enhance

[10] Severe fire frequency which was once every 100–300 years has become once every 10 years.

[11] Fisher, A., M. Hannemann, M.J. Roberts, and W. Schlenker., 2007. Climate change and agriculture reconsidered. Unpublished.

photosynthesis and yield will increase. Wheat will benefit enough to offset some of the negative temperature effects. Increasing carbon dioxide increases the efficiency of water use by rangeland grasses and unfortunately some weeds, which results in more growth for warm-season grasses (Morgan et al., 2011). Corn, which uses a different photosynthetic pathway, may lose efficiency.

As the planet warms, some species that are wanted or liked are likely not to have an important advantage that the unloved have. Insects, diseases, and weeds can and will adapt more easily to warming than large mammals and trees. They evolve quickly and unfortunately for agriculture, many are pests. There is a term for the fate of species that cannot evolve to keep pace with the change: extinction. Gardner (2006) claims climate change is humanity's most urgent environmental challenge because it:

- Is global in scope.
- Has the capacity to remake human civilizations (i.e., ice melting in the north could shut down the Gulf Stream; Europe's temperature would plummet).
- Brings a cascade of unpredictable effects. Of special importance to agriculture are the unknown effects on pests.
- Is occurring faster than predicted.
- Could soon become irreversible.

Berry's (1999, p. 104) always wise counsel is that human "ethical traditions know how to deal with suicide, homicide, and even genocide; but these traditions collapse entirely when confronted with biocide, the extinction of vulnerable life systems of the Earth, and geocide, destruction of the Earth itself." Climate change challenges our ethical foundation in both respects. Berry claims that the danger to and misuse of the earth stem from deficiencies in the "spiritual and humanist traditions of western culture." Both are primarily or exclusively committed to human domination of the earth and its resources.

## Pollution

Soil erosion depletes agriculture's ultimate resource and lost soil pollutes water. Erosion increases the amount of dust carried by wind, which acts as an abrasive and can carry about 20 human infectious disease organisms, including anthrax and tuberculosis. About 60% of eroded soil ends up in rivers, streams, and lakes, making them more prone to flooding and contamination from applied fertilizer and pesticides. Fertilizer in soil leads to eutrophication of rivers, streams and lakes, loss of biodiversity, groundwater and air pollution, and soil and water acidification. Between 30% and 80% of applied nitrogen is lost to the environment (Conway and Pretty, 1991). The relationship between fertilizer use, soil erosion, nitrogen runoff, and the dead zone in the Gulf of Mexico was mentioned in Chapter 1. Harm to nontarget species from pesticide use in production agriculture is a major ecological concern. Global pesticide use has increased from almost none prior to 1950 to 4.7 billion tons per year. The 3 million cases of pesticide poisoning in the world each year (WHO, 1990) mean that, on average, six people are poisoned by pesticides somewhere in the world each minute. Of those poisoned, 220,000 die, mostly in the world's developing countries (WHO cited by Pimentel and Greiner, 1997, p. 52).

To double food production to meet expected demand will exacerbate these problems. Doubling with present technology will require increasing application of nitrogen and phosphorus by more than 2 times. Humans already release more nitrogen and phosphorus to terrestrial ecosystems than all natural systems combined. Doubling food production will increase eutrophication of marine ecosystems, loss of biodiversity, and groundwater and air pollution (Myers, 2009).

## *Loss of Farmers*

Soil is agriculture's most important productive resource, but farmers are agriculture's primary knowledge resource. Most nonfarming people regard farming and ranching as a routine, humble, nonintellectual activity performed by people (usually, it is assumed, by men) who are fundamentally, poorly educated hicks. They farm because they chose to or could not make it in a more challenging career. Nothing could be further from the truth. Farmers and ranchers are the custodians and stewards of the world's productive land and they are disappearing rapidly. We may be compelled to decide that lots more of them are needed. We may decide they should be paid to be stewards of the land and ecological care takers as well as producers. Those who derive their primary income from farming were fewer than 960,000 in 2002, less than 1% of the US population. Forty percent were over 55 and twenty-six percent were over 65. The number of full-time farmers less than 35 has steadily declined to 5.6%. Thus, the few who know how to care for the world's most important resource, the land, is declining as their average age increases. Average farm size was 140 acres in 1910, 216 in 1950, 464 in 1992, followed by a decline to 418 in 2009. There were 6.5 million farms in US in 1935; about 2.4 million in 1980 (Gardner, 2006) and 2.2 million in 2009. While the number of farms and farmers has declined, production per farm and farmer has steadily increased, labor required has steadily decreased, yield per acre has increased, and the cost of food to consumers has steadily declined as the farmer's share of the food dollar decreased (Gardner). Of the remaining farms, 61% of sales is captured by 163,000 large, industrial farms and most of these are contractually tied with a corporation in a prescribed value chain that obligates them to produce for and sell to the corporation and thus they have ceased to be traditional family farms (Kirschenmann, 2000a) where the farmer owns the land, makes the management decisions, and provides most of the labor.

**Highlight 11.3**

Visit http://www.kansasfreeland.com (accessed 2005 and August 2011) to learn that if you are willing to move to Plainville, KS or one of 10 other cities, you can obtain a building lot for free. Plainville (population just over 2000) in Rooks County (population 5,800) of northwest Kansas, is offering free 143 by 175 foot lots for the construction of new homes. The North Town Addition project will give people a chance to build a home and live in a small-town atmosphere, and at the same time, have big-city conveniences not far away. In May 2005, four building lots were available on a first-come, first-served basis. Several communities in Kansas are offering free land and other incentives. The goal is to keep rural areas viable and promote economic growth.

In the first edition of this book, Chugwater, WY had a similar program. In 2000, the population was 244, with a median annual income of $23,750. The average temperature was 46.7 in January and 69.4 in July. The wind blows most of the time. Beginning in May 2005, but no longer, Chugwater granted newcomers a 100 by 120 foot city lot, if the applicant agreed to build a house and live in it for 2 years. If you like peace and quiet, Chugwater may still be your place. In 2000, the town had no policeman, no traffic light and not much traffic, no grocery store, and no bar, but it did have a soda fountain with 48 flavors of milkshakes (see *Denver Post* April 24, 2005, p. 1a and 8a). In 2011, there is a sheriff in town, still no traffic light, not much traffic, and no grocery store, but there is a bar at the Buffalo Lodge and Grill. The Soda Fountain still offers lots of choices for malts and shakes, and if one wants peace and quiet, it is a great place to live.

Rural America is emptying. Almost 700 rural US counties lost more than 10% of their residents between 1980 and 2000. Most, but not all, are in the Central Great Plains. In 1900, 60% of the US population was engaged in some kind of agriculture; today less than 2% is and their number is declining. Fewer than 4% of US farms produce about 56% of all agricultural sales. In Colorado, nearly 6 million acres were "developed" from 1992 to 1997; more than double the conversion rate from 1982 to 1992. There are a few exceptions, but in most US states, census data show that, the number of farmers declined from 1940 to 2000 and the size of farms increased. The average age of US principal farm operators was 55.3 years in the 2002 census and has increased in each census since 1978.

Use of these data often elicits an accusation of nostalgia for the good old days. I am guilty. I am nostalgic for what I know I have lost. The challenge then goes on to assert that society does not have any obligation to preserve or to save what someone may love. The corner gas station is gone. The Mom and Pop grocery store has disappeared in most places, and few lament their passing. Are there any rural blacksmiths left? It is progress, and one gets in its way at one's peril.

Small farms are economically (they make little, if any, profit) and productively inefficient. Yields are frequently high per unit area, but total production is low. Our economic and production system compels small farmers to use technology that they may know is not sustainable and is not compatible with being a good farmer. Survival has a higher value than environmental correctness. Small family farms may not be good stewards of the land. Without thinking about it, the American public has tacitly agreed with the political decision to let small farms disappear. They are small, after all, and they disappear quietly without political turmoil. Those affected don't demonstrate in city streets, they don't riot or cause riots; they just go away quietly.

What we are headed toward is a food system that is supplied by 50,000, or so, large farms and ranches each of which will be efficient and productive. That is what will be gained. What is lost? When we lose farms and ranches where a family owns a majority of the capital resources, makes the important management decisions, and provides most of the labor, have we lost anything else? We have lost a group of people who have a daily, personal contact with nature. People who create, populate, and assure the continuance of rural communities with a social contract that works. These folks are in it for the long haul. They create sustainable human and agricultural systems. Their communities have a tightly knit social fabric that seems to be the antithesis of the alienated urban centers where most Americans live. We have also, in a very real sense, lost our seed stock. Those who teach in land-grant agricultural universities can teach students a lot about farming and ranching but to really learn how to farm or ranch one must walk the land with a farmer or rancher. That knowledge base is disappearing. We always think we know what we are doing and what we are gaining. Dwelling on what we are losing, as we gain is not just foolish nostalgia.

So what? We no longer need a large number of automobile producers to make all the vehicles we need. Why do we need all those farmers and ranchers? Consolidation has been beneficial in most manufacturing industries and, it is assumed, it ought to be in agriculture as well. With ever-improving agricultural technology small, inefficient producers are simply not needed. We need production, and if it can be done best (i.e., at lowest cost) by a few producers, then it should be, even if that means moving much of our food production outside US. That is the nature of the capitalistic enterprise. Capitalism, a process of creative destruction, has winners and losers. The latter, ideally, are absorbed by the winners, the more efficient enterprise. However, as family farms are lost, we will lose the rural communities that the farms and ranches created and sustained. That loss is also regarded by many as a loss that removes a problem, but does not create one. No one, it is claimed, wants to live in rural backwaters that have few of what many assume are the required amenities of modern life (convenient entertainment, places to buy almost anything, fast food, convenient coffee, and so on). However, if these places are the source of important American values and if the people who inhabit them take care of the land, we may lose those things as well. Economists and consumers understand what cheap food costs but it is much more difficult to place costs on qualitative things: the lives of farm families that are destroyed when the farm is lost, the loss of communities, the loss of heritage. We do not understand these costs, because we calculate only what can be quantified. We don't know how to calculate the costs of fundamentally qualitative things. It is worth thinking about.

After spending my career teaching agriculture in a university, it is hard to admit, but undeniably true, that one cannot learn how to farm or ranch in a university. One can learn a lot about farming or ranching and about techniques and technology. But if one wants to learn how to farm or ranch, one must ask (must study with) a farmer or a rancher. They are the best teachers and they are disappearing. According to the US Dept. of Agriculture,[12] Gardner (2006), and Dimitri et al. (2005), the number of farmers has been declining in nearly all US states. We do not understand what the costs of the loss of their experiential knowledge may be.

Some countries have capital to export, but must import food because their land resource is not adequate to feed a large, growing population. Food production is being outsourced to countries that need capital and have abundant land. Capital-rich nations acquire the right to produce food elsewhere. They fulfill their obligation to provide food for the people and thereby recognize the absolute necessity of land. Rich importing nations are acquiring vast tracts of farmland in poor countries. Supporters of these arrangements claim the rich, importing country provides new seeds, technology, and money for agriculture, which poor nations do not have, even though agriculture has been the basis of their economy for decades. The projects claim they will improve agriculture. Opponents say the projects are primarily land grabs and argue that poor farmers will be pushed off the land and the people or the country will not be helped because all the food is exported. There are reasons for skepticism, but it is too early to tell if these programs will reverse the decline of farming in poor countries. The point is that rich countries with minimal land at least tacitly admit that if food is to be produced, land and farmers are the essential resources.

### Population

The world's population is still growing, and barring a major disaster (earthquake, nuclear war, massive flooding, worldwide disease epidemic, and so on), it will continue to grow for the next few decades, but the growth rate (1.4% per year, World Bank, 2002; 1.2%, World Bank, 2010) will continue to decline. One cannot blame those who practice agriculture for population growth but agriculture's role is clear. Without production increases, it will not be possible to feed the expected increase in population. There will not be enough food. Agriculture's practitioners have always claimed credit for feeding people, therefore they must share at least some of the blame for population growth. Most of the problems enumerated herein were, at least partially, enabled by agriculture's adoption of practices that increased production, while creating societal externalities. As mentioned previously, the human ecological footprint (see Wackernagel and Rees, 1996) has grown due to the increased wants of the rich and to the sheer increase in the number of people the planet must support.

## Dominant Scientific Myths

Kirschenmann (2000b), and other thoughtful commentators on agriculture know the agricultural myth that production is all that matters must be abandoned. Production

[12] www.census.gov/population.cen2000. Accessed November 2004.

does matter, but it is not all that matters. We must acknowledge that agricultural practice has caused real, enduring harm to the environment and to people (e.g., migrant labor and small family farmers who have been driven off the farm). As we dispense with pervasive myths about agriculture, we must also dispense with the scientific myths that pervade agricultural and general science. They have been described by Sarewitz (1996, Chapters 2–6) and cited by Busch (2000, pp. 66–67).

When science began, the intent of most scientists was to explore and understand nature's complexity. This was the only course available, because scientists had not yet developed the ability to command or attempt to dominate the natural world. Early agricultural scientists and farmers might have wanted to dominate and subdue nature, but they could not. Humans were dependent on and subject to the natural world. Farming often failed due to bad weather, poor fertility, lack of water, or pest outbreaks that could not be controlled. As science developed, efforts were more and more directed toward developing "technologies that could extract economic benefits from nature" (Kirschenmann, 2002). For example, in weed science the emphasis nearly from the beginning has been on ways to kill weeds selectively in crops. Only recently has it turned toward developing an understanding of the complex biological systems in which weeds occur and often dominate. Weeds were regarded as inevitable companions of growing crops. They were the inevitable outcome of the way crops were grown. They were not seen as problems of the production system that, if modified, might be diminished. That view is changing.

It is a certainty that over time, agricultural scientists have developed myths (stories to explain a phenomenon of nature) that guide the conduct of the science. The dominant and commonly accepted myths about science govern not only the science that is done but also its public acceptance and social consequences. Every scientist brings a conception of science to a problem or a new field. There is no such thing as a scientist with a clean slate (Larson, 2004, p. 165). The following dominant myths are unavoidable (Sarewitz, 1996; Busch, 2000).

## *The Myth of Infinite Benefit*

This asserts "if more science and technology are necessary for a better quality of life, then the more we spend on research the better our quality of life will be" (Sarewitz, 1996, p. 19). Thus, more science and more technology will always yield more public benefits. The scientific enterprise is seen as separate from society and in a pure utilitarian sense it "exists to provide a constant flow of benefits to all" (Busch, 2000). Sarewitz says many scientists hold that "the more innovation we have, the more competitive we will be as an economic entity, and the healthier we'll be as a nation." Science is to be regarded as, and scientists often think of themselves as, people engaged in an activity that provides the greatest benefit to the greatest number of people. This attitude ignores two things: the benefits of science are usually most readily available to the rich, and agricultural research and the resultant technology very often create new, unforeseen problems that have to be solved. The benefits are accepted and credit is sought for solving some problems. Unsolved and new agricultural problems are commonly externalized and become social costs.

Kirschenmann (2002) pointed out that this utilitarian science has at least three unintended consequences. The first is that the scientists often misapprehend the true nature of the problem—why weeds exist in crops may be a more important long-term question than how to control them annually? The dominant question for most agricultural scientists has been how to increase production. This resulted in acceptance of the productionist agricultural ethic defined by Thompson (1995, p. 48), which accepts that any behavior and any technology is good as long as it increases productivity. There was only one imperative: to produce as much as possible, regardless of the ecological costs and perhaps even if it was not profitable to the producer (Thompson). The productionist ethic has become the dominant ethic, because those who practice agriculture have always believed that hard work is followed by an accumulation of wealth, which is morally acceptable. High agricultural production is a sign that the producer has been favored by God's grace. Thompson suggests that the ethic has dominated because it is believed that to leave land alone is to squander resources provided by God for our use. We are the designated stewards and producing more is the best sign of our stewardship, because production benefits all. Second, Kirshenmann suggests utilitarian science has separated us from nature. Utilitarian science believes that nature is to be used by humans but use has led to abuse. Humans believe that they have been selected to have dominion over and to subdue the natural world to provide the greatest good for the greatest number of people. We are not part of nature, it is a place from which we extract benefits; not something to which we belong. While abuse may not be the intent, it is the inevitable result of modern agricultural practice. The experts who conduct agricultural research and those who apply the resultant technology to produce food have not paid much attention to the long-term ecological and social effects of the enterprise because the immediate utilitarian benefit of production has been apparent and welcomed.

The productionist ethic is bankrupt (Kirschenmann, 2004), because it fails to prescribe any standard for agriculture that views nature as anything other than a static, mechanistic structure that can be and ought to be controlled by humans with technology. It assumes nature is stable and largely immune from harm and it assumes that agriculture operates in an economy where value is solely determined by price (Kirshenmann). Because the evidence is clear that the productionist ethic has led to more harm than good (Green et al., 2004), a new ethic is demanded that guides an agriculture that does not ruin the ecological and social communities on which its success and future are dependent. The history of agriculture is replete with examples of ecological failure in single fields and for entire civilizations (Thompson, 1995, p. 76; see McNeill, 2000) but agriculture's practitioners, in their quest for greater production have ignored their own history. In Logan's (1995) view, ignorance is at the base of the problem. "More technology, greater planting rates, more intensive use (of soil), greater dependence on larger holdings, and fewer farmers are supposed to save the day. Instead they hasten decline."

## *The Myth that Science and Scientists are Value Free*

When I was a student, I don't recall any professors who made the explicit claim that science was value free. Of course, one might argue that I don't recall much of what I learned during my student days. However, I did learn or acquire the fact that science is value free and presumably free of the constraints of ethics; an unexplored topic. I and my student colleagues learned that science and scientists value objectivity over subjectivity. Subjectivity, I learned, is the realm of values. Objectivity, the scientific approach, is a "value-free" position. This illustrates what Rollin (2011, p. 91) identifies as an "ideological ubiquitous denial of the relevance of values in general, and of ethics in particular, to science." The denial blinds scientists to moral issues, which are often at the heart of societal concern about science (Rollin). The argument that science is value free is one of a number of value claims and ethical issues that are fundamental to the scientific enterprise. Science includes a mind-set that assumes certain value claims. Its foundation stands within the arena of values and ethics. For example:

- Objectivity is valued over subjectivity.
- Because science is value free, considering the ethical implications of scientific research is not necessary.
- Human life is of greater value than other forms of life.
- Only humans deserve moral consideration.

Claims that value judgments, ethical arguments, and moral questions stand outside science betray a naiveté regarding the assumptions on which science rests.

I and my student colleagues learned that the difference between empirical (scientific) facts and value judgments is that only the former are meaningful. We learned we were logical positivists although none of our teachers ever used or defined the concept. Scientific education and subsequent thought was guided and perhaps distorted by denial of the meaningfulness and relevance of value questions. Science dealt with facts. Ethics deals with what ought to be, with values. One should not, indeed one cannot, get values from facts. Facts do not provide reasons for action. Facts help us make sensible decisions, but no amount of facts can make up my mind about what I ought to do (Singer, 1981, p. 75).

Opponents of scientific technology deserve to be heard, as they raise issues that ought to be addressed by the scientific community. Those who take different positions on the proper value orientation of (pesticides, biotechnology, or sustainability) should not be dismissed because, it is claimed, they simply muddle scientific debate with irrelevant ethical arguments and moral claims, which, of course, is a position with a moral foundation, albeit an unrecognized one.

## *The Myth of Unfettered Research*

This myth asserts that any scientifically reasonable basic research—the study of fundamental natural processes—will yield social benefits, ought to be permitted, and be publicly funded. Scientists are well-educated people whose specialized training demands they be detached from the concerns of daily life—a value—so they can pursue scientific interests that will advance the frontiers of knowledge and improve human life (Busch, 2000). In fact, "researchers motivated by curiosity about nature have produced a great abundance of startling, unexpected and marvelous discoveries over the past fifty years" (Sarewitz, 1996, p. 48). This myth is related to the belief that the scientist, qua scientist, engaged in research using the scientific method is and should be unhindered by values. This is patently false. "Political and historical milieus strongly influence the course of basic research" (p. 39) in all scientific fields. Agricultural science, like all science, is controlled by the constant, required quest for funding. Legislators and funding agencies have priorities that value some lines of research more than others. Therefore, agricultural research and the scientists who conduct it are not unfettered. They are tightly bound in a vortex of largely unexamined and unquestioned values.

## *The Myth of Accountability*

This claim is that peer review of scientific results prior to publication and the necessity of repeatability of conclusions are sufficient to maintain the intellectual integrity and ethical responsibility of scientists. This neat locution which says—trust me, I am a scientist—leaves out the public, which is asked to fund the work and ignore its consequences. If the research meets the criteria of high intellectual integrity and established scientific standards for performance, then society must be satisfied, even if, as has been the case for much agricultural research, it may have undesirable ecological or social consequences.

This myth is explored in Dürrenmatt's (1964) play "The Physicists," in which he asks several relevant questions:

- Is it always best to seek to know everything?
- Who is to be held accountable for the wrongs science commits: those whose work leads to discoveries that harm ecological and social relationships *or* those who use the work that others have done to cause the harm?
- Can anyone be held accountable for the moral aspects of science?

The play asks the audience to consider, when can one be sure they are doing the right thing, and how does one decide what the right thing is? Dürrenmatt's work raises important questions of accountability for all scientists. It is reasonable to postulate that science is essential to the solution of many of the world's problems. It follows then that it is vital that the public's current high esteem (Sarewitz, 1996, p. 58) for and trust in science must be maintained. The integrity of science does not and cannot end with delivery of a product that is quality controlled and intellectually sound according to science's internal criteria. The scientific honor code, in Sarewitz's view (p. 59), is not just about the conduct of science. It must also be about the "ethics and values of science as a component of society."

## *The Myth of Authoritativeness*

The assertion is that science can provide a rational, objective basis for creating political consensus by separating fact from perception. In fact, the opposite seems to be the case: "political controversy seems uniformly to inflame and deepen scientific controversy" (Sarewitz, 1996, p. 77).

Scientists often believe science is objective and value free (see 2 above, Hollander, 2000); therefore, one need only examine the data to know what to do. The falsity of this claim should be immediately obvious. It has been the most contested areas of science that have been the most vigorously debated in the political realm. In agriculture, for example, pesticide use is highly regulated, but there is little political or scientific consensus on the effects of pesticides on human health. Animal rights are also prescribed in law, but there is still great controversy about animal treatment. Most people are meat eaters, but prefer not to know too much about how the animals they eat are treated (e.g., see Schlosser, 2002).

Busch (2000) points out that when scientific consensus emerges, public and political consensus quickly follows. Global warming is a good example of a problem referred to the scientific community by politicians, and, as scientific consensus emerged, international willingness to confront the issues also emerged. But we are a long way from consensus on the use of pesticides in agriculture, confinement rearing of animals, or the value of the productionist paradigm.

### *The Myth of the Endless Frontier*

New scientific knowledge generated at the cutting edge of basic science is or ought to be free of careful consideration of its moral and practical consequences because it will be transformed into new technologies that will be benefit all. Basic science therefore should not be subject to careful scrutiny for its potential consequences because they cannot be known in advance. "Fundamental scientific knowledge is a thing apart, accumulating as if in a reservoir, from which it can later be drawn by applied scientists" (Sarewitz, 1996, p. 98) … who create products and processes. It is the consequences of applied science (technology) that should be of concern and good technological advances are dependent on basic science. Busch (2000) suggests the division is false because basic science and technology are inextricably linked in several ways. Scientific problems emerge from new technologies and most scientific work is dependent on technology developed by science. For example, many university research scientists owe the existence of their position to early observations that some chemicals could be used to selectively control some agricultural pests. Further exploration of pesticides is highly dependent on new chemical analytical technology developed to find pesticide residues.

Sarewitz (p. 103) points that the rise of the environmental movement in the industrialized world marked the end of the myth of the endless frontier. People began to recognize that the conquest of the frontier enabled "liberation from elemental want" and a steadily rising standard of living, but carried with it an "acceleration of exploitation, modification, and despoliation of nature." Moral and practical consequences became more or equally as important as material benefits.

## Production and Ethics

In spite of the apparent railing herein against the productionist ethic, production is essential. Production of sufficient, high-quality food and fiber is the only viable way to feed the world's people (Rist, 1988). However, one must ask production for what and by whom? Agriculture should not abandon its quest to improve and ethically justify production (Kirschenmann, 2004). However, production should not have primacy over everything else. What is needed is public participation in development of an ethical foundation that considers the priority of production in comparison to ensuring the need of all humans for food is met (Rist, 1988). It is an important issue that ought to be part of discussions of agriculture's ethical horizon, but it has not come to the fore. It is readily acknowledged that agricultural systems must be highly productive and sustainable. That affirmation is consistent with agriculture's moral obligation to feed the world. But it has not been accompanied by debate about whether or not there is a human right to food. The Scandinavian nations have done more per capita than others to fulfill the duty to give food and other types of aid to developing countries. Food aid is a priority for international aid agencies (especially the World Food Program) (Thompson, 2010). But in spite of their commendable efforts, a human right to food has not been affirmed. I recommended that agricultural organizations should formulate and prominently include in their mission and objectives a statement of their ethical position that recognizes the obligation to conduct agriculture "in a manner that makes a decent life for humans possible while, at the same time, retaining the ecological dynamics that sustain life on the planet" (Kirschenmann, 2004). It must be an ethic that is human oriented but acknowledges the ecological relationships that make farming possible. Part of achieving food for all is ensuring that markets function to achieve distribution to the one billion people who are not fed well. Markets are essential to the task. To achieve this requires that those involved in agriculture cooperate with all members of the general society. Creating a sustainable agriculture is not and cannot be just an agricultural responsibility, it is a social responsibility (see Chapter 7). It is an agricultural task in that those who practice agriculture must change some of their practices. Some changes that should be considered include (Rist, 1988):

- Reducing losses. Losses during harvest, postharvest storage, and processing should be reduced. Agriculture could lead the way toward a true recycling economy, where the waste from one enterprise becomes the feed stock for another.
- Ending wasteful habits. There are clear, well-reasoned arguments concerning the moral status of animals (e.g., see Cavalieri, 2001; Rollin, 1989, 1992; Singer, 1977, 2002; and others cited in Chapter 10). There are equally clear, well-reasoned arguments that animals are essential to a truly integrated, sustainable agriculture (von Kaufmann and Fitzhugh, 2004; Smil, 2000; Chapter 10). This debate must be resolved and part of the resolution will be a diminution of the excessive consumption of meat by the rich, which is harmful to human health and wasteful of resources (e.g., land and grain) that could be used to feed people.
- Ending pollution. Pesticide use, especially prophylactic use, and excessive fertilizer use will have to be diminished. Erosion of soil must decrease to protect our most valuable environmental resource and to diminish water and air pollution.
- Policy changes. If the public wants farmers to conserve energy, reduce pollution, and promote ecological stability, policies that reward farmers for adopting such practices will have to be developed. Most farmers know how to farm in ways that prevent soil erosion, do not mine water, reduce pollution, promote animal welfare, and achieve ecological harmony. Present government and market policies that reward only production must be changed so desirable practices are rewarded.

The preceding pages have emphasized some of the harms agriculture has done in its endless quest for more production. Those harms have created agriculture's public image, which is not favorable. Kirschenmann (2000a) proposes that those who practice agriculture must change its public image, essentially because agriculture has lost its connection to the public. Most people in developed countries are not farmers or ranchers and have no connection to the source of their food. People eat but do not know how their food is produced. Kirschenmann said it well:

> *If agriculture is purely an industrial act whose only purpose is to manipulate the technologies required to produce some wheat* with which we have no connection, *that is ground into flour in some distant factory* with which we have no connection, *made into frozen bread dough in some warehouse-like bakery* with which we have no connection, *and placed in to a microwaveable plastic container* with which we have no connection—*and all the while the process may be harming monarch butterflies, or rendering our water unfit to drink, or killing off the fish in our favorite streams*—how could we expect the public to support agriculture?

## The Imperative of Responsibility

To create a new, widely accepted public image, those who practice agriculture, those who study it, and all who benefit from it because they eat should consider adopting, as a general standard, what Jonas (1984) calls an ethic of responsibility. The responsibility is to future people, a philosophically debatable proposition because no one knows what the future holds. We cannot know the people who will inhabit the distant future (100 years hence), what their situation will be, or the kind of world they will inhabit, therefore, we cannot assume we are obligated to them. We cannot "catch" them and won't ever "meet" them because we won't be there. Jonas strongly suggests that even though these things are true, we ought to accept an obligation to the future. He suggests present humans do not want to (will not) accept an obligation to assure the happiness of future generations if the price is unhappiness or even death of some present humans. Given this position, it is not logically inconsistent, in Jonas' view, to posit that the price of the happiness and well-being of present humans should not be bought at the cost of the existence or happiness of future generations. The difference in the two cases is that in the first case, the well-being of future humans is ensured, albeit, perhaps, in diminished circumstances, while in the second case, future humans may be eliminated. Sacrifice of the future for the present is logically in Jonas' view "no more open to attack than the sacrifice of the present for the future." The imperative thus becomes— "Act so that the effects of your action are compatible with the permanence of genuine human life." Or, in a negative expression, "Act so the effects of your action are not destructive to the future possibility of such life." In Jonas's view, we are obligated to try, to the best of our ability, to create "the conditions for an indefinite continuation of humanity on earth." Why? Because anyone may choose to risk or end one's own life, but no one has the right "to choose or even risk, nonexistence for future generations to assure a better life for the present one" (Jonas). It is a compelling moral claim.

**Highlight 11.4**

According to the World Bank, almost one-half of the 6.4 billion people on earth live on less than US $2 per day and more than 1 billion live on less than US $1 per day. "Some of this misery is caused by incompetent and rapacious governments in the developing world, but not all of it. For more than two decades, dozens of impoverished countries have been forced to spend more money servicing loans outstanding to wealthy foreigners than on hospitals and schools. In many cases, the governments that took out these loans no longer exist, but their successors are shackled by onerous interest payments."

When one attempts to apply the imperative of responsibility to agriculture, the essential human activity, it is clear that the obligation is a collective not just a personal one. Agriculture is a private enterprise with large public consequences and therefore, the public must act to help create the kind of agriculture that assures "the indefinite continuation of humanity on earth." To illustrate the point that the achievement of a sustainable agriculture is not just an agricultural responsibility one need only look at some trends in US agriculture.[13] From 1900 to 2000, average farm size more than tripled, the number of farms declined by about one-third, while the number of acres farmed remained about the same. The percentage of the US work force in agriculture declined from almost 40% of the population in 1900 to less than 1% in 1990. The US population steadily increased; farm population steadily declined as did the percentage of the population living on farms. Total farm output declined slightly; required inputs and productivity per worker increased dramatically. The market value of agricultural production became concentrated on fewer farms because of the combined effects of the increased capital requirements in farming, higher levels of costly technology, and higher government price supports. Farming and ranching became more efficient in terms of production per worker, more costly, and less profitable for farmers. Improved technology increased production and created the environmental and social problems that have been mentioned. The ethics of agriculture and the ethical dimension of its many problems have not been of concern within agriculture as long as production increased. The USDA web site (footnote 10) claims there is a recognition among US citizens that "families involved in farming and the diversity of farm operators are important to the cultural identity of our country. The farming and ranching lifestyle is still believed to be an important and virtuous endeavor, worthy of continued support." The statistical evidence does not lead to the conclusion that those engaged in this virtuous endeavor have received much support to continue or even to survive in agriculture.

Many will argue that the observed trends in agriculture are to be expected. They simply follow the trends of consolidation and promotion of production efficiency in all important industries. If US can be fed by 10 (or some small number) large, highly efficient, well-managed farms and ranches, that will be good for all. The argument is a clear

[13] http://www.usda.gov/nass/pubs/trends/farmpopulation.htm. Accessed February 2004.

economic one and is persuasive. It is not a moral argument and as Busch (2000) says, "ultimately moral suasion is more effective for most adults than incentives." But agriculture has not engaged in moral suasion. Those engaged in agriculture have ignored moral arguments because they have thought they had already won the moral case by feeding people, a morally acceptable act, and ignoring the harms done. No one in agriculture has tried to learn if the public really wants more production if the environmental and social costs remain high. The Green Revolution helped to feed the poor especially in Asia but not in Africa because the successful high-yield crops (wheat and rice) are less widely planted in Africa. The clear environmental and social costs (massive, often inappropriate, pesticide use, loss of genetic diversity, disruption of stable rural cultures, development of a system that favored large farms, small farmers being driven from the land, and development of genetic crop monocultures with increased disease and insect susceptibility) have been high and largely ignored by agricultural scientists and agribusiness people. These costs have been regarded as the price of bounty. Such unexamined, yet certain, moral positions are potentially dangerous because, if they persist, the argument about agriculture's future may be lost without being engaged.

There is competition and conflict within agricultural science for research funds and attention from the media. There is no concern about agriculture's moral status because it is not debated. Agricultural science, similar to other sciences, has its scientific facts in order although its underlying theories, which determine what facts are acceptable, may be less certain (Barker and Peters, 1993). Agricultural science suffers in its quest for funding and in its public image because there is not one scientific reality but several, which is as it should be and is not unique to agriculture (Barker and Peters). Scientific advice on public policy issues should, at its best, be conflicting because the social and physical worlds are so. Science in agriculture, or in any other discipline, cannot and should not attempt to give the final definitive answer on what ought to be done in public policy. The task is to interpret scientific findings with all their uncertainty, but not to provide definitive answers to complex social and environmental questions. However, those who give advice will be more certain of their answers and advice when they rest on a firm, well-articulated ethical foundation. A firm ethic is a moral theory in which considered intuitions have been brought into equilibrium with moral principles and scientific knowledge (facts) (Comstock, 1995). If science promotes its technology, as it has in the past, in the absence of moral scrutiny, the results are likely to be technologically successful, as they have been, but the social and ecological effects may discredit rather than honor the scientific developments and the entire scientific enterprise (Wright, 1990, p. 236).

Wildavsky (1995, pp. 439–441) is correct in his assertion that many of the public's fears about the harm caused by agricultural science have a moral foundation. The fears explain people's risk perception and may determine government policy, but they are out of place in determining risk consequences. A fear based on a perceived moral wrong (e.g., it is morally wrong to use pesticides that harm humans, nontarget species, or the environment) and the perception of harm, is not the same as the presence of harm. Wildavsky advocates "citizenship in science" as a prerequisite to moral outrage and demands to stop an action or "to get rid of the stuff," regardless of the cost. What is wrong in Wildavsky's view is that moral outrage has been allowed to lead policy in spite of clear scientific evidence, which although the moral outrage is clear, does not support the claim that harm to anything has been caused or is likely to result from continuation of the practice. Those who practice agriculture, those who do agricultural science, and those who raise moral issues and complaints must all be responsible morally and scientifically.

## Finding Partners

As those engaged in agriculture expand the realm of inquiry about agriculture, they may find it interesting to know who is asking the same questions and with whom it may be good to form partnerships to raise and discuss agriculture's ethical and other issues. Zimdahl and Speer (1998)[14] examined mission statements of agricultural producer groups and asked if they shared missions and objectives with environmental groups and agribusiness companies. They asked which of these might be the best source of intellectual and moral support as land-grant universities strive to fulfill their mission.[15]

When discussing interpretation of scientific results with students and colleagues, a common approach is to examine the data. Show me the data is a prominent research theme and pedagogic technique. What the data reveal when expressed quantitatively helps guide one toward the meaning and conclusions of an experiment. Scientists prize and teach students to prize conciseness. When a large truth can be expressed with simplicity and brevity, it approaches scientific truth and perhaps beauty (Krauthammer, 1997). They believe in the wisdom of Occam's razor: When confronted with two or more explanations for a phenomenon, the simpler, less complicated one is most likely to be correct. The goal is to find simple explanations supported by the data.

Agricultural scientists are continually challenged and frustrated by questions based on feelings or opinions, but not on, or in ignorance of, the data. These may come from colleagues in nonscientific fields or from the general public. They are questions that cannot be answered by the data or by attempts at elegant simplicity, because they originate outside the established bounds of scientific procedure. Questions may be about what the data mean. Often they are about what one intends to do because of the data, or about why such work was done at all. They probe the process and goals of science, but they are not empirical, narrow scientific questions that can be answered by appeals to the data. Ultimately, they are questions about the nature or acceptability of the mission of agricultural science and the techniques used to accomplish the mission.

The purpose of Zimdahl and Speer's (1998) paper is to examine divergent views of agriculture and its mission. Publicly available mission statements or statements of objectives from 16 agricultural businesses, 22 agricultural producer and allied groups, and 25 environmental groups were examined (Table 11.1). In 2011, mission

[14] Much of the following is reproduced with permission from Zimdahl, R.L. and R.L. Speer., 1998. Agriculture's mission: finding a partner. Am. J. Alt. Agric. 16:35–46.

[15] See Zimdahl, R.L., 2003. The mission of land-grant colleges of agriculture. Am. J. Alt. Agric. 18(2):103–115.

**Table 11.1** The Agribusiness Companies, Agricultural Producer and Allied Groups and Environmental Groups Surveyed in 1998/99 and 2003

| **Agribusinesses (13)** | **Producer and Allied Groups (21)** | **Environmental (25)** |
|---|---|---|
| AgrEvo (only 98) | Ag in the classroom | California Food Policy Advocates |
| American Cyanamid (only 98) | Agricultural Women's Leadership Network | Californians for Pesticide Reform |
| Archer Daniels Midland | Agriculture Council of America | |
| BASF | | |
| Bayer | | Campaign for Food Safety |
| Cargill | American Agri-Women | Center for Food Safety |
| Conagra | American Egg Board | Center for Rural Affairs |
| Continental Grain (98) ContiGroup (2003) | American Soybean Council | Center for Science in the Public Interest |
| Dow Agro-Sciences | Animal Industry Found. | Consortium for Sustainable Agriculture Res. and Education |
| DuPont | Council for Agric. Sci. and Technol. (CAST) | |
| Farmland Industries, Inc. | Dairy Management | Consumers Union |
| Monsanto | Farm Bureau | Environmental Defense Fund |
| Mycogen | H.A. Wallace Institute for Alt. Agric. | Food Research and Action Center |
| Novartis | | |
| Rhône Poulenc (only 98) | National Agric. Inst. for Alt. Agric. | Greenpeace International |
| United Agri Products | National Agric. Center and Hall of Fame | Interfaith Center on Corporate Responsibility |
| | National Cattlemen's Beef Assoc. | Izaak Walton League of America |
| | National Corn Growers Assoc. | Loka Institute |
| | National Cotton Council | Organic Consumers Association |
| | National Council of Farmer Cooperatives | National Audubon Soc. |
| | National Farmers Union | National Coalition Against Misuse of Pesticides |
| | Future Farmers of America | National Resources Defense Council |
| | National Grange | |
| | National Pork Producers Council | Pesticide Action Network |
| | | Resources For the Future |
| | | Rural Advancement Foundation. Int. |
| | | Sierra Club |
| | | The Nature Conservancy |
| | | Union of Concerned Scientists |
| | | Worldwatch Institute |

statements for 15 agricultural businesses, 21 agricultural producer and allied groups, and 24 environmental groups were reviewed. A few reviewed in 2003 had merged or ceased to exist. Although each group is involved in agriculture, it was assumed that their separate mission statements would create differing views of what agriculture's mission is and perhaps what it ought to be. There was no attempt to pick all possible representatives of each group.

The intent was to ascertain if the mission statements of agribusinesses, agricultural producer, environmental groups, and Land-Grant Colleges of Agriculture demonstrated shared goals. A hypothesis was that colleges of agriculture and environmental groups might share goals even though these groups frequently regard each other as adversaries. A second hypothesis was that agribusiness companies do not share missions or operational objectives with agricultural producer groups. Common interests are rare, although they usually regard each other as allies. Zimdahl and Speer (1998) discussed these hypotheses. That discussion continues herein together with a comparison with the mission statements of 50 land-grant colleges of agriculture. A major limitation is confounding of statements. One must assume that members of each group use accurate, descriptive words they believe will convey the intended message to the public.

## Mission Statements (Summary—Tables 11.1 and 11.2)

Some in each of the four groups proclaim that promotion of the public good is their highest goal. It is an important standard by which they wish to be judged. Agricultural producers and allied groups contribute to the public good by producing food, feed, or fiber. Agribusiness does this by creating the technology for high yields and adding value to farm products. Environmental groups serve the public by working to protect and preserve the environment. Some, but surprisingly very few, mission statements suggest that members of each group agree that environmental integrity is the *sine qua non* for life on the planet.

Although few mission statements say so, it can be assumed, with confidence, that some members of each group recognize the value of good science as a basis for agricultural policy and practice. Science is among the primary tools needed to determine what policy and practice should be. A great deal of agricultural research is done by agribusiness companies. Much (not all) of it is proprietary, which means that, in some cases, neither the process nor the results are published in open, peer-reviewed scientific journals, and are therefore unavailable for use in determining agricultural policy and practice. The results are confidential and used to further the company's interests, rather than to build the corpus of general scientific knowledge or to serve as a basis for public decision-making. This is not *a priori* ethically objectionable; it is good business practice. The frequently closed scientific community of agribusiness does not have great concern about the importance of public participation and evaluation of their business, but they cultivate a favorable public image. This acknowledged good business practice may not be ethically objectionable, but there is a rising belief in US that companies owe stakeholders (whose number is larger than the number of stockholders and includes customers, employees, activist groups, agricultural producers, and other citizens) an accurate, complete reporting of their scientific activities, including both positive and negative findings (Grose, 1999).

The mission statements of US state- and federally-supported land-grant universities were studied in 2003 and in 2011. These institutions receive some funding from their state legislature and funds allocated by the Morrill Acts of 1862 and 1890. There are 105 land-grant universities (Rahm, 1997), at least one in each US state or territory. The number includes 29 Native American Tribal Colleges established in 1994. Not all land-grant schools have a College of Agriculture (e.g., New Jersey, Rutgers School of Environmental and Biological Sciences; the University of Rhode Island, College of Environmental and Life Sciences).

Because land-grant universities receive governmental support, it follows that in the democratic tradition, those affected by the consequences of any activity and those who pay have a right to a voice in decisions about the activities supported by their tax dollars (Sclove, 1998). In short, because of the nature of these institutions, in contrast to agribusiness companies, there is an obligation to be open and public about scientific research and its meaning. Further, it should not be assumed that scientists, professors, and administrators of science programs in universities neither have a monopoly on expertise, nor any claim to a privileged ethical view from which to evaluate the declared scientific, social, or environmental missions that affect scientific research, science policy, the use of scientific research, or the public (Sclove, 1998). Based on their 2011 mission statements, none favor public participation and evaluation of their work.

Zimdahl and Speer (1998) suggested that agribusinesses claim that new agricultural technology and current agricultural practices inevitably support the public good by increasing production of abundant, high-quality food, feed, and fiber. Their evidence supports the claim. Environmentally based objections to this view derive primarily from disagreement about all effects of the technology required or advocated to accomplish the goal of feeding the world's growing population, and secondarily, disagreement on the perceived imperative for increased production to feed a growing world population. Environmental groups often suggest that insider-only approaches to science policy and practice are antithetical to open, vigorous, creative public debate on which democracy and good science thrive (Sclove, 1998), but the mission statements of most environmental groups do mention public participation. Several include the importance of communities and quality of life to their mission. Environmental groups argue (Smith, 1997) that technology developed by agribusinesses in capitalist societies tends "to further social inequality, undermine popular sovereignty, generate environmental crises, and colonize every nook and cranny of everyday life with corporate propaganda." An agricultural example is the rise and ubiquity of herbicide-resistant crops and their advocacy in print media by manufacturers. A large percentage of US corn and soybeans acreage is now planted with seeds genetically modified to be resistant to one or more herbicides and/or be tolerant of an insect. Questions about the effect of this technology on small farmers and rural communities are asked but largely ignored in promotional material.

Resistance to purchasing genetically modified products in many European countries has been recognized by involved agribusinesses (see Chapter 8). The concern, initially dismissed, was to be overcome through education by company marketing and advertising groups. This tactic, regarded as corporate propaganda (see Kroma and Flora, 2003; Chapter 3), was dismissed by the public. Environmental problems (some anticipated and others unknown) will likely follow large-scale planting of herbicide-resistant crops (e.g., resistant weeds). Creation of such problems seems to be antithetical to achievement of human-oriented goals that all groups share, and to agribusinesses' goal of improving business profitability.

Many members of each group, and many scientists, claim that science is value free, a claim that has been dealt with earlier. Science has never been value free. The logic, practice, and results of science are moderated by social and ethical concerns. Rollin (1996) provides examples of the influence of ethics on science, and similar agricultural examples are plentiful. Within agriculture, the massive public and agribusiness support provided for pesticide research versus the comparatively minimal support for organic or alternative agriculture demonstrates the social determination of the subjects that agricultural science investigates. Similarly, there have been few publicly supported investigations of the effects of modern agricultural technology on the survival of small farms and farming communities (see Goldschmidt, 1998.) Second, Rollin suggests that social control of scientific methods is demonstrated by the fact that biomedical (and pesticide) research is done on rats, who as subjects have no choice, rather than mentally deficient children, who could not express a choice. At present, pain felt by the rats must be controlled, yet it was not too long ago in biomedical and pesticide research that pain in test animals was not even considered (Rollin, 2011). Most humans recoil at the suggestion that mentally deficient children, who surely would be more appropriate subjects to determine human effects, could be used to test the potential for human harm from agricultural pesticides. It is beyond the pale to even consider such thoughts, and our social ethic tells us so. Finally, the degree of statistical reliability demanded when a new pesticide is being evaluated, versus testing a new survey instrument to determine social opinion, shows the influence of our social ethic. Statistical estimates of performance and safety are demanded and accepted when an agrichemical manufacturer proposes new pesticide chemistry or a new use for an old pesticide to the US Environmental Protection Agency (EPA) or to a state regulatory agency. On the other hand, when one proposes a new ethical standard, statistical probability is not even considered. For example, who would accept the statement: It is likely that in 5% of the cases (95% probability) the public will reject any justification for preservation of family farms. Such a claim accompanied by a probability statement is outside the bounds of moral philosophy. Science, as suggested above, is regarded as value free and statistical probability is appropriate. Ethical matters, on the other hand, are regarded as value laden. Therefore, there is no clear standard of proof because we tell ourselves it is all just a matter of opinion. In science, one can prove within acceptable statistical limits what the facts are. Science, we often think, reveals the truth, unencumbered by value considerations. Ethics, it is thought, cannot reveal the truth because there is no ultimate truth in matters that rest on opinion rather than fact.

Agribusiness does many things superbly, but its marvelous successes in bringing new products to market, satisfying consumer needs (and wants), and creating wealth should not lull us into believing that the ethics of agribusiness (or any other kind of business) are equally applicable to all realms of life. The market, the *sine qua non* of modern business, deserves a place, and democratic institutions frequently, but not always, provide controls that keep it in its place (Okun, 1975; Kuttner, 1997). Unfettered markets are regarded as "the essence of human liberty and the most expedient route to prosperity" (Kuttner, p. 3). However, everything must not be for sale (p. 363). The value of business, even in a capitalistic society, is only instrumental, it has no intrinsic worth. According to George Soros, an extraordinarily successful capitalist entrepreneur, "economic theory is an axiomatic system: as long as the basic assumptions hold, the conclusions follow" (Soros, 1997). The predominant words in mission statements (Table 11.2)

**Table 11.2** Words and Their Frequency in the 2011 Mission Statements of 13 Agribusinesses, 21 Agricultural Producer Allied Groups, and 25 Environmental Organizations and 50 Land-Grant Universities.[16] Numbers in Parenthesis are the Percent Responding

| | Word Frequency | | | |
|---|---|---|---|---|
| | **Agribusiness** | **Agricultural Producer** | **Environmental** | **Land-Grant Agricultural Colleges** |
| <u>Human-oriented goals</u> | | | | |
| Education | | 6 (29) | 6 (29) | 33 (66) |
| Land-grant mission | | | 2 | 8 |
| Pursuit of knowledge | | 1 | 1 | 41 (82) |
| Human progress | | | | |
| Health, well-being | 4 (30) | 2 | 2 | 9 |
| safety, nutrition, nurture | 6 (46) | 2 | 3 | 24 (48) |
| Social goals, justice, Social responsibility | | | 2 | 36 (72) |
| Quality of life | 4 | 2 | 3 | 24 (48) |
| Community | | | | |
| International service | | | 1 | 19 (38) |
| Feed the world | 2 | | | |
| <u>Environmental goals</u> | | | | |
| Sustain agric, conserve restore, preserve env. | | | 10 (40) | |
| Sustain Inc/Agric. production enhance improve agric. | 1 | | 2 | 37 (74) |
| Stewardship | 1 | 1 | 5 (20) | 28 (56) |
| Protect nature, preserve respect, biodiversity | 1 | | 5 (20) | |
| Science-based Res/Info. | | | | 6 |
| <u>Business goals</u> | | | | |
| Profit, value, growth | 8 (62) | 6 (28) | | 24 (48) |
| Economic well-being | | 3 | | |
| Ethical standards | 2 | 2 | | |
| Ethically responsible | 2 | | | |
| Pursue strategic opp. | 3 | 2 | | 2 |
| Influence agric. policy | | 3 | 5 (20) | |

[16]Numbers in each column are the number of members of each group that used the term.

show that consideration of ethics is included by two agribusinesses but not by any of the other three groups. Moral questions are not addressed by any group's mission statement. Agribusinesses understandably emphasize profit, growth, and market principles. Table 11.2 shows some, but far from all, of the members of each group include human and environmental concern. Some agribusiness mission statements mention, but few emphasize, the environment or nature. Mission statements of agricultural producer groups rarely include words that emphasize improving environmental quality. All use human-oriented words. Ensuring food supply receives more emphasis than social goals (e.g., sustaining communities, justice, public participation). Promotion of the public good is not to be a high goal. Perhaps the most interesting thing that can be learned from study of the mission statements of a diverse set of organizations is that not much can be learned about the purpose, objective, and overall mission. Mission statements are useful but not conclusive evidence of purpose.

## The Role of the University

Universities in general, and those with colleges of agriculture in particular, frequently see themselves trapped among the competing interests (Table 11.2) and demands of the three groups: agribusiness, agricultural producer, and environmental. Well-funded agribusiness is eager to benefit from the intellectual and technological resources of the university to fulfill its research and technology development needs. Most agricultural producer and allied groups regard the university, especially the Cooperative Extension Service of land-grant universities, as an unbiased source of technological and production information about what agribusiness offers. They also fund research, but not at the level agribusiness does. Environmental groups are not notable sources of research funding. They act as a public conscience and are frequent critics of university-based research.

In 2006, private sector funding for US agricultural research and development expenditures was $2.8 billion, federal funding $3 billion, and state and private funding for state experiment stations $2 billion. Gardner (2002, p. 182) showed a steady increase (approx. $40 million per year) for "real" public spending on agricultural research from 1950 to 1990. "In recent years, State funds have declined, USDA funds have remained fairly steady (with changes in the composition of funding), but funding from other Federal agencies and the private sector has increased. Efforts to increase competitively awarded funds for research have fluctuated over time, as have special grants (earmarks). Along with shifts in funding sources, the proportion of basic research being undertaken within the public agricultural research system has declined.[17] " In 2012, USDA funding declined 17%.[18] With declining public funding, university scientists have been compelled to find other sources to support

[17] *Source*: Economic Research Service of the USDA Web site: US Public Agricultural Research: Changes in Funding Sources and Shifts in Emphasis, 1980–2005. Accessed July 17, 2011.

[18] *Source*: Economic Research Service of the USDA Web site: Agricultural Research Funding in the Public and Private Sectors. Accessed July 17, 2011.

research and graduate programs. Agribusiness is an obvious and willing source. Those able to attract such support find acclaim within the university. When such funding is sought and accepted, intellectual leadership may pass from the university scientist to the funding source. Thus, land-grant colleges of agriculture that were conceived and designed to serve agricultural producers find themselves compelled to accept funds and intellectual leadership from agribusinesses, whose logical interest in sales and market share may run counter to the best interests of producers. Private R&D funds are understandably commercially oriented. Companies, which must hold down costs, concentrate funds on research that is likely to result in sales and profits, and lead to intellectual property that can be protected by patents. No company is interested in supporting research that will benefit competitors. For example, more than 40% of private agricultural research support is invested in product development, compared with less than 7% of public agricultural research.

The university also finds itself succumbing to demands for service to its "customers," formerly called students. Now instead of only the student cafeteria, we find McDonalds or Carl's Jr. in the student center near the flower store, beauty parlor, branch bank, and ATM machine. The library has a coffee and snack shop. Each is regarded as a gain without much thought about what has been lost. We risk losing the university as the locus of intellectual culture which exceeds the sum of its mechanically acquired parts (Readings, 1996, pp. 74–75). We risk losing the essence of the university: a place where thinking is a shared process, a place where one thinks and learns how to think. One could argue that McDonalds and ATMs make allocation of time and money more efficient. Once efficiency is invoked, arguments about what may have been lost fade quickly. It is said that efficiency is one of the things the university must increase or intensify in its quest to be excellent. Readings (p. 119) tells us that the omnipresence of the criterion of excellence in modern universities merely "brackets the question of value in favor of measurement and replaces questions of accountability or responsibility with accounting solutions."

The university is confronted with instrumental decision-making within an imposed system that ignores the needs and values we live by and want to live by (Readings, p. 94), so the institution can become as efficient as business. The bottom line of a university's research program should not be measured in dollars or technological advances, but rather in ideas and intellectual creativity (Mac Lane, 1996). Close university/agribusiness partnerships are neither antithetical to ideas and intellectual creativity nor inevitably against the university's best interest and incompatible with the public good. However, if the liaison becomes too close or if the university becomes too dependent on agribusiness, the central locus of investigation may shift from ideas and intellectual creativity toward the university becoming one site among many where judgment is held open (Readings, p. 120).

Although not unanimous, environmental groups, in general, support Smith's (1997) strong accusation that "The university is in danger of becoming like the muscle-bound freak with tremendously developed biceps who has lets the rest of his mind and body atrophy. Corporate funds are the steroids." Corporate funds could also be regarded as the narcotics that lull their recipients into the belief that pursuit of corporate interests (i.e., commercializable technologies) will achieve the greatest good for the greatest number. It is argued that the pursuit maximizes public good, a stated goal of some agribusinesses. I understand these accusations neither achieve the reader's acceptance of the premise that agricultural producer's best friends may be found among environmental groups, nor build consensus for future action. As a strategy to build consensus, Daly (1996) reminds us that "it is probably good to keep the most controversial issues for last, even if they are ultimately the most important. But it would be quite dishonest not to bring them up at all."

A related and equally controversial issue is the nature of the university's primary task, which is to take the long view (Mac Lane, 1996). It is important to maintain the essence of the place where one thinks and learns how to think. This view does not emphasize satisfying the wants of the customer of the moment, but rather the legitimate needs of society for years ahead. Universities have existed longer than any modern industrial corporation. They are one of the best ways societies have devised to accomplish the difficult task of discovering and evaluating ideas and transmitting them and the process to new generations (Mac Lane). Of 50 land-grant colleges of agriculture, 33 identified education and 41 the pursuit of knowledge as primary goals. Only agricultural producers (29%) came close to similar emphasis (Table 11.2). Education—transmission of knowledge—is slow and unpredictable and does not fit the corporate competitive model (Mac Lane), which is often the apparent standard for judgment in the modern university. How big one's grant is seems more important than how big one's ideas are. If the long view is the correct view of the university's mission and therefore of the mission of a college of agriculture, one must ask if that view is most compatible with the stated missions of agribusiness, producer, or environmental groups. Which of these groups will be the best sources of intellectual and other forms of aid to the university as it strives to fulfill its mission of discovery, evaluation, and transmission of ideas through teaching, scholarship, and service? Similarly, are the goals of either agribusiness or environmental groups more compatible with the goals of agricultural producers, and how do these groups support producers? Table 11.2 provides some clues but does not answer the questions.

Orr (1994) in his discussion of the problem of education begins by citing a bit from an unpublished paper by Elie Wiesel,[19] who noted that the designers of the Holocaust were the heirs of Kant and Goethe and were widely believed to be the best educated people on earth. Wiesel described what was wrong with their excellent education:

> *It emphasized theories instead of values, concepts rather than human beings, abstraction rather than consciousness, answers instead of questions, ideology and efficiency rather than conscience.*

Orr (1994) argues that the same can be said of modern environmental (and by implication) agricultural education: it emphasizes theory, concepts, abstraction, answers (the right ones) and efficiency. There are correct answers to all questions. Technical efficiency is paramount. Orr argues that education, even a lot of it, is no guarantee

[19] Wiesel, E., 1990. Unpublished remarks before the Global Forum held in Moscow.

of "decency, prudence, or wisdom." He does not advocate ignorance but a different kind of education that emphasizes standards of decency and human survival. Sir Francis Bacon, the founder of modern science, told us that with increasing knowledge, we would gain power over nature (Busch, 2000). But, is power the proper goal? Orr (p. 8) says that we gain insight into what is wrong with modern education and our culture from the characteristics we know or ought to know well. Each of these shows the madness of the drive to dominate nature that typifies modern agriculture. Marlowe's Faust trades his soul for knowledge and power, Shelley's Dr. Frankenstein refuses to take responsibility for the monster he created (an externality), and Melville's Captain Ahab said "all my means are sane, my motive and my object are mad."

Our modern educational system teaches us that all problems are solvable and even ignorance, which may be part of the human condition, is correctable (Orr). I recall learning as a student that metaphorically speaking, science was able to shine light on human problems and solve them. I learned that as the area of light expanded, it indicated that we knew more and more problems were solved. However, as the area of light grew, the area of darkness surrounding it grew more. It seemed incongruous, but as knowledge grew, ignorance grew even more. But that is how the world works. Education teaches us what we don't know.

Orr also suggests we suffer from the dangerous and false myths that as human knowledge and technology increase we will know better how to manage the earth, human goodness will increase, and we will repair or restore that which has been damaged through human ignorance. I learned that we, the educated, given enough time and money, will be able to fix any problem. The results and recommendations of agricultural research and technology are, after all, "science based," which implies that the scientists know all that needs to be known. Questioning, especially moral questioning, is not required. Further explanation is not needed and, in fact, is likely to be counter-productive (Kirschenmann, 2009).

The data, in each case, incorrectly deny that bad things have happened. Increased agricultural knowledge has led to increased dominance of the natural world, more human misery, and almost no repair of damage. Our cleverness has increased, but our wisdom has not. Agricultural education cannot ignore the necessity of facts. Students must know about the laws of probability, plant and animal physiology, chemistry, and so forth. However, agricultural education must also teach students to think about what to do and why some things should be done and others ignored. Students must somehow learn about the facts and how to deal with what James (2003) identifies as Type I and Type II ethical problems (see Chapter 4). Type I problems are important because of difficulty in deciding what ethical norm should apply. Type II dilemmas, common in agriculture, occur when the general social consensus on what ought to be done is combined with incentives to violate the societal consensus. Presently such things are not an integral part of agricultural education, which has an excess of how to and a paucity of why to. Students arrive with a set of personal and social ethical standards. They will learn some professional ethical standards. It is not as likely that they will leave with a greatly different set of personal or social ethics than those they had when they arrived. It is highly likely that they will leave the university without a firm moral foundation that will guide them as they engage in the practice of agriculture. In that sense, their professors have failed.

## Sustainability as a Goal

Bandwagons come and go (Simmonds, 1991); some pass quickly while others endure and the words associated with them become part of the lexicon. Sustainability is a popular bandwagon term. It is too early to tell if it will endure. Because of its current popularity in agriculture, one might expect it to appear frequently in mission statements, but with the exception of land-grant colleges of agriculture (37 of 50), it does not. The word is common in agricultural publications and in academic discourse. As discussed in Chapter 7, there is no agreed definition; the word means what the user wants it to mean. A simple definition of sustainable agriculture is, "Farmers should farm so they can farm again" (Wojcik, 1989). Harwood (1988, p. 4) suggested: "[An] agriculture that can evolve indefinitely toward greater human utility, greater efficiency of resource use, and a balance with the environment that is favorable both to humans and to most other species." There are many other definitions and each is value laden with words such as human utility, efficiency, balance, and favorable. Most definitions are from Western, developed countries and overlook the fact that a sustainable agriculture in a developing country may be that which increases food production to sustain a growing population (Gressel and Rotteveel, 1999). Most definitions derive from an *ego*centric ethic of management in which the land is an instrument to achieve human ends. They are not based on an *eco*centric ethic in which the land has inherent value (Merchant, 1990). In an ecocentric ethic, the land and its needs are regarded as coincident with human needs. Both are sustained; neither is consistently dominant. Sustainable agriculture demands a shift from an anthropocentric or egocentric to an ecocentric ethic. The former view, often called ethical egoism, is a normative theory about how one ought to behave. It says we have no moral duty except to do what is best for ourselves; self-interest rules (Rachels, 1986). It is the ethical equivalent of Adam Smith's "invisible economic hand." In this view, what is to be sustained is production of abundant food or fiber, and profit for the producer and for those who supply the resources (inputs) required to produce. The primary problem with this view is its failure to internalize the inevitable externalities (Merchant, 1990). Sustaining production is a good thing, but it surely cannot always be the only or the highest value. When the technologies required to sustain production pollute water, harm nontarget species, or contaminate food, it is hard to support the claim that production and profit should always be the highest goals. Thus, producers who are dependent on the land ought to ask if the missions of agribusiness and environmental groups sustain or mitigate against the sustainability of the land, the producer's primary resource. The evidence (Table 11.2) from the mission statements studied, once again, does not answer the question. While it is obvious that brief mission statements cannot include everything, it is equally obvious that what is omitted may be as important as what is included.

**Highlight 11.5**

In 1997, I was invited to spend two months as a Visiting Professor in the Institute of Plant Production and Agroecology in the Tropics and Subtropics of the University of Hohenheim, Stuttgart, Germany. My host, Professor Herr Dr. Werner Koch, enjoyed walks in the countryside and invited me to accompany him. On one occasion, as we walked over a slight rise, I was most impressed with the agricultural vista ahead. There were many small fields; Some bare soil, some green with a winter crop or hay, and some with stubble from the last crop. The large area was laced with cement paths on which people were strolling, jogging, biking, pushing children in strollers, or roller-blading. There were no obvious farm buildings. I commented to Prof. Koch that it was nice of the government to pave all the paths so the public could enjoy the countryside. Prof. Koch kindly and firmly informed me that the paths were built by the Government, but not for the uses I observed. They were built so the farmers, who lived in nearby villages, could get equipment to their fields. He then went on to explain what he called the German Landscaping Program. A major purpose of the program was to keep small farmers in farming. The Government offered farmers the option of accepting some or all of the following conditions for five years: no herbicide or growth regulator use, the inter-row distance in cereals was greater than 17 cm, a cover crop would be kept on the land between crops, 20% less fertilizer than normal would be used, and the land would not be plowed. When a farmer agreed to abide by one or all of these conditions, the Government offered a subsidy of points per hectare, which were converted to a payment of Deutschmarks at the end of the cropping season. Each of the stated conditions lowered yield.

Germany is a rich country and does not need all of its farmers to produce more food. Public policy did not favor increasing production but did favor keeping farmers in business. Farmers were valued because they maintained the land and they maintained villages, the center of valued German culture. Maintaining farmers, Professor Koch, assured me, also meant that Germany would look well when tourists visited, a lesser but important goal of the program.

The German state of Baden-Württemberg had a similar program in 2005, Marktentlastungs und Kulturlandschaftsausgleich (MEKA), which roughly translates to a program to reduce production while preserving and improving the quality of the cultivated area. The program included:

Regular soil analyses to assure appropriate fertilizer use.

Preservation of vineyards on steep slopes to prevent soil erosion. Old supporting walls are maintained.

Preservation of land races of economically useful animals (especially cattle).

No use of pesticides, fertilizers, and growth regulators.

Reduction of use of nitrogen fertilizer on arable land by 20%.

Inevitable decreases in crop yield are compensated by Government subsidies to maintain farmer's income. The program is used most by small farms. Other German states and European Countries have similar programs tailored to their area and specific needs. Each program is part of the agriculture-environment initiative of the European Union and is financed by the European Union.

The programs accomplish four desirable agricultural goals (see Lehman 1995; Chapter 10). Safe food is produced, resources are conserved, and the practices are environmentally benign or friendly. Profit is assured by Government subsidies. The programs achieve desirable elements of agricultural sustainability, including protecting producers but they are not sustainable economically, without public subsidy. American agricultural subsidies, it is worth noting totaled more than $300 billion between 1978 and 2002, while small farmers disappeared and the environment was not favored.

The US Department of Agriculture distributes between $10 billion and $30[20] billion in cash subsidies to farmers and owners of farmland each year. The amount depends on market prices for crops, the level of disaster payments, and other factors. More than 90% of the subsidies go to farmers of five crops—wheat, corn, soybeans, rice, and cotton. More than 800,000 farmers and landowners receive subsidies, but the payments are heavily tilted toward the largest producers.

[20] http://www.downsizinggovernment.org/agriculture/subsidies. Accessed August 25, 2011.

Lehman (1995) claims that a sustainable agricultural production system ought to be one that conserves resources, achieves relatively high energy outputs given its energy inputs, and provides sufficient, safe food for a community of people. Such a system might not yield income in excess of costs. That is, the system may not be profitable for the farmer. The words used in mission statements show that members of all three groups agree with Lehman concerning conservation, energy, food sufficiency, and food safety. Environmental groups use environmental words much more frequently than either of the other groups and therefore have objectives coincident with those of producers. The mission statements provide little evidence, but suggest that members of each group will strenuously object to the thought that any nation could endure an agriculture that was not profitable to producers. Profit is important to agribusiness and producers if they are to survive and meet the continuing demands for change. The evidence suggests it is not important to environmental groups. People in agribusiness believe that agriculture is a business and must respond to the same profit demands and follow or be susceptible to the same economic and market rules that govern any other business.

In spite of the lack of a precise definition, and disagreement over the role and necessity of profit, sustainability is in vogue. Where it does not appear it is implied by the use of words such as environment, conserve, restore, preserve, and stewardship, but these words do not appear in all mission statements. Sustainability is something one might assume all groups support. It might be one of the things that all could agree is good. But in spite of its current popularity, it is not included in most mission statements. It seems obvious that farmers/ranchers want to sustain their farms/ranches. It is not illogical to conclude that environmental groups may be the best allies in their quest.

Mission statements may reveal a great deal about what an organization is about, and still not reveal everything. One must assume they reveal what their creators want to reveal. It is tempting, but not acceptable, to read other meanings into such statements, to read between the lines or try to evaluate what is not included. Mission statements must be accepted for what they say. Coletti (1999) notes that codes of ethics come in glossy brochures, and environmental and social duties are featured in annual reports and presumably in mission statements. However, chief executives do not put such things at the top of their respective agendas. Environmental and social issues ranked only 13th among 16 marketplace challenges in a survey of 656 CEOs worldwide (Coletti, 1999). The top five issues were downward pressure on prices (mentioned by 48% of CEOs), changes in type or level of competition (43%), industry consolidation (41%), changing technology (25%), and increasing innovation (24%). Environmental, health, and safety issues were mentioned by only 46 (7%) of the CEOs. What is omitted may be as important as what is said, and what is said may not be what is meant when the going gets tough. Those in search of partners ought to be aware of both the stated and, in so far as possible, the unstated goals of other organizations. An appropriate question is—Are the words in mission statements reflective of public relations efforts or an accurate reflection of intent?

A test of Zimdahl and Speer's (1998) first hypothesis, that agricultural producers share missions and objectives with environmental groups and that their mission statements demonstrate their shared goals, is not a simple objective exercise. It is not obvious from the mission statements that these groups share missions or objectives with each other or with land-grant universities. It is not obvious that they regard each other either as allies or as adversaries. Nor do the mission statements immediately reveal clear objective information on the second hypothesis, that agricultural producers do not share missions or operational objectives with agribusiness companies, and their mission statements demonstrate the lack of common interests.

Harwood (1988, p. 5) said: "In the early 1900s, popular thinking among farmers had led to the rejection of the portion of Jeffersonian thought that held individualism to be supreme." This led to establishment of farmer organizations such as the Farm Bureau and Grange. Farmers became convinced they could not stand independently of their neighbors, and their knowledge, equipment, and ideas needed to be shared if all were to succeed. Harwood (1988) cites Marcus (1985) to describe two sources of agricultural knowledge. Systematic agriculturalists looked to the developing agricultural support industries as their model and guide about how agriculture should be practiced. These industries included farm machinery as exemplified by the cotton gin, reaper, combine, and steel moldboard plow. The fertilizer industry led the way to the chemicalization of agriculture, and although pesticides existed prior to

World War II, their rapid development was a postwar phenomenon. Agribusiness has been, and continues to be, the source of numerous innovations and the technology for rapid increases in crop yield. Agricultural industry was widely regarded within and outside farming as progressive and forward-looking. New products and new ways led to greater production and profit.

This view was opposed (and still is) by scientific (Harwood, 1988), or as I prefer, natural agriculturalists who look to nature as their model of how agriculture ought to be practiced. The central idea is that nature is the best teacher and its workings can be rationalized and formalized into proper agricultural practice. Farmers were regarded as natural historians whose knowledge of place and process would create good, environmentally benign agricultural systems. The twentieth century exponents of this view included Robert Rodale and Louis Bromfield, and more recently Wendell Berry, Wes Jackson, Miguel Altieri, and Francis Moore Lappé.

The view of the systematic agriculturalists is exemplified by the work of Avery (1995, 1997) and Waggoner (1994). In this view, human population growth is regarded as inevitable. There is general agreement among systematic and scientific agriculturalists that the UN median projection of 9–10 billion people by 2050 is reasonable. Those who will create the children are already here. The systematic agriculturalists assume that food demand will exceed supply. Avery (1997) concludes that by 2040, the world must once again achieve a tripling of yields on existing farmland. If that is not accomplished we will lose millions of square miles of presently wildlands and many now endangered species. Land that should not be farmed will be farmed. The fundamental claim is that one of two things must happen:

1. The same amount of land must become three times as productive or
2. Three times as much land must be brought into production.

It is likely that neither will happen but they set the boundary conditions for the future. Avery (1997) claims that the "world has only one proven, effective strategy for protecting its wildlands and endangered species in the 21st century: getting higher yields of crops and livestock from the land we're already farming." Farmers, in this view, work at the hub of sparing land for nature (Waggoner, 1994). Farmers, enabled by modern technology, can raise more crops or animals per unit area of land. This, in Waggoner's (1994) view, helps keep food prices low and spares land for nature that would have to be used to produce food if yields are not raised. Avery (1995, 1997) suggests that crop protection technology and all of modern agriculture should be seen for what they are, "one of mankind's greatest environmental achievements, in the most conservation-minded era of human history." This view of agriculture is the view that agribusiness supports even though the evidence for this conclusion cannot be obtained from mission statements.

Those with the opposite view, what Harwood (1988) citing Marcus (1985) called the scientific view and I call the natural view, hold that modern/industrial agriculture views only human beings as having inherent worth. The rest of nature has only instrumental value as a resource for humans. This is what Merchant (1990) called the egocentric position. The evolution to modern, capital, energy, and chemically intensive agriculture was not done because of rational ecological analysis, but because of scientific contributions that made it feasible, and low-cost energy that made it possible to use nature as an instrument to produce food (Altieri, 1985). Excessive dependence on the technology that characterizes modern agriculture is not sustainable because of two inherent technological problems (Ausubel, 1996). First, technology's success is self-defeating. It has made the human niche elastic (Ausubel, 1996). It enabled us to overcome the limits of the natural world and impose our will upon it. Dominating and subduing are easier. But technology solves and creates problems at the same time. A good example is insect, weed, and disease resistance to pesticides. In the early days of pesticide development, resistance was not a problem. Now it is a huge problem that is dealt with, in part, by continuing to create and use more of the same technology that creates resistance. Resistance management is now part of pest management science. This is precisely the kind of problem solving that alternative agriculturalists deplore. *Second* (Ausubel) is the "paucity of human wisdom." Technology creates the ability to kill and cure, destroy and create, do good or evil, and to sustain or harm the earth. Few set out to do evil but few can see all consequences of any technology. As agriculture changed rapidly after World War II, agricultural science focused on what could be done: the soil's natural fertility was replaced by adding fertilizer, high-yielding plants were developed, pests were controlled with pesticides. The dominant question was, Can we do it? But unbridled use of agricultural technology has not increased the well-being of all members of society and has hurt some.

I do not argue, as some might suspect, that therefore we ought to stop science or control it more carefully. Scientific freedom is a great virtue. There is no question that the scientific advances that have led to modern agriculture have created more human pleasure than pain. The abundance of a modern grocery store is evidence of the achievements of agriculture. But, more and more, we are faced with a moral question: What ought to be done? What should be done? This means that what was once just a technical question (Can we do X?) is now also a moral question (Ought we do X?).

As those engaged in agriculture begin to ask what ought to be done they may also ask specific questions:

1. What groups in our society seek answers to the same questions?
2. What groups think as we do?
3. With whom should we try to partner and form working relationships?
4. With what groups do we share common goals and methods?

It was assumed that answers or clues to answers to these questions could be found in the words used in the mission statements of agribusiness, producer, environmental groups, and land-grant colleges of agriculture. One must conclude that mission statements are not particularly revealing of an organization's purpose or methods. As any group looks for partners, it is unrealistic to expect that all divisions will disappear as common interests are discovered. It has happened in other areas that groups that were at odds have found common ground and new alliances have been formed (Wilkinson, 1999). Environmentalists and loggers in Pacific Northwest timber towns have found that economy and ecology share more than the same prefix (Wilkinson, 1999). Their alliance has come about because of the marginalization of the labor and environmental movements by corporations. Agricultural producer and allied groups that value small farms and rural communities and the sustainability they imply may want to seek similar alliances. Analysis of mission statements is a place to begin to learn about those with whom one must work or one may choose to work. But it is only a beginning. Behavior and actions, as they always have, speak more loudly than words.

## Conclusion

The preface of this book says a primary goal is to continue the discussion of agricultural ethics begun by others. The task was to explore ethical positions in agriculture or the lack thereof using the metaphor of agriculture's horizon: the boundary that separates and delineates one's outlook and knowledge. The book praises agriculture's myriad accomplishments that have vastly increased food and fiber production and the efficiency of that production per acre and per animal. It has, while acknowledging some opposing arguments, been unrelentingly critical of the fact that consideration of the ethics of agriculture has been lacking and that lack has limited agriculture's horizon and has created some of the public's negative view of agriculture—the essential human activity.

A related task was to demonstrate that underlying each set of views on important agricultural issues there is always an ethical position. I conclude that this is true. However, to demonstrate the correctness of the conclusion one must illustrate it with ethical positions held by those in agriculture and here the book fails. I and others cited herein who have explored the ethics of agriculture more carefully, have concluded that agriculture has only one dominant ethic, which is not openly debated. It is accepted. It is the ethic mentioned above and best described by Thompson (1995): there is only one imperative—to produce as much as possible, regardless of the environmental/ecological costs and perhaps even if it is not profitable to the producer. Therefore, it is incorrect to argue that agriculture has no ethical standard at all. That is not true. The argument should be that the dominant ethical standard is unexamined and should not remain so. Agricultural science and technology have been major contributors to the liberation from elemental want of most who reside in the world's developed countries. Scientific advances and especially agriculture's achievements have been central to attaining a standard of living for many that was beyond human imagination in the mid-nineteenth century (Sarewitz, 1996, p. 103). However, as Sarewitz points out, a parallel consequence has been "an unprecedented acceleration in the exploitation, modification, and despoliation of nature."

Those who practice agriculture cannot escape responsibility for its effects on nature but the dominant ethic ignores such effects. The effects are not a cost but a set of problems to be solved through more science and better technology. They are the price of bounty. Environmental effects are also evidence of human mastery over nature which, we are wont to assume, has been subdued by science (Sarewitz). The highest priority for agricultural research is to continue to produce through domination of nature.

It is logical to conclude that agriculture's practitioners believe they have been extraordinarily successful and therefore deserve praise not criticism. Raising questions about agricultural practice and results is to miss the point. Agriculture is about results. What matters is whether or not people are fed. Ethical questions just get in the way. If one believes there is no objective truth in ethics, then it follows that a search for objective moral truth, "ethical facts," is futile. Objective ethical truth, given this view, is just a clever philosophical illusion.

Agriculture's practitioners including agricultural scientists clearly care about scientific truth. It exists and part of the scientist's task is to discover what is true. Lynch (2004), in a perceptive essay, points out that "caring about truth means that you have to be open to the possibility that your own beliefs are mistaken." It is mistaken beliefs about ethics that inform agriculture and need to be changed. Debates about the ethics of agriculture are not trivial but essential to progress just as the search for scientific truth is. No scientist will hold to a scientific belief that is patently false. Similarly no one should hold fast to an ethical view or a view of ethics that is unexamined. There is objective truth in ethics and knowing the ethical foundation for action is just as essential as knowing the scientific hypotheses that support experimentation. Because agriculture is the essential human activity, it is essential that it rest on a firm ethical foundation. Agriculture is not just about results—principles matter for they determine what truths are sought.

> *Forgive my vehemence, which has deep causes in my hope for the future. This is my subject. I know, or partly know, what I want. I know, and clearly know, what I fear.*
> *John Maynard Keynes' letter to Dean Acheson, August 1941*

## References


Altieri, M. (1985). Ecological diversity and the sustainability of California agriculture. *in Sustainability of California Agriculture: A Symposium*. Davis, CA, University of California, p. 106.

Anonymous. (2004). Matters of scale. *Worldwatch Magazine* March/April:36.

Ausubel, J. (1996). Can technology spare the planet? *American Scientist* 84:166–178.

Avery, D. (1995). *Saving the Planet with Pesticides and Plastic: The Environmental Triumph of High-Yield Arming*. Indianapolis, IN, Hudson Institute.

Avery, D. (1997). Saving the planet with pesticides, biotechnology and European farm reform. *Brighton Crop Prot. Conf.* Pp. 3–18.

Bailey R. (ed.). 1995. *The True State of the Planet: Ten of the World's Environmental Researchers in a Major Challenge to the Environmental Movement*. New York, The Free Press.

Barker, A. and B.G. Peters. (1993). Introduction—Science Policy and Government. Pp. 1–16 *in The Politics of Expert Advice: Creating, Using and Manipulating Scientific Knowledge for Public Policy*. Pittsburgh, PA, University of Pittsburgh Press.

Baskin, Y. (1997). *The Work of Nature*. Washington, DC, Island Press.

Berry, T. (1999). *The Great Work*. New York, Bell Tower.

Birch, B.C. and L.L. Rasmussen. (1978). *The Predicament of the Prosperous*. Philadelphia, The Westminster Press.

Busch, L. (2000). *The Eclipse of Morality: Science, State, and Morality*. Hawthorne, NY, Aldine de Gruyter.

Carson, R. (1962). *Silent Spring*, 25th Anniversary Edition. Boston, MA, Houghton Mifflin Co.

Cavalieri, P. (2001). *The Animal Question—Why Nonhuman Animals Deserve Human Rights*. New York, Oxford University Press.

Coletti, E. (1999). Ethics rank low among CEOs. *Christian Science Monitor* (July 12): 12.

Comstock, G.L. (1995). Do agriculturalists need a new, an ecocentric ethic? *Agric. and Human Values* 12:2–16.

Conway, G.R. and J.N. Pretty. (1991). *Unwelcome Harvest. Agriculture and Pollution*. London, Earthscan Publications, Ltd.

Daly, H.E. (1996). *Beyond Growth*. Boston, MA, Beacon Press. P. 23.

Deschênes, O. and M. Greenstone. (2007). The economic impacts of climate change: Evidence from agricultural output and random fluctuations in weather. *American Economic Review* 97(1):354–385.

Dimitri, C.A., Effland, and N. Conklin (2005). The 20th century transformation of US agriculture and farm policy. *ERS Electronic information bulletin* No. 3, June.

Douglass, G.K. (1994). The meanings of agricultural sustainability. Pp. 3–29 *in* G. Douglass (ed.). *Agricultural Sustainability in a Changing Word Order*. Boulder, CO, Westview Press.

Dürrenmatt, F. (1964). The Physicists (translated from the German by J. Kirkup). New York, Grove Press, 94 pp.

Economist. (2011). The 9 billion people question. Feb. 26:16.

Gardner, B.L. (2002). *American Agriculture in the Twentieth Century*. Cambridge, MA, Harvard University Press.

Gardner, G.T. (2006). *Inspiring Progress—Religion's Contributions to Sustainable Development*. New York, W.W. Norton & Co. New York.

Goldschmidt, W. (1998). Conclusion: The urbanization of rural America. Pp. 183–198 *in* K.M. Thu, E.P. Durrenberger (eds.). *Pigs, Profits, and Rural Communities*. Albany, NY, State University of New York.

Green, R.E., S.J. Cornell, J.P.W. Scharlemann, and A. Balmford. (2004). Farming and the fate of wild nature. *Science Express* Dec:23.

Gressel, J. and T. Rotteveel. (1999). Evaluation of the genetic and agro-ecological risks from biotechnologically-derived herbicide resistant crops (BD-HRC), with decision trees for less biased, regional, risk assessment. *Plant Breeding Reviews* 18:251–303.

Grose, T.K. (1999). Called to Account. *TIME* October 4.

Harwood, R.R. (1988). A history of sustainable agriculture. Pp. 3–19 *in* C.A. Edwards, R. Lal, P. Madden, R.H. Miller, G. House (eds.). *Sustainable Agricultural Systems*. Ankeny, IA, Soil and Water Conservation Society.

Hollander, R. (2000). Scientific research, ethics, and values in science. Pp. 1041–1047 *in* T.H. Murray, M.J. Mehlman (eds.). *Encyclopedia of Ethical, Legal and Policy Issues in Biotechnology* Vol. 1: New York, J. Wiley & Sons, Inc.

James, H.S. (2003). On finding solutions to ethical problems in agriculture. *J Agric. and Env. Ethics* 16:439–457.

Jonas, H. (1984). *The Imperative of Responsibility: In Search of an Ethics for the Technological Age*. Chicago, IL, University of Chicago Press.

Kirschenmann, F. (2000a). Challenges facing philosophy as we enter the 21st century: Reshaping the way the human species feeds itself. The Eddy Lecture, Colorado State University. September 28. Available from Colorado State University Department of Philosophy.

Kirschenmann, F. (2004). Ecological morality: A new ethic for agriculture. Pp. 167–176 *in* D. Rickerl and C. Francis (eds.). Agroecosystems Analysis. No. 43 in the series. Agronomy. American Soc. Agronomy, Crop Sci. Soc. America, and Soil Sci. Soc. America. Madison WI.

Kirschenmann, F. (2000b). Questions we aren't asking in agriculture: Beginning the journey toward a new vision. http://www.leopold.iastate.edu/fredspeech.html (accessed October 23, 2000).

Kirschenmann, F. (2009). The dangers of too much certainity. *Leopold Letter* 8.

Kirschenmann, F. (2002). What constitutes sound science? http://www.leopold.iastate.edu.

Krauthammer, C. (1997). Make it snappy essay. *TIME* July 21:84.

Kroma, M.M. and C.B. Flora. (2003). Greening pesticides: A historical analysis of the social construction of farm chemical advertisements. *Agric. and Human Values* 20:21–35.

Kuttner, R. (1997). *Everything for Sale: The Virtues and Limits of Markets*. New York, A.A. Knopf.

Larson, E.J. (2004). *Evolution: The Remarkable History of a Scientific Theory*. New York, Modern Library.

Lehman, H. (1995). *Rationality and Ethics in Agriculture*. Moscow, ID, University of Idaho Press.

Logan, W.B. (1995). Dirt—The Ecstatic Skin of the Earth. New York, The Berkeley Publishing Group.

Lomborg, B. (2001). *The Skeptical Environmentalist—Measuring the Real State of the World.* Cambridge, UK, Cambridge University Press.

Lynch, M.P. (2004). Who cares about the truth? *The Chronicle of Higher Education* Sept. 10:B6–B8.

Mac Lane, S. (1996). Should Universities Imitate Industry? *Amer. Scientist* 84:520–521.

Marcus, A.I. (1985). *Agricultural Science and the Quest for Legitimacy: Farmers Agricultural Colleges, and Experiment Stations, 1870–1890.* Ames, IA, Iowa State University Press.

McNeill, J.R. (2000). *Something New Under the Sun. An Environmental History of the Twentieth-Century World.* New York, W. W. Norton & Co.

Merchant, C. (1990). Environmental ethics and political conflict: A view from California. *Environ. Ethics.* 12:45–68.

Morgan, J.A., D.R. LeCain, E. Pendall, D.M. Blumenthal, B.A. Kimball, Y. Carrilo, D.G. Williams, J. Heisler-White, F.A. Dijkstra, and M. West. (2011). C grasses prosper as carbon dioxide eliminates dessication in warmed semi-arid grassland. *Nature* 476:202–205.

Myers, S.S. (2009). *Global environmental change: The threat to human health.* Worldwatch Report 181. Washington, DC, Worldwatch Institute. Several sources are cited in end-note 8 on page 42 of the report.

National Geographic. (2010). *Water: Our Thirsty World.* Washington, DC, The National Geographic Society.

Okun, A. (1975). *Equality and Efficiency: The Big Tradeoff.* Washington, DC, Brookings Institution Press.

Orr, D.W. (1994). *Earth in Mind: On Education, Environment, and the Human Prospect.* Washington, DC, Island Press.

Paddock, W. and P. Paddock. (1967). *Famine—1975 America's Decision Who Will Survive.* Boston, MA, Little, Brown and Company.

Pimentel, D. (1995). Environmental and economic costs of soil erosion and conservation benefits. *Science* 267:1117–1123.

Pimentel, D. and A. Greiner. (1997). Environmental and economic costs of pesticide use. Pp. 51–78 *in* D. Pimentel (ed.). *Techniques for Reducing Pesticide Use. Economic and Environmental Benefits*. New York, J. Wiley and Sons.

Pimentel, D. and A. Wilson. (2004). World Population, Agriculture and Malnutrition. *World-Watch* 17(5):22–25.
Postel, S. (1999). *Pillar of Sand; Can the Irrigation Miracle Last?* New York, W. W. Norton & Company, Inc.
Pretty, J.N. (1995). Sustainable Agriculture in the 21st Century: Challenges, Contradictions and Opportunities. British Crop Protection Conf. *Weeds.* 3:111-120.
Rachels, J. (1986). *The Elements of Moral Philosophy*, 2nd Ed. New York, McGraw-Hill, Inc. P. 77.
Rahm, D. (1997). *The Land-Grant University Mission. The Ag Bioethics Forum.* Ames, IA, Iowa State University Bioethics Program, 9(1): 5–6.
Readings, B. (1996). *The University in Ruins.* Cambridge, MA, Harvard University Press.
Rist, M. (1988). Future tasks for agriculture. *J Agric. Ethics* 1:101–107.
Rollin, B.E. (1992). *Animal Rights and Human Morality.* Buffalo, NY, Prometheus Books.
Rollin, B.E. (1996). Bad Ethics, Good Ethics and the Genetic Engineering of Animals in Agriculture. *J. Animal Sci.* 74:535–541.
Rollin, B.E. (2011). *Putting the Horse Before Descartes: My life's Work on Behalf of Animals.* Philadelphia, Temple University Press.
Rollin, B.E. (1989). *The Unheeded Cry: Animal Consciousness, Animal Pain, and Science.* Ames, IA, Iowa State University Press.
Salinger, J.D. (1951). *The Catcher in the Rye.* Boston, MA, Little, Brown Co.
Sarewitz, D. (1996). *Frontiers of Illusion: Science, Technology, and the Politics of Progress.* Philadelphia, Temple University Press.
Schlosser, E. (2002). *Fast Food Nation—The Dark Side of the All-American Meal.* New York, HarperCollins Publishers.
Sclove, R.E. (1998). Editorial—Better approaches to science policy. *Science* 279:1283.
Sen, A. (1981). *Poverty and Famines: An Essay on Entitlement and Deprivation.* Oxford, UK, Clarendon Press.
Simmonds, N.W. (1991). Bandwagons I have known. *Tropical Agriculture Assoc. Newsletter* Pp. 7–9.
Singer, P. (2002). *All Animals are Equal.* New York, Oxford University Press.
Singer, P. (1977). *Animal Liberation.* New York, Avon Books.
Singer, P. (1981). *The Expanding Circle: Ethics, Evolution and Moral Progress.* Princeton, NJ, Princeton University Press.
Singer, P. (2009). *The Life You Can Save: Acting Now to End World Poverty.* New York, Random House.
Smil, V. (2000). *Feeding the World: A Challenge for the Twenty-First Century.* Cambridge, MA, The MIT Press.
Smith, T. (1997). *Some Remarks on University/Business Relations, Technological Development, and the Public Good.* The Ag Bioethics Forum. Iowa State University 9(1):6–9.
Soros, G. (1997). The capitalist threat. *Atlantic Monthly* February:48.
Thompson, P.B. (2010). Food aid and the famine relief argument (brief return). *J. Agric. and Environmental Ethics* 23:209–227.
Thompson, P.B. (1995). *The Spirit of the Soil.* New York, Routledge.
Voltaire. (1759). *Candide, ou l'optimisme.*
von Kaufmann, R.R. and H. Fitzhugh. (2004). The importance of livestock for the world's poor. Pp. 137–159 *in* C.G. Scanes, J.A. Miranowski (eds.). *Perspectives in World Food and Agriculture.* Ames, IA, Iowa State University Press.
Wackernagel, M. and W.E. Rees. (1996). *Our Ecological Footprint—Reducing Human Impact on the Earth.* Stony Creek, CT, New Society Publishers.
Waggoner, P. (1994). How Much Land Can Ten Billion People Spare For Nature? *Council Agric. Sci. Technol. (CAST). Task Force Rpt.* 121. Pp. 64.

WHO (World Health Organization). (1990). *Tropical Diseases News*. 31:3 WHO Press Release. March 28.

Wildavsky, A. (1995). *But Is It True? A Citizens Guide to Environmental and Health and Safety Issues*. Cambridge, MA, Harvard University Press.

Wilkinson, T. (1999). Environmentalists discover a curious ally. *Christian Science Monitor* December 13:3.

Williams, T. (2010). *The Corruption of American Agriculture. Americans for Democratic Action Education Fund.* Washington, DC.

Wojcik, J. (1989). *The Arguments of Agriculture: A Casebook in Contemporary Agricultural Controversy*. Purdue University Press. P. x.

World Bank. (2002). World Development Report – 2002, Building Institutions for Markets. See Table 1. Key Indicators of Development, p. 232.

World Bank. (2010). World Development Report – 2010, Development and Climate Change. See Table 1. Key Indicators of Development, p. 378.

Wright, A. (1990). *The Death of Ramón Gonzales: The Modern Agricultural Dilemma*. Austin, TX, University of Texas Press.

Zilberman, D., X. Liu, D. Roland-Holst, and D. Sunding. (2004). The economics of climate change. *Mitigation and Adaptation Strategies for Global Change* 9:365–382.

Zimdahl, R.L. (2003). The mission of land grant colleges of agriculture. *Am. J. Alt. Agric.* 18(2):103–115.

Zimdahl, R.L. and R.L. Speer. (1998). Agriculture's mission: finding a partner. *Am. J. Alt. Agric.* 16(1):35–45.

Printed in the United States
By Bookmasters